T0180838

Lecture Notes on Mathematical Modelling in the Life Sciences

The rapid pace and development of new methods and techniques in mathematics and in biology and medicine creates a natural demand for up-to-date, readable, possibly short lecture notes covering the breadth and depth of mathematical modelling, mathematical analysis and numerical computations in the life sciences, at a high scientific level.

The volumes in this series are written in a style accessible to graduate students. Besides monographs, we envision the series to also provide an outlet for material less formally presented and more anticipatory of future needs due to novel and exciting biomedical applications and mathematical methodologies.

The topics in LMML range from the molecular level through the organismal to the population level, e.g. gene sequencing, protein dynamics, cell biology, developmental biology, genetic and neural networks, organogenesis, tissue mechanics, bioengineering and hemodynamics, infectious diseases, mathematical epidemiology and population dynamics.

Mathematical methods include dynamical systems, partial differential equations, optimal control, statistical mechanics and stochastics, numerical analysis, scientific computing and machine learning, combinatorics, algebra, topology and geometry, etc., which are indispensable for a deeper understanding of biological and medical problems.

Wherever feasible, numerical codes must be made accessible.

Founding Editors:

Michael C. Mackey, McGill University, Montreal, QC, Canada

Angela Stevens, University of Münster, Münster, Germany

More information about this series at http://www.springer.com/series/10049

Hong Qian • Hao Ge

Stochastic Chemical Reaction Systems in Biology

Springer

Hong Qian
Department of Applied Mathematics
University of Washington
Seattle, WA, USA

Hao Ge
Beijing International Center
for Mathematical Research
and Biomedical Pioneering
Innovation Center
Peking University
Beijing, China

ISSN 2193-4789   ISSN 2193-4797   (electronic)
Lecture Notes on Mathematical Modelling in the Life Sciences
ISBN 978-3-030-86251-0   ISBN 978-3-030-86252-7   (eBook)
https://doi.org/10.1007/978-3-030-86252-7

This Springer imprint is published by the registered company Springer Nature Switzerland AG
The registered company address is: Gewerbestrasse 11, 6330 Cham, Switzerland

For Professor Min Qian (钱敏)
  Father and Mentor,

To Mom 张锦炎, who raised me,
  Madeleine 董玥, who believes in me, and
  Isabelle 钱雨蘅, who humors me.
  – H. Q.

To my father 葛云保 and mother 丁梅华,
    who love me without limit,
  my son 葛一川, who inspires me with
      his smiles.
  – H. G.

Preface

In physics, following Lagrange's analytic tradition, a set of differential equations based on Newton's laws of motion, $\ddot{x}_k = m_k^{-1} F_k(x_1, \cdots, x_n)$, $(1 \leq k \leq n)$ for n point masses is considered perfectly acceptable as a representation of mechanical movements of objects in reality. While engineers do discuss extensively realistic forms of the force functions $F_k(\mathbf{x})$ used in the equation, *e.g.,* constitutive relations in mechanical engineering and molecular force fields in computational biochemistry, one has a sense of confidence that whatever he or she learns from solving this mathematical equation is a study of the motion of objects in the real world, idealized as point masses. *Even when $F(\mathbf{x})$ is not realistic, it could be in a designed laboratory experiment.* Unfortunately, in mathematical biology, one rarely has such confidence that a set of equations is a legitimate representation of reality. There, one worries about imperfect models instead of believing an idealized representation. This difference is significant: for a mathematical model without *a priori* validity to be meaningful, its value has to rely on fitting some data from real laboratory measurements; this often relegates mathematical investigations to an auxiliary importance.

Is there also a mathematical theory that could rise above the level of models *a posteriori* for real data and become a nonmechanical representation of biological reality? In the present text, we present a coherent theory for the kinetics of a population of individuals, each with infinitely many degrees of freedom within, and each situated in a complex environment that consists of others. Chemistry has taught us that even though each and every individual molecule has its own "individualism",[1] they can and should be "grouped" into "chemical species" and studied statistically. Indeed, the field of macroscopic chemical kinetics became a part of *exact science* precisely because the law of large numbers, *e.g.,* Avogadro's 6.022×10^{23}, is at play.

Probability is the new logic of science;[2] there are profound consequences of this realization. Statistics is a tool for making discoveries in terms of probability from data, and **dynamics** is a language for narrating exact science with causality: the Newtonian world view is a clockwise machine. When a many-body system became too complex, fluid mechanists stopped following the position of each and every point mass in aerodynamics and hydrodynamics, and instead resorted to just "counting"

---

[1] P. G. de Gennes (1997) Molecular individualism, *Science*, **276**, 1999–2000.

[2] E. T. Jaynes (2003) *Probability Theory: The Logic of Science*, 1st ed., Cambridge Univ. Press.

the number of point masses at different locations and with certain velocities. Similarly, physicists also invented the notion of "heat" to represent the motions of point masses in probabilistic terms. Through the conception of heat, Newtonian mechanical motions are no longer conservative but become ***dissipative***, which predicts convergence with attractors in dynamical terms. This physicist's worldview, however, remains in sharp contrast with the idea of ***biological diversity***. The readers should keep these important issues in mind at all times while working through this book.[3]

Mathematical biology is a very active and fast-growing interdisciplinary area in which mathematical concepts, techniques, and models are applied to a wide variety of problems in biological and medical sciences. Many biological structures and processes are intrinsically stochastic. This text introduces its readers to a coherent mathematical representation and a set of probabilistic techniques for analyzing stochastic dynamic models in biology and medicine. The biological background for each problem will be described, and mathematical models will be developed following a unified set of principles and then analyzed. Finally, the biological implications of the mathematical results will be interpreted. Biological topics will include gene expression, biochemistry, cellular regulation, and population biology. The mathematical skills are based on differential equations, nonlinear dynamical systems theory, and Markovian stochastic processes, with either discrete or continuous state spaces. The latter is also known as a Feller process.

Stochastic dynamics is actually a new way of seeing and thinking of the world; it has wide applications in all fields of science, engineering, and even life. Different from the traditional deterministic approach, the goal of stochastic analyses is to obtain useful information from seemingly random behavior, and stochastic models provide insights by establishing relationships between intrinsic randomness within the tendency of individuals and the emergent behavior of a population or interacting populations.

The readers are expected to understand the basic concepts of the theory of probability. This text will teach advanced mathematical techniques and introduce several classes of important stochastic processes. These processes will apply to various biological problems. The presentation is focused almost exclusively on chemical and biochemical kinetics, but the same mathematical theory and representation can be applied to most other areas that involve the dynamics of interacting populations of individuals.

Indeed, the notion of "identity" has become one of the most significant concepts in postmodern society. In a sense, chemists have been the masters of "identity politics".

———————— * ———— * ———— * ———— * ————

The mathematical statements in the current book are not always rigorous according to the standard of pure mathematics. Certain technical or specific conditions are sometimes omitted. However, there are always rigorous mathematical theories behind all the mathematics we use and the results we state in the book. We have striven to strike a balance: on the one hand, we try to avoid mathematical incor-

---

[3] J. J. Hopfield (1994) Physics, computation, and why biology looks so different, *Journal of Theoretical Biology*, **171**, 53–60.

rectness, while on the other hand, we try not to get lost in rigor that could distract from the truly important applied nature of the subject. We hope this book will open a window for scholars and students with rigorous mathematical taste to see how stochastic mathematics, as a growing component of applied mathematics, plays a central role in understanding chemistry and biology.

The materials covered in the present text are quite unique. Thus, there are many different ways the book can be used. While it was written as a text for graduate students with a strong mathematical bent, some of the contents are sufficiently new as a monograph. The chapters from Parts I and III had been used as complementary material to standard text [185] for a mathematical biology course at the University of Washington (UW) and as a reference for courses on the mathematical theory of cellular dynamics taught by both authors at UW and Peking University. We expect that the book can also be included as teaching material for *Biophysical Chemistry* or *Systems Biology* or be useful for teaching *Theoretical Biophysics*, supplementing for example [187, 26] in a physics department.

Hong Qian
Department of Applied Mathematics
University of Washington, U.S.A.

Hao Ge
Beijing International Center for Mathematical Research (BICMR)
and Biomedical Pioneering Innovation Center (BIOPIC)
Peking University, China

Acknowledgments

A significant portion of this book was adapted from the joint work of the two authors' close collaborations since 2008, after the late Professor Min Qian of Peking University had introduced us to each other. We dedicate this book to him as a living memory. Both of us have also been extremely fortunate to have a longtime friendship with Sunney Xie, who shared his wisdom and provided encouragement, both intellectually and careerwise. We thank Michael Mackey for being a persistent as well as an understanding editor who has guided us through the writing process.

H. Q.: I would like to express gratitude to Professor Elliot Elson, my dissertation advisor, for showing me the stochastic fluctuating world he had pioneered within biophysics. I am grateful to David Eisenberg and Ka-Kit Tung, who nurtured me during critical times in my career, David for taking me into his research group at UCLA with funding, while at the same time giving me the freedom to pursue my own research interests; and K. K. for first reaching out and then providing vast amounts of valuable advice as the department chair at UW. I acknowledge the comradeship of Ping Ao, Shu-Nong Bai, Sui Huang, Jie Liang, Yuhai Tu, Jin Wang, Wenning Wang, and Michael Q. Zhang, through challenging discussions and occasional coauthorships, in sustaining a difficult intellectual enterprise in the gray zone between biological physics and physical biology. I thank all my friends and colleagues in the field of mathematical biology, too many to list them all, from whom I have learned the trade. Finally, I am grateful to Attila Szabo for his friendship and healthy skepticism and for repeatedly telling me to write a book; I hope this product is not a disappointment.

Madeleine Yue Dong, my wife and professor of History, is responsible for opening my eyes to see the world outside the narrow scientific perspectives of physics and biochemistry. Learning through her made me keenly aware of the need to understand the logic of biology and human societies beyond the traditional Newtonian perspective. It is her confidence in me, her shouldering of the lion's share of the burden of a family in a modern society, and countless support, emotionally and in other unthinkable ways, that has made the best of my work possible. Our daughter, Isabelle Yu-Heng, who entered college in the midst of a worldwide emergency, has been a constant source of inspiration and joy in the family. Her independence, curiosity, and convictions on all things that matter have provided her immigrant father a true liberal education about both the Old and New Worlds.

H. G.: Professor Min Qian, my doctoral thesis advisor, showed me a passion for science and led me into the fields of stochastic thermodynamics and mathematical biophysics. I would like to express my gratitude to Professors Xiaoliang Sunney Xie, Gang Tian, Weinan E, Pingwen Zhang and Dayue Chen for their continuous support during the past ten years since I was recruited at Peking University with a joint position in BICMR and BIOPIC. I thank Professors Minping Qian, Guanglu Gong, Yanxia Ren, Yong Liu, Daquan Jiang, and Fuxi Zhang in the Department of Probability and Statistics, who had been my teachers and are now my colleagues, for their advice and friendship. I am also grateful to my former colleagues at Fudan University for their help and nurturing at the very beginning of my career. Finally, I thank all my collaborators and close colleagues, particularly Fan Bai, Shunong Bai, Ken Dill, Jianfeng Feng, Yiqin Gao, Yanyi Huang, Tiejun Li, Chen Jia, Wei Lin, Jianfeng Lu, Steve Pressé, Xiaodong Su, Yujie Sun, Fuchou Tang, Lei Zhang, Yunxin Zhang, Zhihua Zhang, Xinsheng Zhao, Ziqing Zhao and Da Zhou, among many others.

I'm deeply grateful for everything that my father, 葛云保, and my mother, 丁梅华, have done for me. They raised me and taught me the basics as well as the principles of life. Daily life discussions with my parents, especially my father, inspired my interests in science and the spirits of critical thinking and dialectical reasoning. My eight-year-old son, 葛一川, constantly brings me joy, happiness, and responsibility.

Contents

Acronyms

ADP	adenosine diphosphate
ATP	adenosine triphosphate
cdf	cumulative distribution function
CME	chemical master equation
CSTR	continuous stirred-tank reactor
DNA	deoxyribonucleic acid
DGP	Delbrück–Gillespie process
EV	expected value
FTNS	fundamental theorem of natural selection
LDT	large deviations theory
LMA	law of mass action
MD	molecular dynamics
MFPT	mean first-passage time
MM	Michaelis–Menten
NESS	nonequilibrium steady (or stationary) state
NET	nonequilibrium thermodynamics
NSDS	nonlinear stochastic dynamical system
ODE	ordinary differential equation
PDE	partial differential equation
pdf	probability density function
pmf	probability mass function
RNA	ribonucleic acid
RV	random variable

SDE	stochastic differential equation
WKB	Wentzel–Kramers–Brillouin

Chapter 1
Introduction

1.1 Why and What is a Mathematical Science

Mathematics is used as a tool for quantifying observations as well as the language for narrating the laws of physical world. The successes of mathematical theories in astronomy and classical physics and for describing subatomic and cosmological phenomena are so impressive that they have become the envy of many other intellectual endeavors, such as economics [180] and sociology [260]. One of the reasons that biological research is particularly challenging and exciting is precisely because it stands at the border between the hardcore, exact sciences and the elusive sciences of human activities. Molecular cellular biology, furthermore, is characterized as resembling both the physical sciences, via its biochemistry aspects, and the social sciences, through its evolutionary and population biology aspects.

There are two very different modes for applying mathematics in molecular cellular biology. First, one can build mathematical models for biomolecular systems and cellular processes based on very concrete *laws* of physics and chemistry, e.g., the law of conservation of energy/mass, Boltzmann's law of statistical equilibrium, Gibbs' theory of thermochemistry, and the law of mass action for reactions in ideal solutions, and perform mathematical analyses deductively. Second, one can attempt to develop mathematical formulae based on measurements inductively, often with the only justification being the ability of these formulae to fit the experimental data parsimoniously. Both modes are crucial aspects of the mathematical sciences, and clearly they are intimately related and mutually beneficial: a new formulation that fits a great deal of data will eventually becomes a *nature law*; the discovery of a

H. Qian, H. Ge, *Stochastic Chemical Reaction Systems in Biology*,
Lecture Notes on Mathematical Modelling in the Life Sciences,
https://doi.org/10.1007/978-3-030-86252-7_1

concrete mechanism is based on existing natural laws combined with some additional assumptions that explain certain behaviors; and the observations of deviations between data and mathematical predictions leads to the development of *novel mechanisms* for a phenomenon within given laws or a clarification of the *conditions* for a given law and the promotion of the birth of new laws.

Mathematical rigor allows one to provide precise prediction(s). Comparing mathematical predictions with experimental data is key to developing and improving models of physical and biological processes. In mathematical modeling of high-throughput (big data) biology, the validity of a mathematical model is judged by its predictive power. In fact, this is the nature of most statistical modeling. However, mathematical models of cellular processes and biomolecular systems can also be developed based on the well-established laws from physics and chemistry. Such models, therefore, have certain validity *a priori*. Although usually such models do not fit the particular data exactly, and of course the detailed assumptions for many specific expressions in the models cannot be completely correct, physiochemical-based models deserve analysis in their own right. A mechanistic model usually provides qualitative behaviors, such as bifurcation patterns, rather than a very precise matching to the quantitative output of real systems: it explains a *phenomenon*. After this, the parameters in the model are fine tuned to fit the observed data. More ideally, based on the predicted, qualitative relations, one should be able to design a laboratory system to match the mathematical model. Keep in mind that even when a mechanistic model does not fit the particular set of data at hand, *one could nevertheless still use it to learn about the real world*. This last mode of thinking is precisely what lies behind the recently emerging area called *synthetic biology*. It is this unique feature of mathematical modeling in molecular cellular biology that distinguishes our approach from many other texts on mathematical biology and modeling.

The two very different modes of mathematical modeling discussed above, (*a*) representing scientific data in terms of mathematical formula or equations, and (*b*) describing/explaining a system's behavior (natural or engineered, physical or biological, electronic, chemical, economical, social, etc.) based on existing, established formula and equations, can be fittingly called **data-driven modeling** and **mechanistically derived modeling**, respectively; the dichotomy can be traced back to the Kantian synthetic-analytic distinction [33]. According to Karl Popper (1902–1994) and his philosophy of science, the only legitimate scientific activity is the falsification of a hypothesis, that requires first the formulation of the hypothesis, which often involves just looking for patterns in the data, e.g., mode (*a*), and sometimes involves proposing a mechanism (mode *b*). Then, (*b*) requires a mathematical step to derive rigorous predictions from the mechanism, which is a form of logical, or mathematical, deduction.

1.2 Mechanistic Models Constitute Understanding

Readers might have already noticed that data-driven modeling, the inductive formulation of mathematical descriptions of the real world, constitutes the chief objective of statistical inference. In contrast, mechanistically derived modeling, the deductive derivation of mathematical mechanisms of either observed or anticipated phenomena, is one of the main activities of traditional applied mathematics, *à la* Chia-Chiao Lin (1916–2013) [171, 172]. The present text in part follows the latter tradition. We shall elaborate a bit further the significance of this approach.

1.2.1 The Role of Natural Laws and Scientific Hypotheses

Establishing an equation should never be taken lightly. Note that $1.0001 \neq 1.0000$! Therefore, writing down an equality has to be based on either known natural laws or a precise scientific hypothesis. When an assumption is made, it eventually requires verification through satisfactory predictions. We leave it open for the question that what a natural law is, see [72] for more discussion.

Let us take a problem from mechanics as an example. We all learned about harmonic oscillation from high school physics and freshman calculus:

$$m\frac{d^2x}{dt^2} = -kx, \tag{1.1}$$

where m is the mass of an object, x is its position, defined precisely as the co-ordinates of its center of mass, and $-kx$ represents the restoring force from a spring. One should be consciously aware that this equation is really a combination of two different physical laws: Newton's second law of motion, which states that $md^2x/dt^2 = F(x)$, where $F(x)$ is the total force acting on the object, treated as a point mass, when it is located at position x, and Hooke's Law, $F(x) = -kx$, which relates the force and the length of a simple spring.

There is an essential difference between Newton's Law and Hooke's Law. The difference is that the former is applicable under a very wide range of conditions, and the latter has only very narrow applicability. In fact, one can almost say that "the former is always true while the latter is never true." Therefore, modelers working on mechanics, including almost all mechanical engineers, are constantly changing and improving the right-hand side of Eqn. (1.1) to fit the real world. We also note two important features: first, Newton's equation alone, though very important, does not really tell us much about how an object moves; second, for different materials and circumstances, the right-hand side of Eqn. (1.1) will be different. It has to be studied case by case.

With this newfound understanding of the simple mathematical model in Eqn. (1.1), we realize that all mathematical models in ecology and population dynamics are also based on two parts: a balance law for counting the number of individuals in each species, and a relation between birth (death, and migration) rates of the species and their population sizes.

The mathematical models for cellular biochemical reaction systems follow the same spirit: they are based on the number of molecules of each and every biochemical species and the *rate laws* that relate the rates of chemical reactions and the concentrations of the species. The law of mass action is one widely used example of a rate law. When the spatial movements, e.g., transport, of the biomolecules are also important, then laws that relate transport fluxes and the spatial distributions of concentrations will also be included.

Scientific discoveries are often based on experimental observations and measurements. However, discovery does not have to be based only on experimental science. In the framework of mathematical modeling we discussed above, mathematical science leads to discoveries by the creation of novel formulae that fit large sets of data, the proposal/derivation of novel mechanisms based on existing laws of nature, or the discovery of unknown conditions for an existing natural law. One can learn about nature through mathematical studies.

1.2.2 Dynamics and Causality

The notion of "dynamics" holds a prominent place in science because of its close relationship to *causality*. The present text focuses on the dynamic aspects of molecular cellular biology. There are many other areas, such as macromolecular structures, physiology, ecology, and evolutionary biology, that all involve mathematical theories and modeling: see [240, 17, 144, 185, 151, 66].

Understanding a biological system in molecular terms is in the domain of biochemistry and molecular biology. To understand biochemistry and molecular biology in mathematical terms first requires knowing the fundamental laws and equations that govern the behavior of molecules and chemical reactions. There are essentially two major areas of concern: (1) the biomolecular dynamics and functions of individual molecules (often macromolecules such as DNA, proteins, and their complexes) in terms of their atomic structures and the laws of molecular and submolecular physics; and (2) the dynamics of a system of biochemical reactions in space and time in terms of chemical transformations and interactions between molecules. Generally, a complete understanding of the latter requires a rather thorough understanding of the former. Amazingly, however, one can actually understand the latter nearly independently of the many details of the former *phenomenologically*. This observation is known as the hierarchical structure of science [7], or modular cell biology [113].

We shall not concern ourselves with (1), which is the subject of *molecular biophysics*, for which there are many excellent texts [14, 49, 240]. To approach (2), we consider biochemical systems with spatial homogeneity; in chemical engineering terms, the reaction tank is under rapid stirring. In this case, the fundamental equations are based on two laws: the conservation of atoms and the law of mass action or its variations. We shall illustrate all this starting with the simplest chemical reactions possible.

1.3 What Do We Know About Biological Cells?

1.3.1 Basics of Chemical Reactions

An elementary reaction, as the name implies, is the simplest reaction mechanism in which several chemical substances directly interact with each other through collisions and form a product, or products, with a single recognizable event. The elementary reaction is the most basic step that constitutes a reaction; all nonelementary reactions can be understood as combinations of multiple elementary steps. In biochemistry, reaction schemes generally represent the molecular relationship between reactants and products, and they are not always expressed in elementary reactions, such as with photosynthesis

$$6CO_2 + 6H_2O \rightarrow C_6H_{12}O_6 + 6O_2.$$

Elementary reactions can be divided into three categories based on the number of molecules in the reactants: unimolecular reactions, bimolecular reactions and trimolecular reactions. Most elementary reactions are uni- or bimolecular reactions, and no elementary reaction with more than three reactant molecules has been observed. The number of reactant molecules defines the order of the chemical reaction. For example, unimolecular reaction $X \rightarrow Y$ is first-order, and $A + B \rightarrow C$ is a second-order bimolecular reaction.

In principle, all chemical reactions are reversible, i.e., the reactions in the two opposite directions are both possible. Most of the biochemical reactions in a living body are extremely slow, and they can only occur in a reasonable time under the catalysis of enzymes. Thus they are called catalytic reactions.

1.3.2 Cells: The Proteins, DNA and RNA Within

The basic unit of life is the cell. While cells of lower organisms have no nuclei (prokaryotes), the cells of higher organisms are eukaryotic, containing a nucleus, cytoplasm, and lipid bilayer membrane. There are many organelles, such as the endoplasmic reticulum (ER), ribosome, Golgi apparatus, mitochondria, chloroplast, etc., in the cytoplasm. Organisms in the bacterial and archaeal domains are com-

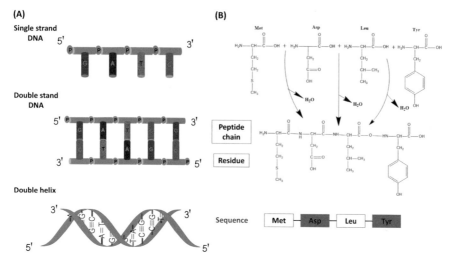

Fig. 1.1 (A) DNA is a double-stranded helical molecule, with each strand consisting of a polynucleic acid chain that is formed from a sequence of nucleotides. There are only four naturally occurring nucleotides with different bases in DNA: adenine (A), cytosine (C), guanine (G) and thymine (T). The two strands in opposite directions are held together through the pairing of A to T and of G to C. (B) A protein is a polypeptide chain that is formed from a sequence of amino acids. There are many other molecular players involved in this chemical process in a cell. A ribosome is the macromolecular machinery that catalyzes protein synthesis, also called mRNA translation. There are twenty naturally occurring amino acids in cells. See the text for more details.

posed of prokaryotic cells; protists, fungi, plants and animals are composed of eukaryotic cells.

The fundamental substances in cells are proteins that consist of polyamino acids, polynucleic acids that include deoxyribonucleic acid (DNA) and ribonucleic acid (RNA), lipids, polysaccharides, and many other kinds of molecules. As biopolymers, protein polypeptide chains are folded into well-defined, often unique three-dimensional structures at room temperature and physiological pH and salt concentrations. There are 20 types of naturally occurring amino acids that form the different residues along a peptide backbone. The three-dimensional structure is believed to be essential in the biological function of a protein. However, in recent years, a growing list of *intrinsically disordered proteins* has been discovered [53].

Both DNA and RNA are composed of nucleotides, and there are five kinds of nucleotides, called bases: A (adenine), T (thymine), C (cytosine), G (guanine), U (uracil), with the last only appearing in RNA, replacing the T in DNA. Within the two strands of nucleic acid in a DNA double helix, the corresponding nucleotides are paired, A with T and C with G, and held together by hydrogen bonds. The fact that chemical structure of base pairing within the double helix forms the mechanism of the genetic heredity of life is one of the greatest discoveries in science. In eukaryotic cells, the DNA double helix is typically wrapped around histone proteins, forming nucleosomes. The nucleus of each eukaryotic cell contains several chromosomes, composed by chains of nucleosomes. At different stages of a cell's life, the spatial arrangement of the chromosomes can be either tight or open.

1.3.3 The Central Dogma of Molecular Biology

One of the greatest achievements in the field of biology in the 20th century was the discovery of the central dogma of molecular biology, following the discovery of the DNA structure discussed above. It indicates the direction and way by which genetic information is transmitted in molecular terms. This central dogma can be simply expressed as follows: The processes by which a cell produces the protein that a gene encodes is called the expression of the gene; that is, "DNA dictates the synthesis of RNA, RNA dictates the synthesis of proteins, proteins in turn are the essential players of the previous two processes and are important in the self-replication of DNA", or simply, "DNA→ RNA → protein". The entire affair is divided into three major parts: transcription, translation and DNA replication.

The RNA transcribed from DNA is also called messenger RNA (i.e., mRNA). The nucleotide base sequence of mRNA encodes the amino acid sequence of the corresponding protein molecule. In its coding region, every three adjacent bases form a "codon" for a specific amino acid. For example, AUG is the codon for me-thionine, and CUA, CUT, CUC, CUG are the codons for leucine. Next, with the participation of ribosomes, mRNA is translated into the corresponding polypeptide chain, which is folded and assembled into a functional protein molecule. After the proteins are synthesized, they are transported to where they are needed: in cytosol, organelles, or the nucleus, etc.

1.3.4 Cellular Regulation

Through the intricate molecular interactions between DNA, or RNA, and the protein complexes that are involved in transcription or translation, a genome also contains regulatory information at the level of gene expression, *e.g.*, the amount of mRNA and corresponding protein being made, and on the timing of the gene expression. The binding of different subsegments of a gene sequence with different proteins or noncoding RNAs form regulatory combinations that control gene expression. The proteins involved often need to undergo various chemical modifications, catalyzed by other enzyme molecules that activate or inhibit themselves. The most common protein modifications are phosphorylation, dephosphorylation, acetylation, ubiquitination, and methylation.

The phosphorylation process, catalyzed by a specific protein kinase, transfers a phosphate group (P) from ATP to proteins; another enzyme called protein phosphatase is responsible for catalyzing the dephosphorylation of proteins. The phosphorylation, or more generally, any chemical modification, of a protein determines the structure and thus the activity of the protein, affects the process of intracellular information transmission, and allows the cell to respond appropriately to external stimuli.These macromolecules, through a network of biochemical reactions, become the "players" of a living community.

Unicellular organisms utilize a gene-protein interaction network and various feedback regulations to sustain the activities of life and adapt to changes in the environment. Multicellular organisms are no different, but involve an additional, higher-level organization of different cell types into "cellular societies", which also rely on communication and signaling between cells through mechanical, chemical, and electrical means to coordinate the behavior of individual cells in the organisms.

The main regulatory activities in the cell are:

(1) Cell cycle: DNA replication-related gene expression and cell entrance into the stages of division and proliferation;

(2) Cell differentiation: selective gene expression and irreversible cell differentiation into mature cells with specific phenotypes and corresponding functions;

(3) Cell survival and apoptosis;

(4) Metabolic regulation: control of the metabolism of substances and energy of cells by regulating the activities of enzymes related to metabolism;

(5) Cell functions: muscle contraction and relaxation, immune cell production of antibodies, and epithelial cell release of glandular secretions.

1.3.5 Cell Biology As a System of Biochemical Reactions

The forgoing sections paint a clear picture of biological cells as highly complex systems of biochemical species and reactions. Fortunately, chemistry allows one to present any complex biochemical reaction network uniformly in a conceptual manner: we can use two very large integers to represent all possible N molecular species in a cell, $X_1, X_2, \cdots, X_N$, and all possible M reactions, as

$$v_{\ell 1}^+ X_1 + v_{\ell 2}^+ X_2 + \cdots + v_{\ell N}^+ X_N \longrightarrow v_{\ell 1}^- X_1 + v_{\ell 2}^- X_2 + \cdots + v_{\ell N}^- X_N, \qquad (1.2)$$

in which $\ell = 1, 2, \cdots, M$. The two sets of nonnegative integers v^+'s and v^-'s are called stoichiometric coefficients in elementary chemistry [222]. They represent the number of various molecules involved in a particular chemical reaction. Note that "chemical species" has a very fluid definition: for example, it could represent the "kinase B inside the nucleus", thus describing the spatial information of the molecules.

It is this uniform representation of biochemical reactions in cellular biology that gives the present text the possibility of presenting a unified mathematical treatment of biological subjects.

1.3.6 Chemical Reactions in Terms of Molecular Processes and Beyond

Chemical reactions describe molecular transformations; and molecular changes are usually called chemical reactions. Therefore, it may seem that the title of the present subsection is redundant. Instead, we would like to define *reaction processes* abstractly as the system defined in Eqn. (1.2). Then, in this abstract sense, when the symbols $X_1, \cdots, X_N$ represent molecules, these are chemical reactions; but when symbols represent biological organisms in an ecological system with interacting populations, then (1.2) describe the ecological processes of birth, death, predation, competition, viral infection etc. The system in (1.2) provides a uniform representation not only of chemical reactions but also of many other biological dynamical systems in which the main variables are the number of individuals of various "species". The theory of stochastic *population processes* pioneered by D. G. Kendall (1918–2007), J. E. Moyal (1910–1998), and others [146, 183], as well as the notion of *elementary processes and canonical theory* formulated by J. E. Keizer (1942–1999) [145, 77], can be fittingly applied. The mathematical foundation connecting stochastic population processes with deterministic nonlinear dynamics systems widely employed in mathematical biology can be found in T. G. Kurtz's work [154, 65].

For molecular reaction processes, each transformation depicted in Eqn. (1.2) can be further investigated and understood at the level of atoms and the bonds, covalent or noncovalent, that hold them together. This is the subject of biochemistry proper; the science of complex systems and matters can be classified into a hierarchy [7]. In Chapters 8, 15 and 16, we shall discuss the relevance of the core materials presented in the present text as a possible paradigm that connects different levels in the hierarchy.

Part I
Essentials of Deterministic and Stochastic Chemical Kinetics

Part I mainly discusses mathematical methods. In terms of deterministic mathematics, we expect the readers to have a certain proficiency in calculus, introductory ordinary differential equations (ODEs), linear algebra, and some exposure to nonlinear dynamical systems approaches to ODEs: phase portraits, planar systems, Lyapunov functions, and linear stability analysis. Chapter 2 applies this type of mathematics to setting up rate equations for deterministic chemical reaction kinetics followed by basic analyses. We do not expect the readers to have similar levels of mathematical background in stochastic mathematics, except the knowledge gained from an elementary course. Therefore, Chapter 3 starts almost from the very beginning, although with a rapid pace, and touches upon all the basic concepts that are needed for the rest of the book.

Chapter 4 is somewhat unique: it introduces the notion of large deviations, a rather advanced concept from the theory of probability. In traditional teaching, this material has to come after one has learned the laws of large numbers and the central limit theorems. We are experimenting with a very applied approach to the subject: using a few simple and explicit examples, we first illustrate the idea of exponentially small asymptotics. We then note that the theory of large deviations has a theoretical objective, proving the existence of the exponential asymptotic form, called the large deviations principle (LDP), and a more practical objective of actually deriving the functional form of the asymptotics, known as the large deviation rate function (LDRF). If one assumes the former to be the so-called WKB (Wentzel, Kramers, and Brillouin) ansatz, then the formal computations of the LDRF, or at least derivation of the differential equation satisfied by the LDRF, is a rather straightforward task in applied mathematics. It turns out that one can go quite far with this approach. Section 4.5 contains a very recent result on $\vec{\lambda}$-surgery. This chapter culminates in Kramers' stochastic diffusion theory for the rate of an individual unimolecular reaction: even though a single macromolecular transition in an aqueous solution, e.g., a protein conformational change, is an extremely complex affair in atomic physics, Kramers' theory shows that it can be represented by a single Markov transition at a different time scale with one parameter.

Chapter 5 uses the mathematics from the previous chapters to lay the foundation for a probabilistic description of chemical kinetics, tracking reactions as random events one by one, and counting individual molecules one at a time. In 1940, while

H. A. Kramers (1894–1952) published his stochastic diffusion theory for the rate of an individual reaction [152], M. Delbrück (1906–1981) developed a stochastic formulation for *a system of* many different chemical reactions, each with a given rate law [47]. Section 5.5 presents the Delbrück–Gillespie process (DGP), which represents stochastic chemical reaction systems. Kramers' theory applies to a single molecule in terms of a collection of atoms; the DGP applies to single-cell biochemistry in terms of a collection of stochastic elementary reactions, i.e., a discrete jump with an exponentially distributed waiting time.

We emphasize that a stochastic Markov process is a single mathematical object with two complementary representations by either a differential equation for its probability distribution or a stochastic trajectory. Therefore, a DGP has a chemical master equation as well as a stochastic trajectory that can be expressed in terms of Poisson jump processes and simulated according to the Gillespie algorithm [104]. This is analogous to the Fokker–Planck equation and the stochastic differential equation, both for a given stochastic diffusion process in continuous space.

Through studying this Part, the readers will gradually appreciate one of the most important insights from the understanding of chemical reactions: while molecules involved in a reaction can be highly complex in terms of the atoms within, *a reaction* can be statistically represented by a simple mathematical model, called rate equations, with a few parameters. Therefore, to a large extent, the more detailed submolecular physics does not have a significant role *per se* in understanding cellular behavior; probabilities and statistics help reduce vast details into emergent compact descriptions.

Chapter 2
Kinetic Rate Equations and the Law of Mass Action

In modern biology, one cannot discuss cells without talking about biochemistry and molecular reactions in an aqueous environment. It can be quite crowded inside a cell, but we shall not be too worried about this fact for now. To a casual observer, chemistry is dynamic; furthermore, since all the molecules are moving under the influence of thermal agitations from the temperature of the room or body, chemistry has to be stochastic. Paradoxically, however, to many professional scientists who study biology from a molecular standpoint, a molecule is merely a static collection of atoms with an intricate architecture. Furthermore, the behavior of a cell is determined simply by what is there and what is not in terms of macromolecules, usually in a completely deterministic way without dynamics and without probability.

These assumptions have been challenged in recent years owing to the observations and multiplex measurements on single cells and single molecules. These studies have revealed a highly dynamic and statistical picture of the cellular and subcellular world: the terms *heterogeneity* and *variation* have become widely used in this context.

In a nutshell, a fundamental equation in biology is the balance of the "head counts of individuals within populations", which can be biological organisms in population dynamics, viruses and people in infectious disease epidemics, and cells and molecules in cellular and molecular biology.

© The Author(s), under exclusive license to Springer Nature Switzerland AG 2021
H. Qian, H. Ge, *Stochastic Chemical Reaction Systems in Biology*,
Lecture Notes on Mathematical Modelling in the Life Sciences,
https://doi.org/10.1007/978-3-030-86252-7_2

2.1 Reaction Kinetic Equation: Conservation of Molecules and the Law of Mass Action

We start with two simple examples. First, recall the simplest linear isomerization reactions. In biochemistry, a stereoisomerization reaction of a protein is also called a conformational change:

$$X \longrightarrow Y. \tag{2.1}$$

Second, a nonlinear combination reaction, also called an association reaction or binding in biochemistry, involving at least one protein as one of two reactants:

$$A + B \longrightarrow C. \tag{2.2}$$

Let $J_1(t)$ and $J_2(t)$ be the respective instantaneous fluxes of the two reactions at time t. The flux of a reaction sometimes is also called the instantaneous rate of the reaction. The implicit assumption for the concept of a "flux", or a "rate", is that each reaction is an event that occurs instantaneously in time, leading to a decrease in the number of species X (or A and B) and an increase in the number of species Y (or C) by 1. The flux is the reciprocal of the mean time between two consecutive events or the number of events per unit time. Macroscopically, the flux can be set in units of concentration per unit time. The flux J is the number of occurrences of a reaction, measured in terms of the unit that matches the concentration of the reactants and products of the reaction.

In terms of J, the counting of the number of molecules for the unimolecular reaction in (2.1) yields

$$\frac{dc_Y(t)}{dt} = -\frac{dc_X(t)}{dt} = J_1(t), \tag{2.3}$$

and that for the bimolecular reaction in (2.2) yields

$$\frac{dc_C(t)}{dt} = -\frac{dc_A(t)}{dt} = -\frac{dc_B(t)}{dt} = J_2(t), \tag{2.4}$$

where $c_A(t)$, $c_B(t)$ and $c_X(t)$ denote the concentrations of the chemical species A, B, and X, respectively, at time t.

While Eqs. (2.3) and (2.4) are absolutely correct as long as the concentrations of c can be treated as a smooth function of time t, they are not very useful until J is specified in terms of the concentrations of the chemical species involved. In general, $J_1(c_X(t), c_Y(t), t)$ is a function of the number of all participating chemical species and many other factors; similarly, J_2 is a function of c_A, c_B, c_C, and t. The law of mass action (LMA) states that the instantaneous rate of a reaction, i.e., the number of molecules transformed in unit time and per volume is proportional to the concentrations of all the reactants: $J_1 = k_1 c_X$ for the reaction in (2.1) and $J_2 = k_2 c_A c_B$ for the reaction in (2.2). k_1 has the dimension of $[\text{time}]^{-1}$, and k_2 has the dimension of $[\text{time}]^{-1}[\text{concentration}]^{-1}$. $J_1(c_X)$ is linear and $J_2(c_A, c_B)$ is nonlinear. k_1 and k_2 are called the *rate constants* of the corresponding reactions. For a dimerization reaction $X + X \longrightarrow Z$ with rate constant k_3, the rate of reaction should be $J_3(c_X) = k_3 c_X^2$.

Combining the conservation of molecules with the LMA, we have, for the unimolecular reaction, i.e., isomerization, in (2.1)

$$\frac{dc_Y}{dt} = -\frac{dc_X}{dt} = k_1 c_X \tag{2.5}$$

and for the bimolecular reaction in (2.2)

$$\frac{dc_C}{dt} = -\frac{dc_A}{dt} = -\frac{dc_B}{dt} = k_2 c_A c_B. \tag{2.6}$$

$J_1(c_X)$ and $J_2(c_A, c_B)$ are also called the instantaneous fluxes of the chemical reactions. Hence, the conservation of molecules relates the time derivative of concentration(s) to the reaction flux(es), and the LMA relates the flux(es) to the concentration(s). This establishes a system of ordinary differential equations. For the abovementioned dimerization reaction $X + X \longrightarrow Z$, one has

$$\frac{dc_X}{dt} = -2J_3(c_X) \quad \text{and} \quad \frac{dc_Z}{dt} = J_3(c_X). \tag{2.7}$$

The factor of 2 is a stoichiometric coefficient; it relates the reaction flux J_3 and the rate of change of $c_X(t)$.

A comparison between the chemical kinetic equation with Newton's equation of mechanical motion is in order.[1] We recognize that Eqs. 2.3 and 2.4, *e.g.*, the conservation of molecules, merely provide a definition for the notion of an instantaneous reaction flux. They are like the equation of mechanical motion $d^2 x(t)/dt^2 = a(t)$, which defines the notion of acceleration. However, these equations are not very useful until each and every reaction flux, as a function of its reactants' concentrations,

---

[1] It is interesting to note that the triplet "force, mass, and change in velocity" can be replaced by "flux, stoichiometry, and change in concentration".

is given, just as $d^2x(t)/dt^2 = F/m$ is not very useful until the functional form $F(x)$ is provided. The latter is called the *constitutive relation* in mechanical engineering and the molecular force field in computational molecular dynamic (MD) simulations. We also note that chemical kinetics should be compared with *kinematics* in mechanics: they are simply concerned with, respectively, the change in molecular numbers and the motion of objects in terms of speed without reference to the forces, or energy, that cause the motion.

2.2 Forward and Backward Reaction Fluxes J_+ and J_-, Chemical Potential and Boltzmann's Law

Ordinary differential equations such as (2.5) and (2.6), called *chemical kinetic rate equations*, are also widely found in population dynamics of other biological species. Chemical kinetics, however, owes its unique sophistication to the fundamental relations between kinetics and *chemical thermodynamics*, the theory that provides the notions of energy and force in the world of chemical species and their transformations. As we shall show, it is productive to draw a parallel between the kinematics and dynamics of mechanical motions on the one hand and kinetics and thermodynamics of chemical reactions on the other.

From atomic physics, we know that no chemical reaction is absolutely irreversible at finite temperature. Thus, in general we have

$$X \underset{k_{-1}}{\overset{k_{+1}}{\rightleftarrows}} Y, \tag{2.8}$$

and

$$A + B \underset{k_{-2}}{\overset{k_{+2}}{\rightleftarrows}} C. \tag{2.9}$$

In these cases, the conservation of molecule counts takes the forms of

$$\frac{dc_Y}{dt} = -\frac{dc_X}{dt} = J_{+1} - J_{-1}, \tag{2.10}$$

and

$$\frac{dc_C}{dt} = -\frac{dc_A}{dt} = -\frac{dc_B}{dt} = J_{+2} - J_{-2}. \tag{2.11}$$

If we further assume the LMA, then $J_{+1}(c_X) = k_{+1}c_X$, $J_{-1}(c_Y) = k_{-1}c_Y$, $J_{+2}(c_A, c_B)$ $= k_{+2}c_A c_B$, and $J_{-2}(c_C) = k_{-2}c_C$.

According to the theory of Gibbsian chemical thermodynamics, for molecular species X and Y with concentrations c_X and c_Y, the likelihood of X becoming Y, or Y becoming X, is determined by the *chemical potential* difference between them:

$$\mu_X = \mu_X^o + k_B T \ln c_X, \tag{2.12}$$

$$\mu_Y = \mu_Y^o + k_B T \ln c_Y, \tag{2.13}$$

where k_B is Boltzmann's constant, and T is the temperature in Kelvin (273.15 K $= 0°C$). We see that there are two terms that determine the chemical potential of a species: its concentration, and a term intrinsic to the molecular structure in an appropriate solvent condition, which is related to the internal energy (μ^o). When $\mu_X = \mu_Y$, there will be no net change in concentration for X or Y. This defines a *chemical equilibrium* between X and Y. Thus, we have:

$$\mu_X^o + k_B T \ln c_X^{eq} = \mu_Y^o + k_B T \ln c_Y^{eq}. \tag{2.14}$$

This yields

$$\frac{c_Y^{eq}}{c_X^{eq}} = e^{-(\mu_Y^o - \mu_X^o)/k_B T} = \frac{k_1^+}{k_1^-}. \tag{2.15}$$

For an ideal solution in which solutes are statistically independent, the proportions of the chemical species X and Y, i.e.,

$$\frac{c_X^{eq}}{c_X^{eq} + c_Y^{eq}} \quad \text{and} \quad \frac{c_Y^{eq}}{c_X^{eq} + c_Y^{eq}},$$

can be regarded as the probabilities of chemical states X and Y of a single molecule. Hence, the first equality in (2.15) is known as Boltzmann's law : the probability of a molecular system having energy E in equilibrium is proportional to $e^{-E/k_B T}$. The second equality is from the LMA, connecting equilibrium with kinetics. It also shows that no rate constants can be absolutely zero since there can be no infinite amount of energy.

Finally, it is easy to show that we have a relation, which actually is very general, between the chemical potential difference of a chemical reaction and its forward and backward fluxes [236, 16, 204]:

$$\Delta\mu = k_B T \ln \left(\frac{J_-}{J_+}\right), \tag{2.16}$$

where $\Delta\mu$ is the chemical potential *difference* of a reaction. For the reaction in (2.8), $\Delta\mu = \mu_Y - \mu_X$, and for the reaction in (2.9), $\Delta\mu = \mu_C - \mu_A - \mu_B$. The net flux in the reaction is

$$J = J_+ - J_-. \tag{2.17}$$

Therefore, Eqs. (2.16) and (2.17) together give us

$$\Delta\mu \times J \leq 0, \tag{2.18}$$

and the equality holds true if and only if $\Delta\mu$ and J are both zero: they become zero simultaneously.

What is the physical meaning of Eqn. (2.18)? Let us make an analogy between a chemical reaction system and an electrical circuit. We have that electrical current $\times$ voltage is the electrical power: the amount of energy consumed per unit time. This turns out to be the same for a chemical reaction and its $|\Delta\mu \times J|$.

The inequality in Eqn. (2.18) means that a chemical reaction can only dissipate heat; it cannot absorb thermal energy from a single temperature source and turn 100% of the heat into chemical energy, represented by the values of μ. This is a statement of the second law of thermodynamics.

Like mechanical energy, which consists of potential and kinetic parts, chemical energy is also intimately involved in our daily lives: when one goes to a supermarket and reads the amount of calories indicated for a food, you are learning about its chemical energy.

2.3 General Chemical Kinetic System in a Continuously Stirred Vessel

When a reaction vessel is continuously and rapidly stirred, one can safely assume that the concentrations of each and every chemical species within are spatially uniform. In this case, the concentrations of all the species satisfy the general kinetic equations.

The most general nonlinear chemical reaction system contains N distinct chemical species, $X_1, X_2, \cdots, X_N$, and M reversible chemical reactions:

$$v_{\ell 1}^+ X_1 + v_{\ell 2}^+ X_2 + \cdots + v_{\ell N}^+ X_N \underset{J_{-\ell}(\mathbf{x})}{\overset{J_{+\ell}(\mathbf{x})}{\rightleftharpoons}} v_{\ell 1}^- X_1 + v_{\ell 2}^- X_2 + \cdots + v_{\ell N}^- X_N, \qquad (2.19)$$

in which $1 \leq \ell \leq M$. The stoichiometric coefficients v_{ij}^+ and v_{ij}^- are nonnegative integers. In fact, most of them are zero since usually only a few chemical species are involved in a particular reaction. The stoichiometric coefficients relate species to reactions.

For an aqueous chemical solution in a continuous stirred-tank reactor (CSTR), the concentrations of the species at time t, $x_i(t)$ for X_i, satisfy the system of rate equations, the ordinary differential equations that describe the balance of all the molecular species, , *e.g.*, the conservation of molecules:

$$\frac{dx_i(t)}{dt} = \sum_{\ell=1}^{M} \left(v_{\ell i}^- - v_{\ell i}^+ \right) \left(J_{+\ell}(\mathbf{x}) - J_{-\ell}(\mathbf{x}) \right), \qquad (2.20)$$

in which $\mathbf{x} = (x_1, x_2, \cdots, x_N)$ denotes the concentrations of these chemical species and $J_{+\ell}(\mathbf{x})$ and $J_{-\ell}(\mathbf{x})$ are the forward and backward fluxes for the ℓth reaction.

Recall that from Section 2.1, the functional forms of $J_{\pm\ell}(\mathbf{x})$ are called **rate laws** in chemical kinetics. The LMA is one of the most widely used rate laws. It states that all the fluxes have a particular functional form

$$J_{+\ell}(\mathbf{x}) = k_{+\ell} \prod_{k=1}^{N} (x_k)^{v_{\ell k}^+}, \quad J_{-\ell}(\mathbf{x}) = k_{-\ell} \prod_{k=1}^{N} (x_k)^{v_{\ell k}^-}, \qquad (2.21)$$

in which $k_{+\ell}$ and $k_{-\ell}$ are the forward and backward rate constants for the ℓth reaction. Corresponding to the law of mass action, the chemical potential difference for the ℓth reaction in an ideal solution is

$$\Delta \mu_\ell = \sum_{k=1}^{N} \left(v_{\ell k}^- - v_{\ell k}^+ \right) \mu_k = \sum_{k=1}^{N} \left(v_{\ell k}^- - v_{\ell k}^+ \right) \left(\mu_k^o + k_B T \ln x_k \right), \qquad (2.22)$$

in which

$$\sum_{k=1}^{N} \left(v_{\ell k}^- - v_{\ell k}^+ \right) \mu_k^o = \ln \left(\frac{k_{-\ell}}{k_{+\ell}} \right). \qquad (2.23)$$

Therefore,

$$\Delta \mu_\ell = k_B T \ln \left(\frac{J_{-\ell}(\mathbf{x})}{J_{+\ell}(\mathbf{x})} \right). \qquad (2.24)$$

Even though we only derived this result based on the LMA, it is actually valid for rate laws in general [97, 98, 204].

When a reaction vessel is open to inflow and outflow, the right-hand side of Eqn. (2.20) will have corresponding terms that represent the input and output fluxes. A chemical reaction vessel is said to be "open" when the concentrations of certain chemical species are subject to external control. In that case, the controlled chemical species will have constant concentrations, and their differential equations are eliminated from the system in (2.20). Such an open chemical reaction vessel is called a **chemostat**. If a system under a chemostatic control can reach a steady state, the chemical species in the inflow and outflow have to be balanced: the entire reaction system can be viewed as "a reaction" that transforms the input chemicals into output chemicals. In that case, there is a well-defined overall $\Delta\mu$ for the chemostat, which could be compared with the voltage of a battery for an electrical appliance. Driven by a battery, a radio functions; likewise, driven by such a $\Delta\mu$, a cell lives [113]. There is a great deal to be understood, but nothing is mysterious [242].

2.4 The Principle of Detailed Balance for Chemical Reaction

Now that we have introduced energy, via the concept of chemical potential, into chemical reactions, should conservation of energy not play a role in the dynamics of a chemical reaction system? Is it possible to even include this fundamental law of physics into the the mathematical theory of chemical reaction kinetics?

The answer to this question is that there is still an important missing piece in the mathematical theory of chemical and biochemical reaction systems. The consequence of energy conservation does not appear until one deals with a system of chemical reactions going through a complete **cycle** [117, 157, 23]. To illustrate the role of energy conservation in the context of chemical reactions, let us consider three isomerization reactions in a cycle:

$$A \underset{k_{-1}}{\overset{k_1}{\rightleftharpoons}} B, \quad B \underset{k_{-2}}{\overset{k_2}{\rightleftharpoons}} C, \quad C \underset{k_{-3}}{\overset{k_3}{\rightleftharpoons}} A. \tag{2.25}$$

According to the relation between rate constants and the chemical energy given in Eqs. (2.15) and (2.23), we have

$$\frac{k_1}{k_{-1}} = e^{-(\mu_B^o - \mu_A^o)/k_B T}, \quad \frac{k_2}{k_{-2}} = e^{-(\mu_C^o - \mu_B^o)/k_B T}, \quad \frac{k_3}{k_{-3}} = e^{-(\mu_A^o - \mu_C^o)/k_B T}. \tag{2.26}$$

Therefore,

$$\frac{k_1 k_2 k_3}{k_{-1} k_{-2} k_{-3}} = 1. \tag{2.27}$$

This is an interesting but very stringent constraint on the rate constants. What is the physical interpretation of this? To answer this question, let us solve the mathematical steady state of cyclic chemical reaction systems with three reactions. We find that

$$c_A^{ss} = \frac{k_2 k_3 + k_3 k_{-1} + k_{-1} k_{-2}}{\Delta} c_T,$$

$$c_B^{ss} = \frac{k_3 k_1 + k_1 k_{-2} + k_{-2} k_{-3}}{\Delta} c_T,$$

$$c_C^{ss} = \frac{k_1 k_2 + k_2 k_{-3} + k_{-3} k_{-1}}{\Delta} c_T,$$

in which $c_T = c_A(t) + c_B(t) + c_C(t)$ is a constant for all times t, and the denominator

$$\Delta = k_2 k_3 + k_3 k_{-1} + k_{-1} k_{-2} + k_3 k_1 + k_1 k_{-2}$$
$$+ k_{-2} k_{-3} + k_1 k_2 + k_2 k_{-3} + k_{-3} k_{-1}.$$

Then, we have the steady state flux in the cycle

$$J^{ss} = c_A^{ss} k_1 - c_B^{ss} k_{-1} = c_B^{ss} k_2 - c_C^{ss} k_{-2} = c_C^{ss} k_3 - c_A^{ss} k_{-3} \tag{2.28}$$

$$= \frac{k_1 k_2 k_3 - k_{-1} k_{-2} k_{-3}}{\Delta} c_T.$$

Thus, $J^{ss} = 0$ if and only if Eqn. (2.27) holds true.

Eqn. (2.27), therefore, indicates that in a closed chemical reaction system, a system without chemical energy input, the steady-state can not have a non-zero flux in any one of the reactions; otherwise, there would be a continuous production of heat energy by the reaction system, which violates the law of energy conservation.

Eqn. (2.27) is the mathematical consequence of energy conservation. Systems that satisfy (2.27) are not driven; with time, they spontaneously reach chemical equilibrium, with $\Delta\mu = 0$ for each and every reaction. Furthermore, if Eqn. (2.27) is not true, then we have for each reaction: $\Delta\mu \neq 0$ and $J^{ss} \neq 0$. Hence $\Delta\mu \times J^{ss} \neq 0$. This is the rate of energy dissipation. We will see a more detailed analysis of this in Chapter 10.

The above considerations led to the formulation of the *principle of detailed balance*, also called *the principle of entire equilibrium* by G. N. Lewis [164]. It states that in the thermodynamic equilibrium of a chemical reaction system, each and every reaction has to have its forward flux exactly balanced by its backward flux, thus producing a system with zero net flux and zero chemical potential difference. Therefore, in a chemical reaction system at equilibrium, no matter the complexity of the reaction network, the reactants and products for each and every reaction in (2.19) will satisfy

$$J_{+\ell}(\mathbf{x}^{eq}) = J_{-\ell}(\mathbf{x}^{eq}), \tag{2.29}$$

for each ℓ, in which $\mathbf{x}^{eq}$ are the concentrations of the chemical species in equilibrium. With the law of mass action, we arrive at

$$\prod_{k=1}^{N} (x_k^{eq})^{v_{\ell k}^- - v_{\ell k}^+} = \frac{k_{+\ell}}{k_{-\ell}}. \tag{2.30}$$

The ratio of the forward to backward rate constants is called the *equilibrium constant* of the reaction.

We have seen that the cycle rule in Eqn. (2.27) is intimately related to energy conservation in a closed system. It can be shown that it is also intimately related to the "cycle condition of detailed balance" which is widely used by chemists and physicists. This condition states simply that *for each reaction cycle in a nondriven chemical reaction system, the product of all forward rates is equal to the product of all the backward rates.* In simpler terms, the product of all the equilibrium constants over a reaction cycle has to be 1.

However, for a chemical kinetic system that is under a chemostat, the dynamic steady state, *e.g.*, the fixed point of the kinetic rate equation, need not be a chemical equilibrium. In that case, some reactions can have nonzero fluxes, and the cycle condition is violated. Such a fixed point is called a **nonequilibrium steady state** (NESS) in contrast to an equilibrium steady state.

2.5 Equilibrium and Kinetics of Closed Chemical Reaction Systems

We now apply the principle of detailed balance to prove an important property of chemical kinetics with LMA in a closed system: the system can only go to equilibrium, and all the eigenvalues near equilibrium are real. We prove the first part of the result by the Lyapunov method and the second part by linear analysis. The concept of a Lyapunov function in standard textbooks has a rather strict definition. In the present text, a Lyapunov function is understood in a broad sense, as introduced by J. P. LaSalle [158]: It is a continuous function on $\mathbb{R}^N$, it need not be convex, and it is nonincreasing following $\mathbf{x}(t)$. By this terminology, the Lyapunov function in standard textbooks is called *positive definite* [158]. For recent studies on the Lyapunov function as a nonequilibrium potential function, see [6].

2.5.1 An Example

As an example, first let us consider a system of nonlinear chemical reactions known as the reversible Schnakenberg model, which consists of four species and three reactions:

$$A \underset{k_{-1}}{\overset{k_1}{\rightleftharpoons}} C, \ B \underset{k_{-2}}{\overset{k_2}{\rightleftharpoons}} D, \ 2C + D \underset{k_{-3}}{\overset{k_3}{\rightleftharpoons}} 3C. \tag{2.31}$$

The system of differential equations for the four concentrations is

$$\frac{dc_A}{dt} = -J_1, \ \frac{dc_B}{dt} = -J_2, \ \frac{dc_C}{dt} = J_3 + J_1, \ \frac{dc_D}{dt} = J_2 - J_3, \tag{2.32}$$

in which $J_k = J_{+k} - J_{-k}$, $k = 1, 2, 3$, and,

$$J_{+1} = k_{+1}c_A, \ J_{-1} = k_{-1}c_C, \ J_{+2} = k_{+2}c_B, \ J_{-2} = k_{-2}c_D,$$

$$J_{+3} = k_{+3}c_C^2 c_D, \ J_{-3} = k_{-3}c_C^3. \tag{2.33}$$

It is easy to show that the principle of detailed balance, i.e., $J_{+1} = J_{-1}, J_{+2} = J_{-2}$ and $J_{+3} = J_{-3}$, is satisfied under any steady-state condition of Eqn. (2.32).

Let us assume that the system is in a steady state. If so, then concentrations of all species have to be positive since all the reactions are reversible. Let c_X^* denote the steady-state concentrations, $X = A, B, C, D$. We consider a function of the dynamic variables, i.e., all concentrations:

$$L(c_A, c_B, c_C, c_D) = \sum_X c_X \ln\left(\frac{c_X}{c_X^*}\right). \tag{2.34}$$

We now prove three things:

(1) $L(c_X) \geq 0$ and $L(c_X) = 0$ if and only if $c_X = c_X^*$;

(2) $L(c_X)$ is convex;

(3) $\frac{d}{dt} L[c_X(t)] \leq 0$.

This L is called a Lyapunov function. In addition, c_X^* is an asymptotically stable fixed point if it is isolated. Moreover, if the equilibrium state is unique, then it is globally asymptotically stable.

Proof:

(1) we have $\ln(c_X/c_X^*) \geq 1 - (c_X^*/c_X)$, $\forall$ $(c_X^*/c_X) > 0$. Therefore,

$$\sum_X c_X \ln \left(\frac{c_X}{c_X^*} \right) \geq \sum_X c_X \left(1 - \frac{c_X^*}{c_X} \right) = \sum_X c_X - c_X^* = 0.$$

The inequality holds only when $\frac{c_X^*}{c_X} = 1$.

(2)

$$\frac{\partial^2 L}{\partial c_Y^2} = \frac{1}{c_Y},$$

and all cross-terms $\frac{\partial^2 L}{\partial c_Y \partial c_Z}$ with $Y \neq Z$ are zero. Hence, the function L is convex.

(3)

$$\frac{dL}{dt} = \sum_X \frac{\partial L}{\partial c_X} \frac{dc_X}{dt}$$

$$= -J_1 \ln \left(\frac{c_A}{c_A^*} \right) - J_2 \ln \left(\frac{c_B}{c_B^*} \right) + (J_3 + J_1) \ln \left(\frac{c_C}{c_C^*} \right) + (J_2 - J_3) \ln \left(\frac{c_D}{c_D^*} \right)$$

$$= J_1 \ln \left(\frac{c_A^* c_C}{c_A c_C^*} \right) + J_2 \ln \left(\frac{c_B^* c_D}{c_B c_D^*} \right) + J_3 \ln \left(\frac{c_C c_D^*}{c_C^* c_D} \right)$$

$$= J_1 \ln \left(\frac{J_1^-}{J_1^+} \right) + J_2 \ln \left(\frac{J_2^-}{J_2^+} \right) + J_3 \ln \left(\frac{J_3^-}{J_3^+} \right) \leq 0.$$

The last equation uses the detailed balance relations: $k_{+1} c_A^* = k_{-1} c_C^*$, etc.   □

The linear analysis is carried out by computing the Jacobian matrix near the fixed point $(c_A^*, c_B^*, c_C^*, c_D^*)$:

$$\mathbf{A} = \begin{pmatrix} -k_{+1} & 0 & k_{-1} & 0 \\ 0 & -k_{+2} & 0 & k_{-2} \\ k_{+1} & 0 & -k_{-1} + 2k_{+3} c_C^* c_D^* - 3k_{-3} c_C^{*2} & k_{+3} c_C^{*2} \\ 0 & k_{+2} & -2k_{+3} c_C^* c_D^* + 3k_{-3} c_C^{*2} & k_{-2} - k_{+3} c_C^{*2} \end{pmatrix}.$$

Now if we denote

$$Q = \begin{pmatrix} \sqrt{c_A^*} & 0 & 0 & 0 \\ 0 & \sqrt{c_B^*} & 0 & 0 \\ 0 & 0 & \sqrt{c_C^*} & 0 \\ 0 & 0 & 0 & \sqrt{c_D^*} \end{pmatrix},$$

then

$$Q^{-1}AQ = \begin{pmatrix} -k_{+1} & 0 & k_{-1}\sqrt{\frac{c_C^*}{c_A^*}} & 0 \\ 0 & -k_{+2} & 0 & k_{-2}\sqrt{\frac{c_D^*}{c_B^*}} \\ k_{+1}\sqrt{\frac{c_A^*}{c_C^*}} & 0 & -k_{-1}+2k_{+3}c_C^*c_D^*-3k_{-3}c_C^{*2} & k_{+3}c_C^{*\frac{3}{2}}\sqrt{c_D^*} \\ 0 & k_{+2}\sqrt{\frac{c_B^*}{c_D^*}} & -2k_{+3}c_C^{*\frac{3}{2}}\sqrt{c_D^*}+3k_{-3}c_C^{*\frac{5}{2}}c_D^{*-\frac{1}{2}} & k_{-2}-k_{+3}c_C^{*2} \end{pmatrix},$$

which is symmetric according to the detailed balance relations. Hence, all the eigenvalues of A are necessarily real.

What is the meaning of the function $L(c_X)$? We now show that it is the *total chemical energy* of the system (known as the Gibbs function in thermodynamics). In fact, $k_B T L$ and

$$G = \sum_X c_X \left(\mu_X^o + k_B T \ln c_X \right) \tag{2.35}$$

differ only by a constant. This is because in equilibrium, $\mu_A = \mu_B = \mu_C = \mu_D$, denoted as μ_{eq}, followed by

$$G - k_B T L = \sum_X c_X \left(\mu_X^o + k_B T \ln c_X^* \right) = \mu_{eq} \sum_X c_X. \tag{2.36}$$

2.5.2 Shear's Theorem for General Mass-action Kinetics

Within the framework of general mass-action kinetics with N species and M reactions, D. B. Shear showed that if a detailed balance holds for one fixed point $\{x^{eq}\}$ such that $J_{+\ell}(x^{eq}) = J_{-\ell}(x^{eq})$ for every ℓ, then

$$F(x) = \sum_{k=1}^{N} x_k \ln \left(\frac{x_k}{x_k^{eq}} \right) - x_k + x_k^{eq} \tag{2.37}$$

is a Lyapunov function of the system of kinetic equations [250, 251]. This result establishes that such systems will always approach their equilibrium steady state, which is also stable.[2] Such systems can never exhibit limit cycle behavior.

Proof:

First, $F(\mathbf{x})$ in (2.37) is a nonnegative function of $\mathbf{x}$. Applying the inequality $\ln x \geq 1 - 1/x$:

$$F(\mathbf{x}) = \sum_{k=1}^{N} x_k \ln\left(\frac{x_k}{x_k^{eq}}\right) - x_k + x_k^{eq}$$

$$\geq \sum_{k=1}^{N} x_k \left(1 - \frac{x_k^{eq}}{x_k}\right) - x_k + x_k^{eq} = 0.$$

The equality holds if and only if all $x_k = x_k^{eq}$.

Second, the Hessian matrix for $F(\mathbf{x})$,

$$
\begin{pmatrix}
\frac{\partial^2 F}{\partial x_1^2} & \frac{\partial^2 F}{\partial x_1 \partial x_2} & \cdots & \frac{\partial^2 F}{\partial x_1 \partial x_N} \\
\frac{\partial^2 F}{\partial x_2 \partial x_1} & \frac{\partial^2 F}{\partial x_2^2} & \cdots & \frac{\partial^2 F}{\partial x_2 \partial x_N} \\
\vdots & \vdots & \ddots & \vdots \\
\frac{\partial^2 F}{\partial x_N \partial x_1} & \frac{\partial^2 F}{\partial x_N \partial x_2} & \cdots & \frac{\partial^2 F}{\partial x_N^2}
\end{pmatrix}
=
\begin{pmatrix}
\frac{1}{x_1} & 0 & \cdots & 0 \\
0 & \frac{1}{x_2} & \cdots & 0 \\
0 & \ddots & \ddots & 0 \\
0 & \cdots & 0 & \frac{1}{x_N}
\end{pmatrix},
$$

is positive definite.

Third,

$$\frac{d}{dt}F(\mathbf{x}(t)) = \sum_{k=1}^{N} \frac{dx_k}{dt} \ln\left(\frac{x_k}{x_k^{eq}}\right)$$

$$= \sum_{k=1}^{N} \sum_{\ell=1}^{M} \left(v_{\ell k}^- - v_{\ell k}^+\right)\left(J_{+\ell}(\mathbf{x}) - J_{-\ell}(\mathbf{x})\right) \ln\left(\frac{x_k}{x_k^{eq}}\right)$$

$$= \sum_{\ell=1}^{M} \left(J_{+\ell}(\mathbf{x}) - J_{-\ell}(\mathbf{x})\right) \sum_{k=1}^{N} \left(v_{\ell k}^- - v_{\ell k}^+\right) \ln\left(\frac{x_k}{x_k^{eq}}\right)$$

$$= \sum_{\ell=1}^{M} \left(J_{+\ell}(\mathbf{x}) - J_{-\ell}(\mathbf{x})\right) \ln\left[\frac{\prod_{k=1}^{N}\left(\frac{x_k}{x_k^{eq}}\right)^{v_{\ell k}^-}}{\prod_{k=1}^{N}\left(\frac{x_k}{x_k^{eq}}\right)^{v_{\ell k}^+}}\right]$$

---

[2] D. B. Shear also proved the uniqueness of the equilibrium steady state, which is not described here.

$$= \sum_{\ell=1}^{M} \left(J_{+\ell}(\mathbf{x}) - J_{-\ell}(\mathbf{x}) \right) \ln \left[\frac{\left(\frac{J_{-\ell}(\mathbf{x})}{J_{-\ell}(\mathbf{x}^{eq})} \right)}{\left(\frac{J_{+\ell}(\mathbf{x})}{J_{+\ell}(\mathbf{x}^{eq})} \right)} \right].$$

Since $J_{+\ell}(\mathbf{x}^{eq}) = J_{-\ell}(\mathbf{x}^{eq})$, then

$$\frac{\mathrm{d}}{\mathrm{d}t} F(\mathbf{x}(t)) = \sum_{\ell=1}^{M} (J_{+\ell}(\mathbf{x}) - J_{-\ell}(\mathbf{x})) \ln \left(\frac{J_{-\ell}(\mathbf{x})}{J_{+\ell}(\mathbf{x})} \right) \le 0. \qquad \qquad \Box$$

This theorem can be understood as a thermodynamic result. J. W. Gibbs (1839–1903) was the first to formulate the *free energy function* and showed that the LMA was closely related to a variational principle in terms of that function, connecting thermodynamics with kinetics [249]. In units of $k_B T$ and per unit volume, the Gibbs function for a dilute solution is:

$$G[\mathbf{x}] = \sum_{j=1}^{N} x_j \left(\mu_j - 1 \right), \quad \mu_j = \mu_j^o + \ln x_j, \tag{2.38}$$

in which μ_j^o is a constant associated with the structure of the jth chemical species in an aqueous solution, and the -1 term is the contribution from the solvent.

If all the reaction rate constants $k_{\pm\ell}$ satisfy the cycle condition for detailed balance mentioned at the end of Section 2.4, the chemical potential *difference* at the final equilibrium state with detailed balance is

$$\Delta\mu_\ell[\mathbf{x}^{eq}] \equiv \sum_{j=1}^{N} \left(v_{\ell j}^- - v_{\ell j}^+ \right) \left(\mu_j^o + \ln x_j^{eq} \right) = 0, \tag{2.39}$$

in which $\{x_j^{eq}\}$ is the equilibrium concentration. Then, for a reaction system with a constant volume that is not in its equilibrium at time t,

$$\frac{\mathrm{d}}{\mathrm{d}t} G[\mathbf{x}(t)] = \sum_{j=1}^{N} \frac{\mathrm{d}x_j(t)}{\mathrm{d}t} \left(\mu_j^o + \ln x_j \right) \tag{2.40}$$

$$= \sum_{j=1}^{N} \sum_{\ell=1}^{M} \left(v_{\ell j}^- - v_{\ell j}^+ \right) \left(J_{+\ell}(\mathbf{x}) - J_{-\ell}(\mathbf{x}) \right) \left(\mu_j^o + \ln x_j \right)$$

$$= \sum_{j=1}^{N} \sum_{\ell=1}^{M} \left(v_{\ell j}^- - v_{\ell j}^+ \right) \left(J_{+\ell}(\mathbf{x}) - J_{-\ell}(\mathbf{x}) \right) \left(\ln x_j - \ln x_j^{eq} \right)$$

$$= \sum_{\ell=1}^{M} \left(J_{+\ell}(\mathbf{x}) - J_{-\ell}(\mathbf{x}) \right) \ln \left(\frac{J_{-\ell}(\mathbf{x})}{J_{+\ell}(\mathbf{x})} \right) = \frac{\mathrm{d}}{\mathrm{d}t} F(\mathbf{x}(t)) \le 0. \tag{2.41}$$

Actually, $-\frac{d}{dt}G[\mathbf{x}(t)] = \sum_{\ell=1}^{M}[J_{+\ell}(\mathbf{x}) - J_{-\ell}(\mathbf{x})]\ln[J_{+\ell}(\mathbf{x})/J_{-\ell}(\mathbf{x})]$ is the entropy production rate (epr) of such a system approaching its equilibrium state.

2.5.3 Complex Balanced Chemical Reaction Networks

For open chemical systems that do not reach chemical equilibrium with detailed balance, Horn, Jackson, and Feinberg introduced the notion of *complex balance* at the steady state of a reaction network under the general assumption of mass action in 1972 [124, 125, 67]. This concept is a generalization of both linear reaction networks and kinetics with detailed balance. Complex balanced kinetics can be non-linear in addition to having nonequilibrium steady states.

It also has a deep relation to the topological structure of a reaction network. Recall that for $F(\mathbf{x})$ in (2.37), generally we have

$$\frac{dF(\mathbf{x})}{dt} = \sum_{\ell=1}^{M}\left(J_{+\ell}(\mathbf{x}) - J_{-\ell}(\mathbf{x})\right)\ln\left(\frac{J_{-\ell}(\mathbf{x})J_{+\ell}(\mathbf{x}^{ss})}{J_{+\ell}(\mathbf{x})J_{-\ell}(\mathbf{x}^{ss})}\right), \tag{2.42}$$

in which $\{\mathbf{x}^{ss}\}$ is a stationary state of the system, even without detailed balance.

A chemical reaction system is "complex balanced" if and only if

$$\sum_{\ell=1}^{M}\left(J_{+\ell}(\mathbf{x}^{ss}) - J_{-\ell}(\mathbf{x}^{ss})\right)\left(\prod_{i=1}^{N}\left(\frac{x_i}{x_i^{ss}}\right)^{v_{\ell j}^{+}} - \prod_{i=1}^{N}\left(\frac{x_i}{x_i^{ss}}\right)^{v_{\ell j}^{-}}\right) \tag{2.43}$$

is zero for any $\mathbf{x} = (x_1, x_2, \cdots, x_N)$. This is because any unique multitype-nomial term

$$\prod_{i=1}^{N}\left(\frac{x_i}{x_i^{ss}}\right)^{\xi_i} \tag{2.44}$$

represents a particular "complex" $(\xi_1 X_1 + \xi_2 X_2 + \cdots + \xi_N X_N)$. Therefore, a complex balanced steady state has all the influx to the complex precisely balanced by the outflux of that complex:

$$\left\{\sum_{\ell=1}^{M}\left(\delta_{v_{\ell}^{-},\xi} - \delta_{v_{\ell}^{+},\xi}\right)\left(J_{+\ell}(\mathbf{x}^{ss}) - J_{-\ell}(\mathbf{x}^{ss})\right)\right\}\prod_{i=1}^{N}\left(\frac{x_i}{x_i^{ss}}\right)^{\xi_i} = 0, \tag{2.45}$$

in which $v_\ell^+ = \{v_{\ell j}^+\}$ and $v_\ell^- = \{v_{\ell j}^-\}$. Since the nomials in Eqn. (2.44) for different complexes are linearly independent functions of x, the polynomial in Eqn. (2.43) is zero if and only if each and every nomial is zero, e.g., Eqn. (2.45).

Detailed balance is a special case in which $J_{+\ell}^{ss} = J_{-\ell}^{ss}$ for every ℓ. Detailed balance is a kinetic concept. Complex balance, however, has a network topological implication for a reaction network [199].

2.5.3.1 Lyapunov Function for Complex Balanced Kinetics

For complex balanced kinetics,

$$
\begin{aligned}
\frac{dF(\mathbf{x})}{dt} &= \sum_{\ell=1}^{M} \left[J_{+\ell}(\mathbf{x}) \ln \left(\frac{J_{-\ell}(\mathbf{x}) J_{+\ell}(\mathbf{x}^{ss})}{J_{+\ell}(\mathbf{x}) J_{-\ell}(\mathbf{x}^{ss})} \right) + J_{-\ell}(\mathbf{x}) \ln \left(\frac{J_{+\ell}(\mathbf{x}) J_{-\ell}(\mathbf{x}^{ss})}{J_{-\ell}(\mathbf{x}) J_{+\ell}(\mathbf{x}^{ss})} \right) \right] \\
&\le \sum_{\ell=1}^{M} \left[J_{+\ell}(\mathbf{x}) \left(\frac{J_{-\ell}(\mathbf{x}) J_{+\ell}(\mathbf{x}^{ss})}{J_{+\ell}(\mathbf{x}) J_{-\ell}(\mathbf{x}^{ss})} - 1 \right) + J_{-\ell}(\mathbf{x}) \left(\frac{J_{+\ell}(\mathbf{x}) J_{-\ell}(\mathbf{x}^{ss})}{J_{-\ell}(\mathbf{x}) J_{+\ell}(\mathbf{x}^{ss})} - 1 \right) \right] \\
&= -\sum_{\ell=1}^{M} \left(J_{+\ell}(\mathbf{x}^{ss}) - J_{-\ell}(\mathbf{x}^{ss}) \right) \left(\frac{J_{+\ell}(\mathbf{x})}{J_{+\ell}(\mathbf{x}^{ss})} - \frac{J_{-\ell}(\mathbf{x})}{J_{-\ell}(\mathbf{x}^{ss})} \right) \\
&= -\sum_{\ell=1}^{M} \left(J_{+\ell}(\mathbf{x}^{ss}) - J_{-\ell}(\mathbf{x}^{ss}) \right) \left(\prod_{i=1}^{N} \left(\frac{x_i}{x_i^{ss}} \right)^{v_{\ell i}^+} - \prod_{i=1}^{N} \left(\frac{x_i}{x_i^{ss}} \right)^{v_{\ell i}^-} \right) = 0.
\end{aligned}
$$

Furthermore, since $F(\mathbf{x}) \ge 0$, it is a Lyapunov function for the mass-action kinetics. The convexity of $F(\mathbf{x})$ is easily established: $\partial^2 A / \partial x_i \partial x_j = x_i^{-1} \delta_{ij}$. Therefore, any isolated steady state $\mathbf{x}^{ss}$ of complex-balanced reaction kinetics is stable[3]. This is a well-known result. The existence of Lyapunov function $F(\mathbf{x})$ for kinetic systems with complex balance prompted Horn and Jackson's proposition of "quasithermodynamics" [125].

One might have noticed that the Lyapunov function method lacks rigorous treatment at the boundary of the positive quadrant of $\mathbb{R}^N$. Actually, the *global attractor conjecture* for kinetics with the complex-balanced LMA remains an open problem in mathematical chemistry [4, 106]. For the nonuniqueness of equilibrium states outside the LMA, see [198].

---

[3] Horn and Jackson proved a stronger result, implying the uniqueness of the steady state [125].

2.5.3.2 Null Space of a Stoichiometric Matrix

If we let the $N \times M$ matrix $\mathbf{S} = \{S_{k\ell} = v_{\ell k}^- - v_{\ell k}^+; 1 \leq k \leq N, 1 \leq \ell \leq M\}$ contain all the stoichiometric coefficients, then the system of rate equations in Eqn. (2.20) can be expressed in a very compact form as

$$\frac{d\mathbf{x}}{dt} = \mathbf{S}J(\mathbf{x}). \tag{2.46}$$

A kinetic steady state $\mathbf{x}^{ss}$ will have its corresponding $J^{ss} \equiv J(\mathbf{x}^{ss})$, as a vector in $\mathbb{R}^M$, in the null space of matrix $\mathbf{S}$: $J^{ss} \in \text{null}(\mathbf{S})$.

Assuming there are total of L reaction complexes, matrix $\mathbf{S} = \mathbf{S}_1\mathbf{S}_2$, in which the $N \times L$ matrix $\mathbf{S}_1$ contains all the stoichiometric coefficients that relate the N chemical *species* to the L *complexes* and the $L \times M$ matrix $\mathbf{S}_2$ connects the L *complexes* to the M *reactions*.

A chemical kinetic steady state with detailed balance has $J^{ss} = \mathbf{0} = (0, \cdots, 0)^T$, with the flux in each of the M reactions being zero. A complex-balanced steady state corresponds to $J^{ss} \in \text{null}(\mathbf{S}_2)$, and the most general steady state without complex balance corresponds to $\mathbf{S}_2 J^{ss} \neq \mathbf{0}$, but $\mathbf{S}_2 J^{ss}$ is in the null space of $\mathbf{S}_1$: $\mathbf{S}_1\mathbf{S}_2 J^{ss} = \mathbf{S}J^{ss} = \mathbf{0}$. In summary,

$$\mathbf{0} \in \text{null}(\mathbf{S}_2) \subset \text{null}(\mathbf{S}) \subset \mathbb{R}^M, \tag{2.47}$$

in which the matrix $\mathbf{S}_2$ contains only $0, 1, -1$, and there are only one 1 and one -1 in each column. This matrix corresponds to the incidence matrix of a *simple graph*, with the complexes being its vertices and the reactions being its edges. In contrast, the entries of stoichiometric matrix $\mathbf{S}$ are integers; it cannot be represented by a simple graph. Rather, $\mathbf{S}$ corresponds to a *bipartite graph* [18].

2.5.4 Equilibrium Gibbs Function for General Rate Laws

Beyond the special law of mass action, for general rate laws, we have formulated the detailed balance of a kinetic system mathematically by a *potential condition* [97, 98]: the existence of a $G(\mathbf{x})$ that satisfies

$$\ln\left(\frac{J_{+\ell}(\mathbf{x})}{J_{-\ell}(\mathbf{x})}\right) = -v_\ell \cdot \nabla G(\mathbf{x}), \tag{2.48}$$

for each $\mathbf{x}$, in which

$$\mathbf{v}_\ell = \left(v_{\ell 1}^- - v_{\ell 1}^+, v_{\ell 2}^- - v_{\ell 2}^+, \cdots, v_{\ell N}^- - v_{\ell N}^+ \right).$$

In a stoichiometric reaction network, the right-hand side of (2.48) is the gradient of G, which is defined on the nodes, and the left-hand side is the *chemical force*, which is defined on the edges. $\mathbf{v}_\ell \cdot \nabla$ should be understood as a "gradient operator" on the discrete stoichiometric network. Eqn. (2.48) implies the chemical force has a potential function, the equilibrium Gibbs function.

The consequence of the existence of an equilibrium Gibbs function is clear. If $\mathbf{x}(t)$ is the solution to the rate equation (2.20), and $H(\mathbf{x})$ is any differentiable function of $\mathbf{x}$, then one immediately has

$$\frac{d}{dt} H(\mathbf{x}(t)) = \sum_{i=1}^N \frac{\partial H(\mathbf{x})}{\partial x_i} \left(\frac{dx_i(t)}{dt} \right)$$

$$= \sum_{i=1}^N \sum_{j=1}^M \frac{\partial H(\mathbf{x})}{\partial x_i} v_{ji} \left(J_{+j}(\mathbf{x}) - J_{-j}(\mathbf{x}) \right)$$

$$= \sum_{j=1}^M \ln \left(\frac{J_{+j}(\mathbf{x})}{J_{-j}(\mathbf{x})} e^{\nabla H(\mathbf{x}) \cdot \mathbf{v}_j} \right) \left(J_{+j}(\mathbf{x}) - J_{-j}(\mathbf{x}) \right) \qquad (2.49a)$$

$$- \sum_{j=1}^M \left(J_{+j}(\mathbf{x}) - J_{-j}(\mathbf{x}) \right) \ln \left(\frac{J_{+j}(\mathbf{x})}{J_{-j}(\mathbf{x})} \right). \qquad (2.49b)$$

The nonnegative term in (2.49b) is again the entropy production rate (epr). When $H(\mathbf{x}) = G(\mathbf{x})$, the Gibbs function satisfying potential condition (2.48), then the term in (2.49a) is zero! Therefore, $dG(\mathbf{x}(t))/dt = -\text{epr} \leq 0$, which echoes the second law of thermodynamics in chemical kinetics.

For recent developments in the nonequilibrium thermodynamics of chemical reaction networks, see [222, 205, 206, 231, 204].

2.6 Nonlinear Chemical Kinetics: Bistability and the Limit Cycle

A linear dynamical system can only have a single isolated fixed point: if the right-hand side of $\dot{x} = f(x)$ is a linear function of x and $f'(x) \neq 0$, there will be only one root to $f(x) = 0$. This implies there is only one fixed point to the ordinary differential equation (ODE). Bistability and limit cycle oscillation, therefore, are both nonlinear phenomena. Furthermore, Shear's theorem shows that in the framework of mass-action kinetics, either phenomenon only occurs in a driven chemical reaction system. In this section, we study two canonical examples of **nonlinear, nonequilibrium chemical reaction systems**: The Schlögl system with chemical bistability and the Schnakenberg system with limit-cycle chemical oscillation. The reversible version of the latter has already been introduced in Eqn. (2.31).

2.6.1 Schlögl System and Chemical Bistability

The Schlögl system consists of two reversible reactions [241]:

$$A + 2X \underset{k_2}{\overset{k_1}{\rightleftharpoons}} 3X, \quad X \underset{k_4}{\overset{k_3}{\rightleftharpoons}} B. \tag{2.50}$$

The first of the two reactions is *autocatalytic* since the presence of X aids in the transformation of A to X. According to the law of mass action, the differential equation for the kinetics of X is

$$\frac{dx}{dt} = J_{+1} - J_{-1} - J_{+2} + J_{-2}, \tag{2.51a}$$

in which

$$J_{+1} = k_1 a x^2, \ J_{-1} = k_2 x^3, \ J_{+2} = k_3 x, \ J_{-2} = k_4 b. \tag{2.51b}$$

For simplicity, lowercase x, a, b are the concentrations of chemical species X, A, and B. It is assumed that the chemical species A and B are chemostatic; that is, their concentrations are kept constant. Since the right-hand side of the ODE in (2.51) is a third-order polynomial of x when $k_2 \neq 0$, it is possible to have three steady states for the dynamics in general.

If one combines the two reactions in (2.50), one obtains

$$A + 3X \rightleftharpoons B + 3X.$$

This shows that overall, the reaction system accomplishes transformations between A and B, and the species X is merely an *intermediate species* in this regard.

If the concentrations of A and B satisfy the detailed balance condition for both reactions in (2.50), i.e., $k_1 a (x^{eq})^2 = k_2 (x^{eq})^3$ and $k_3 x^{eq} = k_4 b$, in which x^{eq} is the corresponding equilibrium concentration of X, then $a k_1 k_3 = b k_4 k_2$. Under this condition,

$$\frac{dx}{dt} = f(x) = k_1ax^2 - k_2x^3 - k_3x + k_4b$$

$$= k_1ax^2 - k_2x^3 - k_3x + \frac{ak_1k_3}{k_2} = (k_1a - k_2x)\left(x^2 + \frac{k_3}{k_2}\right). \qquad (2.52)$$

Therefore, the only true fixed point to (2.52) is exactly $x^{eq} = \frac{k_1a}{k_2} = \frac{k_4b}{k_3}$.

The space for parameters k_1a, k_2, k_3, k_4b can be divided into regions in which the ODE has one or three fixed points. The boundary of the two regions can be determined by simultaneously satisfying $f(x) = 0$ and $f'(x) = 0$, which corresponds to $f(x) = 0$ having two roots when $f(x)$ is tangent to the x-axis:

$$k_1ax^2 - k_2x^3 - k_3x + k_4b = 2k_1ax - 3k_2x^2 - k_3 = 0.$$

These two equations yield two aggregated, nondimensional parameters $\sigma = \frac{k_2k_3}{(k_1a)^2}$ and $\chi = \frac{k_1k_4ab}{k_3^2}$, which satisfy $\hat{x}^2 - \sigma\hat{x}^3 - \hat{x} + \chi = 2\hat{x} - 3\sigma\hat{x}^2 - 1 = 0$ and $\hat{x} = \frac{k_1ax}{k_3}$. The boundary is given as a parametric curve in the σ-χ plane

$$\sigma = \frac{2\xi - 1}{3\xi^2}, \quad \chi = \frac{\xi(2-\xi)}{3}, \quad 0 \le \xi \le \infty. \qquad (2.53)$$

There is necessarily a cusp in this curve at $\xi = 1$, corresponding to $\sigma = \chi = \frac{1}{3}$, as shown in Figure 2.1.

Regarding the stability of these steady states (fixed points), usually, a nonlinear kinetic system can have multiple steady states, i.e., $f(u,v) = g(u,v) = 0$ has multiple roots, as is the case for the one-dimensional Schlögl system. For each steady state, one can analyze its *stability* in terms of the Jacobian matrix. One can find this in every textbook on differential equations. Very briefly, if the real part of the two eigenvalues of the matrix in (2.60) is negative, the steady state is *asymptotically stable*; if it is positive, then the steady state is *unstable*. Furthermore, if the eigenvalues have an imaginary part, then the kinetics are oscillatory, which is the case in the following section.

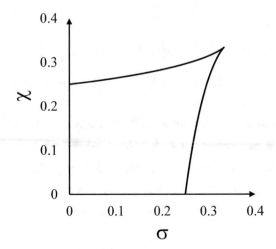

Fig. 2.1 The roots of the third-order polynomial $-\sigma\hat{x}^3 + \hat{x}^2 - \hat{x} + \chi = 0$ form a surface $\hat{x}(\sigma, \chi)$ in σ-χ-$\hat{x}$ space. In a certain region of (σ, χ), the surface has three layers, corresponding to the three real roots to the polynomial. The boundary of the region, given by the parametric equation in Eqn. (2.53), is shown in the figure here. The boundary has a cusp at $(\sigma, \chi) = (\frac{1}{3}, \frac{1}{3})$. Note both $d\sigma/d\xi$ and $d\chi/d\xi$ equal zero at $\xi = 1$, but their ratio $d\chi/d\sigma = 1$.

2.6.2 The Schnakenberg System and Chemical Oscillation

Now we consider the Schnakenberg model:

$$A \underset{k_{-1}}{\overset{k_1}{\rightleftharpoons}} X, \; B \overset{k_2}{\longrightarrow} Y, \; Y + 2X \overset{k_3}{\longrightarrow} 3X. \tag{2.54}$$

Assuming the concentrations of A and B are fixed, This describes an open system. Then, the law of mass action gives

$$\frac{dc_X}{dt} = k_1 c_A - k_{-1} c_X + k_3 c_X^2 c_Y, \tag{2.55a}$$

$$\frac{dc_Y}{dt} = k_2 c_B - k_3 c_X^2 c_Y. \tag{2.55b}$$

In fact, one can always simplify a kinetic model with a reduced number of parameters by simply introducing nondimensional dependent and independent variables. For the system in (12.2), this *nondimensionalization* procedure entails[4]

$$u = \sqrt{k_3/k_{-1}} c_X, \quad v = \sqrt{k_3/k_{-1}} c_Y, \quad \tau = k_{-1}t, \tag{2.56}$$

---

[4] The nondimensionalization procedure is always not unique.

which yields the following nondimensionalized parameters:

$$a = \sqrt{k_1^2 k_3 / k_{-1}^3}\, c_A, \quad b = \sqrt{k_2^2 k_3 / k_{-1}^3}\, c_B. \tag{2.57}$$

Then, Eqn. (12.2) is simplified into

$$\frac{du}{d\tau} = a - u + u^2 v = f(u, v), \tag{2.58a}$$

$$\frac{dv}{d\tau} = b - u^2 v = g(u, v). \tag{2.58b}$$

Note that there are only two parameters now in the system of equations, a and b. It should be noted that nondimensionalization is not unique; there could be other choices in (2.56).

A steady state of a kinetic system is a set of (u^{ss}, v^{ss}) that satisfy $f(u^{ss}, v^{ss}) = g(u^{ss}, v^{ss}) = 0$. Visualizing $(f, g)(u, v)$ as a planar vector field, (u^{ss}, v^{ss}) is also called a *fixed point*. Eqn. (2.58) has a unique steady state at

$$u^{ss} = a + b, \quad v^{ss} = b/(a+b)^2. \tag{2.59}$$

The Jacobian matrix $\mathbf{A}$ at (u^{ss}, v^{ss}) is

$$\mathbf{A} = \begin{pmatrix} \frac{\partial f}{\partial u} & \frac{\partial f}{\partial v} \\ \frac{\partial g}{\partial u} & \frac{\partial g}{\partial v} \end{pmatrix}_{u^{ss}, v^{ss}} = \begin{pmatrix} \dfrac{b-a}{a+b} & (a+b)^2 \\ -\dfrac{2b}{a+b} & -(a+b)^2 \end{pmatrix}, \tag{2.60}$$

which has trace and determinant:

$$\mathrm{tr}(\mathbf{A}) = \frac{b-a}{a+b} - (a+b)^2, \quad \det(\mathbf{A}) = (a+b)^2 > 0. \tag{2.61}$$

For the particular matrix in (2.60), a positive determinant implies that the product of the two eigenvalues λ_1 and λ_2 is > 0. Furthermore, if $\Delta(\mathbf{A}) = \mathrm{tr}(\mathbf{A})^2 - 4\det(\mathbf{A}) < 0$, then λ_1 and λ_2 are a pair of complex numbers with the same imaginary parts. When $\mathrm{tr}(\mathbf{A}) = 0$, the two eigenvalues are purely imaginary. Therefore, when $\mathrm{tr}(\mathbf{A})$ is slightly greater than zero, the steady state is unstable and oscillatory, and a stable limit cycle emerges around the unstable steady state. This is known as a *Hopf bifurcation*.

With the fundamentals of deterministic chemical kinetics presented in the present chapter, the field of nonlinear chemical kinetics can essentially be studied in applied mathematics. There are many excellent chapters in textbooks and monographs, for example, [10, 108, 61]. We shall, however, move in a different direction.

Chapter 3
Probability Distribution and Stochastic Processes

3.1 Basics of Probability Theory

3.1.1 Probability Space and Random Variables

The fundamental objects in the theory of probability are random events that might occur but usually not with certainty. Each of the possible outcomes of a random event, therefore, is assigned a probability. The probability satisfies the rule of addition for nonoverlapping events, the rule of multiplication for *independent events*, and the law of total probability, among others. The total probability of all the possible outcomes that have to be exhaustively declared *a priori* is considered to be 1.

In this setup, experimental observables, e.g., scientific measurements, are defined as functions on the set of all elementary events, Ω, called the sample space. Ω together with the probability measure **P** constitute the probability space.[1] None of the elementary events can be divided; all measurements are perfectly deterministic at the level of elementary events. This sample space Ω is thus independent of measurements! The situation is rather similar to the logic of quantum mechanics: the fundamental object in the latter theory is a *wave function* that "lives" in a Hilbert space; the observations are the linear self-adjoint operators on the Hilbert space. In the theory of probability, an observable is called a *random variable*, which is a *measurable function* in the probability space $X(\omega)$, $\omega \in \Omega$, We shall not devolve

---

[1] Rigorously, the probability space $(\Omega, \mathscr{F}, \mathbf{P})$ also needs the set of all measurable events that form a σ-algebra $\mathscr{F}$. As a matter of fact, the probability **P** is not even defined on Ω; it is a function of $\mathscr{F}$. This implies, according to the mathematics, that probability measures cannot be observed. One can only measure functions defined on Ω.

© The Author(s), under exclusive license to Springer Nature Switzerland AG 2021
H. Qian, H. Ge, *Stochastic Chemical Reaction Systems in Biology*,
Lecture Notes on Mathematical Modelling in the Life Sciences,
https://doi.org/10.1007/978-3-030-86252-7_3

into the precise mathematical notion of measurability, but merely notice the logical relationship between probability and random variables.

Random variables (RVs) are a new concept motivated from our daily life experience. They are not scalar quantities in the conventional, deterministic sense. Because an RV is the outcome of a random event, it does not have a particular value; rather, it can be equal to many different values **with probabilities**. Thus, when measuring such an RV, the results are also random, and repeated measurements give different values. However, the probability of observing a particular value x is deterministic: it is the total probability of those events $\omega \in \Omega$ that gives the particular value $X(\omega) = x$.

There are many familiar situations in the real world that can be modeled by RVs. The number on a rolled dice is the most elementary example. Others include the amount of time one waits at a particular bus stop (assuming the bus is on a regular schedule), the life span of a person in a population, and the velocity of a molecule in a bottle of gas.

There are two types of RVs that are widely studied: discrete and continuous. For a continuous RV $\mathbf{X}$, one cannot talk about the probability that $\mathbf{X} = x$ (note the x is simply a number); it is zero. One has to introduce the concept of probability density. The notion of RVs can certainly be generalized to more complex objects; matrix-valued random variables and graph-valued random variables have become two very popular subjects in recent years, connected to the theory of random matrices and the theory of random graphs.

For clarity, proper notations are very important to the theory of probability. For a discrete random variable $\mathbf{X}$ taking real values x_i, $i \in \mathbb{Z}^+$, we have a *probability mass function* (pmf)

$$p_X(x_i) = \Pr\{\mathbf{X} = x_i\}. \tag{3.1a}$$

For a continuous, real-valued random variable $\mathbf{X} \in \mathbb{R}$, we have a *probability density function* (pdf) such that for each $a < b$,

$$\int_a^b f_X(x)\mathrm{d}x = \Pr\{a < \mathbf{X} \le b\}. \tag{3.1b}$$

Note that the subscript X is very important. It signifies the random variable; the x inside $(\cdot)$ is just a dummy variable for the probability function.

In measure-theoretical language, probability is a measure of a set that consists of elementary random events in Ω, $\{a < \mathbf{X} < b\}$, known as a nonelementary random event, which is the inverse image of $\mathbf{X}$, recalling that a random variable is a function defined on the probability space. In the study of RVs, it is important to identify the "nonelementary event generated by the RV".

3.1.2 Probability Distribution Functions — Mass and Density

The probability distribution of a random variable $\mathbf{X}$, be it discrete or continuous, can also be described by a *cumulative distribution function* (cdf):

$$F_X(x) = \Pr\{\mathbf{X} \leq x\}.$$

Note the three standard notions: p, f, and F. We usually use uppercase F to denote a cdf, lowercase f for a pdf, and lowercase p for pmf. While one needs to have separated discussions of the pmf and pdf for discrete and continuous RVs, respectively, the concept of cdf covers both types of RVs in terms of discontinuous and continuous functions.

The cumulative distribution function is also simply called the distribution function. The properties of a distribution function $F(x)$ include positivity, monotonicity, $F(-\infty) = 0$, and $F(\infty) = 1$. In the case of continuous RVs, the cdf and pdf are intimately related:

$$F(x) = \int_{-\infty}^{x} f(\xi)\,d\xi, \qquad f(x) = \frac{dF(x)}{dx}.$$

For discrete RVs, we have

$$F(x_i) = \sum_{x_i \leq x} p(x_i), \qquad p_X(x_i) = F(x_i) - F(x_i^-),$$

where $F(z^-) = \lim_{\varepsilon \to 0} F(z - \varepsilon)$ with $\varepsilon \geq 0$. In terms of the Dirac δ function, the derivative of the Heaviside step function, one can also express:

$$\frac{dF(x)}{dx} = \sum_{\text{all } i} p(x_i)\delta(x - x_i).$$

3.1.3 Expected Value, Variance, and Moments

The expected value (EV) is the official name, in the mathematical theory of probability, for "average", or "mean". For a discrete RV $\mathbf{X}$:

$$\mathbb{E}[\mathbf{X}] = \sum_i x_i \Pr\{X = x_i\} = \sum_i x_i p_X(x_i). \tag{3.2}$$

For a continuous RV, we have

$$\mathbb{E}[\mathbf{X}] = \int_{-\infty}^{+\infty} x f_X(x)\mathrm{d}x. \tag{3.3}$$

To understand the significance of the EV, let us consider the problem of buying a lottery ticket. Can you explain why statisticians suggest people select numbers larger than 31 when buying tickets?

The EV gives us the weighted average of an RV. To get a sense of how broad the distribution of an RV X is, one computes its variance :

$$\mathrm{Var}[\mathbf{X}] = \mathbb{E}\left[(\mathbf{X} - \mathbb{E}(\mathbf{X}))^2\right]. \tag{3.4}$$

It is easy to show that $\mathrm{Var}[\mathbf{X}] = \mathbb{E}\left[\mathbf{X}^2\right] - \left(\mathbb{E}[\mathbf{X}]\right)^2$.

The probability distribution of an RV is a function. Therefore, in general, it cannot be completely determined just by the expected value and the variance. The mth moment of an RV is another numerical characteristic of an RV, defined as $E[\mathbf{X}^m]$. The expected value is the 1st moment, and the variance is related to the 2nd moment. The 3rd and 4th moments are related to the skewness and kurtosis, respectively, of the distribution.

3.1.4 Function of An RV and Change of Variables

3.1.4.1 Cdf and Pdf of a Function of An RV

For any function $g(x)$, $g(\mathbf{X})$ is a function of the RV $\mathbf{X}$, which is also a random variable, $g(\mathbf{X}(\omega))$.

Suppose $\mathbf{X}$ is a continuous random variable, and let us only deal with a monotonic function $y = g(x)$. Then, the probability density function of $\mathbf{Y} = g(\mathbf{X})$ is

$$f_Y(y) = \frac{f_X(x)}{|g'(x)|}, \tag{3.5}$$

where $x = g^{-1}(y)$.

The proof for Eqn. (3.5) is truly a practice of the notations and definitions of the basics we have learned thus far. By definition,

$$
\begin{aligned}
f_Y(y)dy &= \Pr\{y < \mathbf{Y} \leq y + dy\} \\
&= \Pr\{y < g(\mathbf{X}) \leq y + dy\} \\
&= \Pr\{g^{-1}(y) < \mathbf{X} \leq g^{-1}(y + dy)\} \\
&= \Pr\{x < \mathbf{X} \leq x + |g'(x)|^{-1}dy\} \\
&= f_X(x)|g'(x)|^{-1}dy.
\end{aligned}
$$

In terms of the cdf, the proof is cleaner. Assuming $g(x)$ is monotonically increasing,

$$
\begin{aligned}
F_Y(y) &= \Pr\{\mathbf{Y} \leq y\} \\
&= \Pr\{g(\mathbf{X}) \leq y\} \\
&= \Pr\{\mathbf{X} \leq g^{-1}(y)\} \\
&= F_X(g^{-1}(y));
\end{aligned}
$$

$$f_Y(y) = \frac{dF_Y(y)}{dy} = f_X(g^{-1}(y)) \left(\frac{dg^{-1}(y)}{dy}\right).$$

Example.

(i) If $\mathbf{X}$ is a normal RV, then $e^{a\mathbf{X}}$ has a log-normal distribution.

(ii) Random number generator - the inverse transform method. We want to generate a random variable $\mathbf{X}$ with cumulative probability function $F_X(x)$ and pdf $f_X(x) = dF_X(x)/dx$. We wish to find a function g such that $\mathbf{U} = g(\mathbf{X})$ is a uniform distribution on $[0, 1]$. Since $\mathbf{X} = g^{-1}(\mathbf{U})$, we have:

$$f_U(u) = \frac{f_X(x(u))}{|g'(x(u))|}, \qquad (0 \leq u \leq 1)$$

Therefore, if we choose $g(x) = F_X(x)$, then $g'(x) = f_X(x)$, and

$$f_U(u) = \frac{f_X(x(u))}{|f_X(x(u))|} = \begin{cases} 1 & \text{for } u \in [0,1] \\ 0 & \text{elsewhere} \end{cases}$$

(iii) The Jacobian is also called entropy in coordinate transformation.

3.1.4.2 Expected Value of a Function of An RV

The expected value, or expectation, of RV $Y = g(\mathbf{X})$ is given by

$$\mathbb{E}[\mathbf{Y}] = \sum_i g(x_i) p_X(x_i), \tag{3.6}$$

for discrete RV $\mathbf{X}$ and

$$\mathbb{E}[\mathbf{Y}] = \int_{-\infty}^{\infty} y f_Y(y) \mathrm{d}y = \int_{-\infty}^{\infty} y \, \mathrm{d}F_Y(y) = \int_{-\infty}^{\infty} g(x) f_X(x) \mathrm{d}x.$$

for continuous RV $\mathbf{X}$.

One can write the EV for both discrete and continuous RVs as

$$\mathbb{E}[g(\mathbf{X})] = \int_{-\infty}^{+\infty} g(x) \mathrm{d}F_X(x). \tag{3.7}$$

If $\mathbf{Y} = \mathbf{X}^k$, then $\mathbb{E}[\mathbf{Y}]$ is the kth moment of $\mathbf{X}$:

$$E[\mathbf{X}^m] = \int_{-\infty}^{+\infty} x^m \mathrm{d}F_X(x). \tag{3.8}$$

Homework. If we have a random number generator for RV $\mathbf{X}$ with pdf $f_X(x)$ and distribution $F_X(x)$, what is the function g such that we need to generate RV $\mathbf{Y} = g(\mathbf{X})$ with pdf $f_Y(y)$?

3.1.5 Popular Discrete and Continuous RVs and Their Distributions

In this section, we introduce several widely encountered RVs.

A multinomial distribution for $n_1, n_2, \cdots, n_m$

$$p(n_1,\cdots,n_m) = \frac{N!}{n_1!n_2!\cdots n_m!}\, p_1^{n_1} p_2^{n_2}\cdots p_m^{n_m}, \quad \left(n_k \geq 0, \ \sum_{k=1}^{m} n_k = N\right)$$

has $\mathbb{E}[n_k] = Np_k$, $\text{Var}[n_k] = Np_k(1-p_k)$, and $\mathbb{E}[n_k n_\ell] = N(N-1)p_k p_\ell$ when $k \neq \ell$. For $m=2$, it is known as a binomial distribution.

A Poisson distribution

$$p(n) = \frac{\lambda^n}{n!}e^{-\lambda}, \quad (n = 0,1,2,...)$$

has $\mathbb{E}[n] = \text{Var}[n] = \lambda$.

A uniform distribution on $[0,L]$

$$f(x) = \frac{1}{L}, \quad (0 \leq x \leq L)$$

has $\mathbb{E}[x] = \frac{L}{2}$ and $\text{Var}[x] = \frac{L^2}{12}$.

A Gaussian distribution

$$f(x) = \frac{1}{\sqrt{2\pi}\sigma}e^{-\frac{(x-\mu)^2}{2\sigma^2}}, \quad (-\infty \leq x \leq \infty)$$

has $\mathbb{E}[x] = \mu$ and $\text{Var}[x] = \sigma^2$.

An exponential distribution

$$f(x) = \lambda e^{-\lambda x}, \quad (x \geq 0)$$

has $\mathbb{E}[x] = \lambda^{-1}$ and $\text{Var}[x] = \lambda^{-2}$.

Example.

(i) A Poisson distribution as the limit of a binomial distribution with $N \to \infty$, $p \to 0$ and $Np = \lambda$. This is known as the Poisson theorem (see Section 5.2.3).

(ii) A binomial distribution as the conditioned sum of two Poisson RVs. Let $\mathbf{N}_i$, $(i=1,2)$, represent two Poisson RV with pmfs

$$p_{N_i}(n) = \frac{\lambda_i^n}{n!}e^{-\lambda_i}.$$

Now the conditional probability is calculated as

$$\Pr\{\mathbf{N}_1 = m | \mathbf{N}_1 + \mathbf{N}_2 = M\} = \frac{p_{N_1}(m)p_{N_2}(M-m)}{\sum_{m=0}^{M} p_{N_1}(m)p_{N_2}(M-m)}$$

$$= \frac{M!}{m!(M-m)!}\theta^m(1-\theta)^{M-m}.$$

where $\theta = \frac{\lambda_1}{\lambda_1 + \lambda_2}$.

(iii) The equilibrium distribution for a simple isomerization reaction $A \rightleftharpoons B$. Each molecule is either in state A or state B, and all molecules are independent of each other. Assume that the probability that each molecule is in state A is p_A, while that for state B is $p_B = 1 - p_A$. If the total number of molecules is N, the number of molecules in state A (n_A) follows a binomial distribution with parameters N and p_A. Hence, the expected value of n_A is Np_A and that of n_B is Np_B. The equilibrium constant of the reaction $A \rightleftharpoons B$ is just $\frac{n_B}{n_A} = \frac{1-p_A}{p_A}$. The variance of n_A is $Np_A p_B$, the same as the variance of n_B. Therefore, the standard deviation of n_A, the square root of the variance, divided by the expected value, quantifying the noise level of the reaction, is $\frac{1}{\sqrt{N}}\sqrt{\frac{1-p_A}{p_A}}$, which tends to zero as $\frac{1}{\sqrt{N}}$.

3.1.6 A Pair of RVs and Their Independence

Let us now consider a pair of continuous random variables $(\mathbf{X}, \mathbf{Y})$. The pdf for this pair of RVs is a multivariate function $f_{XY}(x,y)$. Clearly, it has to be nonnegative, and its integration is

$$\int_{-\infty}^{+\infty} \int_{-\infty}^{+\infty} f_{XY}(x,y) dx dy = 1.$$

The function

$$f_X(x) = \int_{-\infty}^{+\infty} f_{XY}(x,y) dy$$

is called the *marginal distribution* of $\mathbf{X}$, irrespective of $\mathbf{Y}$. Clearly, it is also a pdf. Additionally, for a given $\mathbf{Y} = y_0$, the function of x

$$\frac{f_{XY}(x,y_0)}{\int_{-\infty}^{+\infty} f_{XY}(x,y_0) dx}$$

is also a pdf. This is called a *conditional probability* density function. It is the pdf of $\mathbf{X}$ when $\mathbf{Y} = y_0$ is given. It is usually denoted as $f_{X|Y}(x|\mathbf{Y} = y_0)$. In terms of the conditional probability density function, we have

$$f_{XY}(x,y) = f_{X|Y}(x|\mathbf{Y} = y) f_Y(y).$$

If a conditional pdf $f_{X|Y}(x|\mathbf{Y} = y_0)$ is independent of the value of y_0, then we say that RVs $\mathbf{X}$ and $\mathbf{Y}$ are *independent*. In this case,

$$f_{XY}(x,y) = f_X(x)f_Y(y)$$

can be factored into two distributions. This is the multiplicative law of probability. It can also be taken as the definition of the independence of RVs **X** and **Y**.

For independent RVs **X** and **Y**, $\mathbb{E}[\mathbf{XY}] = \mathbb{E}[\mathbf{X}]\mathbb{E}[\mathbf{Y}]$. It is a necessary but not sufficient condition for independence. The two RVs satisfying this condition is called *uncorrelated*.

The EV of the sum of two RVs is always the sum of the EVs. The variance of the sum of two RVs, however, is the sum of the variances only when the RVs are uncorrelated:

$$\mathbb{E}[\mathbf{X}+\mathbf{Y}] = \mathbb{E}[\mathbf{X}]+\mathbb{E}[\mathbf{Y}];$$
$$\mathrm{Var}[\mathbf{X}+\mathbf{Y}] = \mathbb{E}\left[(\mathbf{X}+\mathbf{Y})^2\right] - \left(\mathbb{E}[\mathbf{X}]+\mathbb{E}[\mathbf{Y}]\right)^2$$
$$= \mathbb{E}[\mathbf{X}^2] - (\mathbb{E}[\mathbf{X}])^2 + \mathbb{E}[\mathbf{Y}^2] - (\mathbb{E}[\mathbf{Y}])^2 + 2\mathbb{E}[\mathbf{XY}] - 2\mathbb{E}[\mathbf{X}]\mathbb{E}[\mathbf{Y}]$$
$$= \mathrm{Var}[\mathbf{X}] + \mathrm{Var}[\mathbf{Y}].$$

3.1.7 Functions of Two RVs

Let us assume that we have the joint pdf for **X** and **Y**: $f_{XY}(x,y)$. We further assume that g and h are monotonic. Then, we define

$$\mathbf{Z} = g(\mathbf{X},\mathbf{Y}), \quad \mathbf{W} = h(\mathbf{X},\mathbf{Y})$$

.

Hence, the joint pdf for **Z** and **W**:

$$f_{ZW}(z,w) = \frac{f_{XY}(x,y)}{\begin{vmatrix} \partial g/\partial x & \partial g/\partial y \\ \partial h/\partial x & \partial h/\partial y \end{vmatrix}}$$

In a special case:

$$\mathbf{Z} = g(\mathbf{X},\mathbf{Y}), \quad \mathbf{W} = \mathbf{X},$$

then

$$f_{ZW}(z,w) = \frac{f_{XY}(x,y)}{\left|\frac{\partial g}{\partial y}\right|},$$

and

$$f_Z(z) = \int_{-\infty}^{\infty} f_{ZW}(z,w)dw = \int_{-\infty}^{\infty} \frac{f_{XY}(x,y)}{\left|\frac{\partial g}{\partial y}\right|}dx.$$

Similarly,

$$f_W(w) = \int_{-\infty}^{\infty} \frac{f_{XY}(x,y)}{|g_x'(x,y)|}dy.$$

3.2 Discrete-time, Discrete-state Markov Chains

Thus far we have only dealt with random variables. We are now interested in time-dependent random phenomena. Stochastic processes are introduced to represent dynamic systems that involve randomness. One now describes a *stochastic process* using the **probability** that $X(t)$ is x at time t with probability $\Pr\{X(t) = x\}$. Consider a dynamic system with discrete times $n = 0, 1, 2, \cdots$, for describing an entire trajectory; one might be interested in the joint probability

$$\Pr\{X(0) = x_0, X(1) = x_1, \cdots, X(n) = x_n, \cdots\}. \tag{3.9}$$

3.2.1 Transition Probability Matrix

A Markov chain represents a kind of system whose future dynamics, in statistical terms, are completely determined by the current state, without memory from the past. In terms of conditional probability, this means that for any state $x_0, x_1, \cdots, x_n$, x_{n+1},

$$\Pr\{X(n+1) = x_{n+1}|X(n) = x_n, X(n-1) = x_{n-1}, \cdots, X(0) = x_0\}$$
$$= \Pr\{X(n+1) = x_{n+1}|X(n) = x_n\}. \tag{3.10}$$

With this Markovian property, the probability of the entire trajectory in (3.9) can be simplified to

$$\Pr\{\mathbf{X}(0) = x_0, \mathbf{X}(1) = x_1, \cdots, \mathbf{X}(n) = x_n\}$$
$$= \Pr\{\mathbf{X}(n) = x_n | \mathbf{X}(n-1) = x_{n-1}\} \times \Pr\{\mathbf{X}(n-1) = x_{n-1} | \mathbf{X}(n-2) = x_{n-2}\}$$
$$\cdots \Pr\{\mathbf{X}(1) = x_1 | \mathbf{X}(0) = x_0\} \times \Pr\{\mathbf{X}(0) = x_0\}. \tag{3.11}$$

There are only two things to be specified for (3.11): the initial distribution for the $\mathbf{X}(0)$, and the *transition probability*

$$P_{x_i x_j} = \Pr\{\mathbf{X}(n) = x_j | \mathbf{X}(n-1) = x_i\}. \tag{3.12}$$

Here, we assume $P_{x_i x_j}$ is independent of time n.

If one uses positive integers to denote discrete states, then the transition probability can be written as a matrix P_{ij}, in which

$$P_{ij} \geq 0, \ \sum_j P_{ij} = 1,$$

for each i and j. Such a nonnegative matrix is also called a *stochastic matrix*.

The Chapman–Kolmogorov Equation. Suppose $P_{ij}(n)$ is the probability that the process is in state x_j at time n, given that it starts in state x_i at time 0:

$$P_{ij}(n) = P\{\mathbf{X}(n+m) = j | \mathbf{X}(m) = i\}.$$

Then

$$P_{ij}(m+n) = \sum_{k=0}^{\infty} P_{ik}(m) P_{kj}(n). \tag{3.13}$$

This equation states that to move from state s_i to state s_j in time $m+n$, $X(m)$ moves to some state s_k in time m and then from s_k to s_j in the remaining time n.

3.2.2 Invariant Probability Distribution

Denote the distribution of $\mathbf{X}(n)$ as $\{p_i(n)\}$, then the distribution satisfies

$$p_i(n) = \sum_j p_j(n-1) P_{ji}.$$

Under very mild conditions, it can be proven that the distribution $\{p_i(n)\}$ converges to a limit distribution $\{\pi_i\}$ when n goes to infinity, which satisfies

$$\pi_i = \sum_j \pi_j P_{ji}.$$

This implies that if the initial distribution $\{p_i(0)\}$ is equal to $\{\pi_i\}$, then all the distributions $\{p_i(n)\}$ for each n are equal to $\{\pi_i\}$. That is why $\{\pi_i\}$ is called the invariant probability distribution, and in this case, the stochastic process is said to be stationary.

3.3 Continuous-time, Discrete-state Markov Processes

3.3.1 Poisson Process

A Poisson process is associated with the repeated occurrence of an event along a one-dimensional axis (called time). These events can be represented by points randomly distributed on the time axis with the following properties:

(i) all events are independent in each disjoint interval;

(ii) the probability $P_n(\tau)$ of having n events in a very small interval $(t, t+\tau)$ is

$$P_1(\tau) = \lambda\tau + o(\tau), \quad \text{and} \quad P_0(\tau) = 1 - P_1(\tau) + o(\tau)$$

where λ is called the intensity of the Poisson process. Then, we have:

$$P_k(\tau) = \frac{(\lambda\tau)^k e^{-\lambda\tau}}{k!}.$$

Proof: To show this, we note that

$$P_0(t+dt) = P_0(t)P_0(dt) = P_0(t)\left[1 - \lambda\,dt + o(dt)\right]$$

$$\implies \qquad P_0'(t) = -\lambda P_0(t)$$

and since $P_0(0) = 1$, we have

$$P_0(t) = e^{\lambda t}.$$

For $k \geq 1$,

$$P_k(t+dt) = P_k(t)P_0(dt) + P_{k-1}P_1(dt) + o(dt) = P_k(t)(1 - \lambda\,dt) + P_{k-1}(\lambda\,dt) + o(dt),$$

$$\implies \frac{P_k(t+dt) - P_k(t)}{dt} = -\lambda P_k(t) + \lambda P_{k-1}(t) + o(1),$$

$$\Longrightarrow \frac{dP_k(t)}{dt} = -\lambda P_k(t) + \lambda P_{k-1}(t). \tag{3.14}$$

To solve Eqn. (3.14), we introduce $\psi_k(t) = P_k(t)e^{\lambda t}$, where the equation for $\psi_k(t)$ is

$$\begin{aligned}
\frac{d\psi_k(t)}{dt} &= \frac{dP_k(t)}{dt}e^{\lambda t} + \lambda \psi_k(t) \\
&= -\lambda \psi_k(t) + \lambda \psi_{k-1}(t) + \lambda \psi_k(t) \\
&= \lambda \psi_{k-1}(t)
\end{aligned}$$

with condition $\psi_k(0) = 0$. We therefore have

$$\psi_k(t) = \frac{(\lambda t)^k}{k!}, \quad \text{and} \quad P_k(t) = \frac{(\lambda t)^k}{k!}e^{-\lambda}. \qquad \Box$$

Homework. If $\mathbf{Z}_1, \mathbf{Z}_2, ..., \mathbf{Z}_m$ are independent Poisson random variables, then

$$\mathbf{Z} = \mathbf{Z}_1 + \mathbf{Z}_2 + \cdots + \mathbf{Z}_m$$

is still Poisson.

Homework. Try to use the method of generating functions to show that the random sum of

$$\mathbf{N}_1 + \mathbf{N}_2 + \cdots + \mathbf{N_K},$$

where $\mathbf{K}$ is a Poisson RV with mean λ, and the RVs $\mathbf{N}$ are i.i.d. binary RVs with $\Pr\{\mathbf{N} = 1\} = p$ and $\Pr\{\mathbf{N} = 0\} = 1 - p$, is still Poisson.

3.3.1.1 Waiting-time Distribution

There are two aspects of a Poisson process: the distribution of the corresponding *counting process*, which is exactly $P_k(t)$ defined in the previous section, and the distribution of the times of occurrences, called the *point processes*.

What is the probability distribution of the arriving time of first event? Let us denote the time by an RV $\mathbf{T}_1$, then

$$\Pr\{\mathbf{T}_1 \le t\} = 1 - \Pr\{\mathbf{T}_1 > t\}$$
$$= 1 - \Pr\{\mathbf{N}_t = 0\}$$
$$= 1 - e^{-\lambda t},$$

where $\mathbf{N}_t$ is the number of events before time t (counting process). The pdf for $\mathbf{T}_1$ is obtained by taking the derivative:

$$f_{\mathbf{T}_1}(t) = \lambda e^{-\lambda t}.$$

Hence, the Poisson process is intimately related to the exponential distribution.

Similarly, let us consider $\mathbf{T}_k$:

$$\Pr\{\mathbf{T}_k \le t\} = 1 - \Pr\{\mathbf{T}_k > t\}$$
$$= 1 - \Pr\{\mathbf{N}_t = k - 1\}$$
$$= 1 - \sum_{\ell=0}^{k-1} \frac{(\lambda t)^\ell e^{-\lambda t}}{\ell!}.$$

Again, differentiating with respect to t, we have the pdf:

$$f_{\mathbf{T}_k}(t) = \frac{d}{dt}\Pr\{\mathbf{T}_k \le t\} = -\sum_{\ell=0}^{k-1} \frac{\lambda \ell (\lambda t)^{\ell-1} - \lambda (\lambda t)^\ell e^{-\lambda t}}{\ell!}$$
$$= -\sum_{\ell=0}^{k-2} \frac{\lambda (\lambda t)^\ell e^{-\lambda t}}{\ell!} + \sum_{\ell=0}^{k-1} \frac{\lambda (\lambda t)^\ell e^{-\lambda t}}{\ell!}$$
$$= \frac{\lambda (\lambda t)^{k-1} e^{-\lambda t}}{(k-1)!},$$

which is called *gamma distribution*. When $k = 1$, $\mathbf{T}_k$ is said to be exponentially distributed.

We now show that the gamma distribution of order k is just the sum of k independent exponential distributions:

$$\mathbf{T}_k = \mathbf{T}_1 + (\mathbf{T}_2 - \mathbf{T}_1) + (\mathbf{T}_3 - \mathbf{T}_2) + \cdots + (\mathbf{T}_k - \mathbf{T}_{k-1}),$$

where the waiting times for the ith event $\mathbf{T}_i - \mathbf{T}_{i-1}$ are independent.

Proof: We first determine the characteristic function for the gamma distribution:

$$\int_0^\infty \frac{\lambda(\lambda t)^{k-1}e^{-\lambda t}}{(k-1)!}e^{-ist}dt = -\left(\frac{\lambda}{\lambda+is}\right)\int_0^\infty \frac{(\lambda t)^{k-1}}{(k-1)!}de^{-(\lambda+is)t}$$

$$= \left(\frac{\lambda}{\lambda+is}\right)\int_0^\infty \frac{\lambda(\lambda t)^{k-2}}{(k-2)!}e^{-(\lambda+is)t}dt$$

$$= \left(\frac{\lambda}{\lambda+is}\right)^k.$$

Note that

$$\frac{\lambda}{\lambda+is} = \int_0^\infty \lambda e^{-\lambda t}e^{-ist}dt$$

is the characteristic function of an exponential distribution. □

The mean of the gamma distribution is

$$i\cdot\frac{\partial}{\partial s}\left(\frac{\lambda}{\lambda+is}\right)^k\bigg|_{s=0} = \frac{k\lambda^k}{(\lambda+is)^{k+1}}\bigg|_{s=0} = \frac{k\lambda^k}{\lambda^{k+1}} = \frac{k}{\lambda},$$

as expected. λ^{-1} is the mean waiting time.

3.3.1.2 Uniform Distribution and Poisson Process

If points generated by a Poisson process are labeled on a time axis, what is their distribution? The answer is they are uniformly distributed with density λ. This is not a very intuitive result. The key is that the statement is *conditioned on a fixed total number of events in an interval* !

Let T_1, T_2, ... be the occurrence times in a Poisson process of intensity λ. Conditioned on $N_t = n$, the random variables T_1, T_2, ..., T_n are order statistics of uniformly distributed n random variables on the interval $[0,t)$.

We prove the simplest version of this result: $n = 1$. We know the joint pdf for RVs T_1 and T_2:

$$\Pr\{t\leq T_1 < t+dt, \tau \leq T_2 < \tau+d\tau\} = \lambda^2 e^{-\lambda t}e^{-\lambda(\tau-t)}dt d\tau \quad (\tau > t);$$

therefore,

$$\Pr\{t\leq T_1 < t+dt, T_2 \geq \tau\} = \left(\int_\tau^\infty \lambda^2 e^{-\lambda t}e^{-\lambda(\tau'-t)}d\tau'\right)dt$$

$$= \lambda\left(1-e^{-\lambda\tau}\right)dt, \quad (\tau > t). \tag{3.15}$$

Hence,

$$\Pr\{t \le \mathbf{T}_1 < t + dt | \mathbf{T}_2 \ge \tau\} = \frac{\Pr\{t \le \mathbf{T}_1 < t + dt, \mathbf{T}_2 \ge \tau\}}{\Pr\{\mathbf{T}_2 \ge \tau\}} \quad (t < \tau)$$

$$= \frac{\lambda \left(1 - e^{-\lambda\tau}\right) dt}{\displaystyle\int_0^\tau \lambda \left(1 - e^{-\lambda\tau}\right) dt} \quad (t < \tau)$$

$$= \frac{dt}{\tau} \quad (t < \tau).$$

This is a uniform distribution on $[0, \tau)$. Note that without the condition,

$$\Pr\{t \le \mathbf{T}_1 < t + dt\} = \lambda e^{-\lambda t} dt \quad (0 \le t < \infty).$$

This result shows how important the condition is to a probabilistic problem. Its inclusion or exclusion can change the result!

3.3.2 Continuous-time, Discrete-state Markov Process

Let us consider a continuous-time Markov process $\mathbf{X}(t)$ with a finite number N of possible states. For simplicity we shall label the states $1, 2, \cdots, N$. Define the transition probability matrix $\mathbf{P}(t) = \{P_{ij}(t)\}$, in which $P_{ij}(t) = P(\mathbf{X}(t) = j | \mathbf{X}(0) = i)\}$. Assume that we know the infinitesimal transition rates q_{ij} for the continuous-time Markov process: for a very small t, $P_{ii}(t) = 1 - q_i t + o(t)$ approximately represents the probability that the process has *not* escaped from the state s_i, and similarly,

$$P_{ij}(t) = q_{ij}t + o(t), \; q_{ij} \ge 0,$$

in which $q_i = -q_{ii} = -\sum_{k \ne i} q_{ik} > 0$.

Therefore, according to the Chapman–Kolmogorov equation $\mathbf{P}(t+s) = \mathbf{P}(t)\mathbf{P}(s)$, when h is small,

$$P_{ij}(t+h) = P_{ij}(t)P_{jj}(h) + \sum_{k \ne j} P_{ik}(t)P_{kj}(h)$$

$$= P_{ij}(t)(1 - q_j h) + \sum_{k \ne j} P_{ik}(t)q_{kj}h + o(h),$$

followed by the *Kolmogorov forward equation*

$$\frac{d}{dt}\mathbf{P}(t) = \mathbf{P}(t)\mathbf{Q} \tag{3.16}$$

with initial value $\mathbf{P}(0) = \mathbf{I}$, in which $\mathbf{Q} = \{q_{ij}\}$.

The fundamental solution to (3.16) is $\mathbf{P}(t) = \exp(\mathbf{Q}t)$, which also implies the *Kolmogorov backward equation* $\frac{\mathrm{d}}{\mathrm{d}t}\mathbf{P}(t) = \mathbf{Q}\mathbf{P}(t)$. Strictly speaking, the Kolmogorov forward and backward equations are equations satisfied by the matrix $\mathbf{P}(t)$. We note that if a Markov system has an initial distribution $p_i(0)$, and if one denotes the probability of being in state k at time t as $p_k(t)$, then $p_k(t) = \sum_i p_i(0)P_{ik}(t)$ satisfies

$$\frac{\mathrm{d}}{\mathrm{d}t}p_k(t) = \sum_{i \neq k} p_i(t)q_{ik} - \sum_{i \neq k} p_k(t)q_{ki}. \tag{3.17}$$

This equation has been called a master equation in physics.

If $\{\pi_k\}$ is the stationary probability distribution, when the right-hand side of (3.17) is zero, i.e., the solution to

$$\sum_{k=1}^{N} \pi_k q_{k\ell} = 0, \quad \ell = 1, 2, \cdots, N,$$

under the constraint $\sum_{k=1}^{N} \pi_k = 1$, then $p_i(t)$ converges to π_i for each i under very mild conditions.

Similarly, a differential equation for the vector $\{u_k\}$

$$\frac{\mathrm{d}u_k}{\mathrm{d}t} = \sum_{\ell=1}^{N} q_{k\ell}u_\ell, \tag{3.18}$$

is sometimes also called a *backward equation*. u_k has a very interesting probabilistic meaning: it can be understood as a certain conditional expectation of a random variable $\mathbf{Y}(t) = y(\mathbf{X}(t))$. Then, its conditional expected value $u_k \equiv \mathbb{E}[\mathbf{Y}(t)|\mathbf{X}(0) = k]$ satisfies

$$\frac{\mathrm{d}u_k(t)}{\mathrm{d}t} = \frac{\mathrm{d}}{\mathrm{d}t}\mathbb{E}[\mathbf{Y}(t)|\mathbf{X}(0) = k] = \frac{\mathrm{d}}{\mathrm{d}t}\sum_{\ell=1}^{N} P_{k\ell}(t)y(\ell) = \sum_{\ell=1}^{N}\left(\sum_{j=1}^{N} q_{kj}P_{j\ell}(t)\right)y(\ell)$$

$$= \sum_{j=1}^{N} q_{kj}\mathbb{E}[\mathbf{Y}(t)|\mathbf{X}(0) = j] = \sum_{j=1}^{N} q_{kj}u_j(t). \tag{3.19}$$

The solution to the backward equation $u_k(t)$ has the important property where

$$\sum_{k=1}^{N} u_k(t)\pi_k$$

is independent of time t, e.g., it is a conserved quantity.

The solutions to the master equation have another important property. Let $p_k(t)$ and $q_k(t)$ be two solutions to a forward equation with different initial distributions $p_k(0)$ and $q_k(0)$, respectively. Then,

$$\frac{d}{dt} \sum_{k=1}^{N} p_k(t) \ln \left(\frac{p_k(t)}{q_k(t)} \right) \leq 0. \tag{3.20}$$

One special case of this, which is widely known, is when $q_k(t) = \pi_k$, if $\pi_k > 0$, $\forall k$.

Similarly, the positive solutions to a backward equation, $u_k(t)$ and $v_k(t)$, have the following property:

$$\frac{d}{dt} \sum_{k=1}^{N} \left(\pi_k u_k(t) \right) \ln \left(\frac{u_k(t)}{v_k(t)} \right) \leq 0. \tag{3.21}$$

One special case of this is when $v_k(t) \equiv 1$. The quantity in Eqn. (3.21) is called an H-function; the quantity in Eqn. (3.20) is called relative entropy, Kullback–Leibler divergence in information theory, or free energy in physical chemistry.

3.3.3 Exact Algorithm For Generating a Markov Trajectory

The algorithm for generating a Markov trajectory is based on a well-known theorem describing the statistical properties of trajectories. It can be traced back to at least J. L. Doob (1910–2004) [50], who states that the sojourn time of each state i is exponentially distributed with parameter q_i, and independently, the probability that the trajectory will transit directly from state i to state j immediately after the system leaves state i is $\frac{q_{ij}}{q_i}$.

The algorithm was developed by Boltz, Kalos, and Lebowitz for kinetic Monte Carlo simulations in physics and independently by D. T. Gillespie (1938–2017) for stochastic simulations of reactions in chemistry [28, 103]. For each step where the system is in state i, generate a random number that is exponentially distributed with parameter q_i, determine the time interval, then generate another random number to determine the next state and repeat.

3.3.4 Mean First Passage Time

For any subset A of the state space of a Markov process, e.g., $A = \{i\}$ or $A = \{i,j\}$, we can define the mean time when the set A is first hit starting from any state k, which is also called the mean first passage time and denoted as $\tau_A(k)$. Obviously, $\tau_A(k) = 0$ if $k \in A$.

One can derive a system of equations for $\tau_A(k)$. We know that, starting from state k, the process will first wait for an average exponential time $\frac{1}{q_k}$, then jump to any state j satisfying $q_{kj} > 0$ according to the probability that is proportional to q_{kj}. Hence, $\tau_A(k)$ would be the summation of $\frac{1}{q_k}$ and the probability weighted $\tau_A(j)$ satisfying $q_{kj} > 0$.

Hence, $\tau_A(k)$ satisfies the following equations with boundary $\tau_A(k) = 0$ if $k \in A$:

$$\tau_A(k) = \frac{1}{q_k} + \sum_{j \neq k} \left(\frac{q_{kj}}{q_k} \right) \tau_A(j).$$

Noting that q_k is the sum of all q_{kj} with $j \neq k$, the above equation can be rearranged into

$$\sum_{j \neq k} q_{kj} \left(\tau_A(j) - \tau_A(k) \right) = -1. \tag{3.22}$$

3.3.5 Stochastic Trajectory In Terms of Poisson Processes

One can always use integers $X = 1, 2, \cdots, N$ to label the N states of a discrete-state, continuous-time Markov process. Then $X(t)$ is a function of time with jumps at the times of transition. This stochastic trajectory can be represented in terms of $N \times (N-1)$ independent, unit-rate Poisson processes $Y_{ij}(t)$, $1 \leq i \neq j \leq N$ through an integral equation:

$$X(t) = X(0) + \sum_{i,j=1}^{N} (j-i) Y_{ij} \left(\int_0^t q_{ij} \delta_{i,X(s)} \, ds \right), \tag{3.23}$$

in which δ_{hk} is the Kronecker δ function: $\delta_{hk} = 1$ when $h = k$ and $\delta_{hk} = 0$ when $h \neq k$. The expression in (3.23) is called a *time-changed Poisson representation* of the discrete-state, continuous-time Markov process, which was developed by T. G. Kurtz in 1980 [156].

3.3.6 The Birth-death Process

3.3.6.1 The Pure Birth Process

A birth process is a Markov process $\mathbf{X}(t)$ with trajectories that can be modeled as increasing staircase functions. $\mathbf{X}(t)$ takes values of 1, 2, 3, ... and increases by 1 at the discontinuous point t_i (birth time). The transition rates λ_{ij} are nonzero only if $j = i + 1$ and are denoted

$$\lambda_{i,i+1} = \mu_i,$$

and the diagonal can be represented as

$$-\lambda_{i,i} = \mu_i.$$

The pure birth process is a generalization of the Poisson process where $\mu_i \equiv \mu$.

Just as a Poisson process can be understood from either its counting process or the waiting time intervals, a pure birth process can also be understood by considering the random sojourn time $\mathbf{S}_k$. RV $\mathbf{S}_k$ is the time between the kth and $(k+1)$th birth events. Hence,

$$P_n(t) = \Pr\left\{ \sum_{i=0}^{n-1} \mathbf{S}_i \leq t \leq \sum_{i=0}^{n} \mathbf{S}_i \right\}.$$

It is easy to show that, as in a Poisson process, all $\mathbf{S}_k$ have exponential distributions with respective means $\frac{1}{\lambda_k}$.

A particular birth process that characterizes the growth of a population with identical and independent individuals is the Yule process, in which $\lambda_{n,n+1} = n\alpha$. α is known as the growth rate *per capita* and is the same α in the deterministic exponential growth model $dN(t)/dt = \alpha N(t)$.

3.3.6.2 The Birth-death Process

Similar to the birth process, one can define a death process, which has a nonzero $\lambda_{i,i-1}$. The birth-death process is a generalization of the pure birth and pure death processes. It is the continuous-time analog of a random walk with nonuniform bias.

The stationary solution to a birth-death process, with nonzero $\lambda_{i,i-1}$ and $\lambda_{i,i+1}$, can be explicitly obtained. We note that the stationary probability p_i^s, if it exists, satisfies

$$p_i^s \lambda_{i,i+1} = p_{i+1}^s \lambda_{i+1,i}. \tag{3.24}$$

Therefore, we have

$$p_i^s = C \prod_{\ell=1}^{i} \frac{\lambda_{\ell-1,\ell}}{\lambda_{\ell,\ell-1}}. \tag{3.25}$$

The C in the equation is a normalization factor that can be determined by noting $\sum_{i=0}^{\infty} p_i^s = 1$:

$$C = \left(1 + \sum_{i=1}^{\infty} \prod_{\ell=1}^{i} \frac{\lambda_{\ell-1,\ell}}{\lambda_{\ell,\ell-1}} \right)^{-1}.$$

One specific birth-death process is Kendall's birth-death-immigration process, in which $\lambda_{j,j-1} = j\mu$, representing the rate of death, and $\lambda_{j,j+1} = v + j\alpha$, representing the rate of birth ($j\alpha$) and a constant rate of immigration (v).

The equation for the probability distribution at time t follows the system of equations

$$\frac{dp_0(t)}{dt} = -vp_0(t) + \mu p_1(t) \tag{3.26}$$

$$\frac{dp_j(t)}{dt} = -(v + j\alpha + j\mu)p_j(t) + (v + (j-1)\alpha)p_{j-1}(t) + (j+1)\mu p_{j+1}(t)$$

$$(j = 1,2,...).$$

To deal with the time-dependent solution to this set of equations, we define a generating function

$$G(s,t) = \sum_{j=0}^{\infty} p_j(t)s^j, \quad (0 \le s \le 1), \tag{3.27}$$

and we have:

$$\frac{\partial G(s,t)}{\partial t} = (\alpha s - \mu)(s-1)\frac{\partial G(s,t)}{\partial s} + v(s-1)G(s,t). \tag{3.28}$$

One can show that Eqn. (3.28) is valid even when α, μ and v are functions of time.

If the initial state of the Markov process is in state i, then $G(s,0) = s^i$ corresponding to $p_j(0) = \delta_{ji}$. To obtain the moments for the birth-death-immigration process, the partial differential equation (PDE) can be simplified to an ODE. Taking the mean as an example,

$$m = \left. \frac{\partial G(s,t)}{\partial s} \right|_{s=1}.$$

Hence, differentiating (3.27) with respect to s:

$$\frac{dm}{dt} = \frac{\partial^2}{\partial t \partial s} G(1,t) = (\alpha - \mu)m(t) + v$$

and $m(0) = i_0$. This ODE can be solved, and we have

$$m(t) = i_0 \exp\left[-\int_0^t [\alpha(z) - \mu(z)]dz\right] + \int_0^t v(\tau) \exp\left[\int_\tau^t [\alpha(z) - \mu(z)]dz\right] d\tau.$$

For the special case of constant, time-independent α, μ and v, we have

$$m(t) = i_0 e^{(\alpha-\mu)t} + \frac{v}{\alpha - \mu}\left(e^{(\alpha-\mu)t} - 1\right).$$

When $\alpha < \mu$, i.e., the birth rate is less than the death rate,

$$m(t) \longrightarrow \frac{v}{\mu - \alpha} \quad \text{as } t \longrightarrow \infty.$$

If $\alpha(t) = \mu(t) = 0$, but v is not constant but is a function of time, $v(t)$, the process is a pure birth process, called a nonhomogeneous Poisson process. Then, for initial condition $i = 0$, we have the equation for a nonhomogeneous Poisson process

$$\frac{\partial}{\partial t} G(s,t) = v(t)(s-1)G(s,t),$$

$$G(s,t) = \exp\left(-(1-s)\int_0^t v(\tau)d\tau\right),$$

$$P_{0j}(t) = \frac{1}{j!}\left(\int_0^t v(\tau)d\tau\right)^j \exp\left(-\int_0^t v(\tau)d\tau\right).$$

3.3.6.3 Birth-death Process and Chemical Reactions

Interestingly, the population models for both linear births and deaths, as well as mutations, can be represented by a simple system of chemical reactions. Specifically, let X and Y be two related populations, each with its own birth and death. Moreover, there is a mutation from population X to Y. Now consider a chemical reaction system in which the number of molecules for species X and Y (by abusing the notation intentionally) changes with time according to the reaction scheme

$$A + X \xrightarrow{\lambda} 2X, \quad X \xrightarrow{\mu} B, \quad X + C \xrightarrow{\rho} X + Y, \tag{3.29}$$

in which the parameters λ, μ, ρ are the birth rate, death rate, and mutation rate, respectively. Similarly, for Y, we have

$$A + Y \xrightarrow{\lambda'} 2Y, \quad Y \xrightarrow{\mu'} B. \tag{3.30}$$

The values of λ', μ' might or might not be the same as those of λ, μ, respectively.

It is easy to determine that the models for birth-death processes with mutations are equivalent to the models that we discussed in Chapter 2 and will discuss in Chapter 5 for chemical reaction systems. Here we describe only a linear example; nonlinear examples are equivalent.

3.4 Continuous-time, Real-valued Diffusion Process

This is a vast subject. Since the stochastic models used in this book are mainly based on integer-valued, continuous-time Markov processes, we shall only give a very brief overview of the subject. The readers are referred to [262, 85, 175, 244, 194] and [143] for more information.

Let $X(t)$ be a real-valued, continuous-time Markov process with continuous trajectories, also known as a *diffusion process*. With initial value $X(0) = x_0$, its probability density function at t, $f(x,t|x_0)$, satisfies the partial differential equation

$$\frac{\partial f(x,t)}{\partial t} = D \frac{\partial^2 f(x,t)}{\partial x^2} - \frac{\partial}{\partial x}\left(V(x) f(x,t) \right), \tag{3.31}$$

with initial condition $f(x,0|x_0) = \delta(x - x_0)$. Its stochastic trajectory satisfies the integral equation

$$X(t) = X(0) + \int_0^t V(X(s)) ds + \sigma W(t), \tag{3.32}$$

where $D = \frac{\sigma^2}{2}$ is called the diffusion coefficient[2], $V(x)$ is called drift, and $W(t)$ is standard Brownian motion. The probability density function for $W(t)$ is Gaussian,

$$\Pr\{x < W(t) \leq x + dx\} = \frac{1}{\sqrt{2\pi t}} e^{-\frac{x^2}{2t}} dx, \tag{3.33}$$

which is precisely the solution to the PDE in (3.31) with $D = \frac{1}{2}$ and $V(x) = 0$.

Typically, in rigorous mathematical theory, general diffusion processes that have a nonconstant $D(x)$ needs drift and diffusion to satisfy certain conditions, such as global Lipschitz continuity, twice-continuous differentiability, and uniform ellipticity. Under these conditions, all the results in this section can be rigorously proven.

---

[2] In some literature, σ or σ^2 is also called the diffusion coefficient.

The Kolmogorov forward equation (3.31) and its corresponding stochastic differential equation (3.32) are the diffusion counterparts of Eqs. (3.17) and (3.23), respectively, for discrete state Markov processes.

The conditional probability density function $f(x,t|y)$, as a function of y and t, also satisfies the Kolmogorov backward equation

$$\frac{\partial f(x,t|y)}{\partial t} = D\frac{\partial^2 f(x,t|y)}{\partial y^2} + V(y)\frac{\partial f(x,t|y)}{\partial y}. \tag{3.34}$$

To prove Eqn. (3.34), we start with the Chapman–Kolmogorov integral equation for total probability:

$$f(x,t+s|y) = \int_{\mathbb{R}} f(x,t|z)f(z,s|y)\mathrm{d}z. \tag{3.35}$$

We then have

$$\begin{aligned}
\frac{\partial f(x,t+s|y)}{\partial t} &= \frac{\partial f(x,t+s|y)}{\partial s} = \int_{\mathbb{R}} f(x,t|z)\frac{\partial f(z,s|y)}{\partial s}\mathrm{d}z \\
&= \int_{\mathbb{R}} f(x,t|z)\left(D\frac{\partial^2 f(z,s|y)}{\partial z^2} - \frac{\partial}{\partial z}\Big(V(z)f(z,s|y)\Big)\right)\mathrm{d}z \\
&= \int_{\mathbb{R}} \left(D\frac{\partial^2 f(x,t|z)}{\partial z^2} + V(z)\frac{\partial f(x,t|z)}{\partial z}\right)f(z,s|y)\mathrm{d}z.
\end{aligned}$$

If we let $s \to 0$, since $f(z,s|y) = \delta(z-y)$, we have Eqn. (3.34).

In Eqns. (3.32), (3.31), and (3.34), when $D = \frac{\sigma^2}{2} = 0$, we have

$$\frac{\mathrm{d}X}{\mathrm{d}t} = V(X), \tag{3.36a}$$

$$\frac{\partial f(x,t)}{\partial t} = -\frac{\partial}{\partial x}\Big(V(x)f(x,t)\Big), \tag{3.36b}$$

$$\frac{\partial u(y,t)}{\partial t} = V(y)\frac{\partial u(y,t)}{\partial y}. \tag{3.36c}$$

The two linear PDEs are actually "equivalent", in a certain sense, to the nonlinear ODE. Indeed, one of the methods for solving the latter two PDEs is to construct their *characteristic lines*, which are precisely the solutions to the ODE. In classical mechanics, Eqn. (3.36b) is known as the Liouville equation for the ODE; in modern dynamic systems theory, the solutions to the linear Eqns. (3.36b) and (3.36c) are called the *Perron–Frobenius–Ruelle transfer operator* and the *Koopman operator*, respectively. They are linear representations of the nonlinear dynamics of (3.36a).In stochastic dynamics, the triplet in (3.36) become the SDE and its corresponding Kolmogorov forward and backward equations.

3.5 The Theory of Entropy and Entropy Production

The term "information entropy" has been widely used in engineering in recent years, and it is probably one of the most elusive concepts in science. There have been continuous discussions on more rigorous definitions of *information* and *information entropy*. As we discussed at the beginning of this chapter, any stochastic problem has a probability space, and a laboratory observable $\mathbf{X}$ is defined as a random variable. A cardinal rule in physics and applied mathematics is that only a nondimensionalized, positive scalar quantity can be a legitimate argument of a logarithmic function. With this in mind, *information* is defined as the logarithm of the relative probability of **two** probability distributions μ_1 and μ_2 (called the Radon–Nikodym derivative in mathematics), rendered as a dimensionless random variable as follows:

$$-\ln\left(\frac{d\mu_1}{d\mu_2}(\omega)\right), \tag{3.37}$$

and Shannon's information entropy is the expected value of this unique random variable:

$$H\left[\mu_1\|\mu_2\right] = -\mathbb{E}^{\mu_1}\left[\ln\left(\frac{d\mu_1}{d\mu_2}(\omega)\right)\right], \tag{3.38}$$

in which the expectation is carried out with respect to the probability distribution μ_1.

We shall emphasize this argument further: even for a discrete random variable, the probability mass function has an implicit assumption of using the naive *counting measure* as a reference. For a finite discrete random variable, there is a normalization factor of $\ln\|\Omega\|$; for countable discrete random variables, one needs to find a legitimate second probability distribution in order to define entropy.

In many applications, μ_2 is actually the normalized Lebesgue measure from an observable X in a *compact space*, a random variable defined on the probability space. With respect to such an observable, Shannon's entropy is actually the quantitative characterization of the information that the RV X provides about the probability space. To a laboratory scientist, Y is a higher-resolution observable if

$H[Y] \geq H[X]$. The random variable with the largest entropy is the one whose probability density function is uniform.

Applying the logic to a random variable on a discrete, finite probability space $\Omega = \{1, 2, \cdots, N\}$ with probability measure μ_i, the entropy associated with a random variable X is the Radon–Nikodym derivative (RND):

$$H[X] = -\sum_{g_i} \Pr\{X = g_i\} \ln \left(\frac{\Pr\{X = g_i\}}{N^{-1}} \right), \tag{3.39}$$

in which $\Pr\{X = g_i\} = \sum_{j:X(j)=g_i} \mu_j$.

It is easy to show that $-\ln N \leq H(X) \leq 0$. $H(X)$ for a uniform distribution is 0, the maximum value. If $g_i \equiv c \ \forall i$, that is, X is actually deterministic without uncertainty, then its entropy reaches $-\ln N$, the lower bound. In this case, the random variable X does not at all inform on the probability space and distribution $\{\mu_i\}$. Similarly, for a real-valued continuous random variable X with probability density function $f_X(x)$, if the metric of the whole space is finite, i.e., $\|\Omega\| < \infty$, then we can define

$$H[X] = -\int_\Omega f_X(x) \ln \left(\frac{f_X(x)}{\|\Omega\|^{-1}} \right) dx, \tag{3.40}$$

which satisfies $H[X] \leq 0$. There is no lower bound, however. If $\|\Omega\| = \infty$, then we define the entropy as

$$H[X] = -\int_\Omega f_X(x) \ln f_X(x) dx. \tag{3.41}$$

A deterministic continuous random variable has zero variance. Consider a Gaussian random variable with variance σ^2; the entropy

$$-\frac{1}{\sqrt{2\pi\sigma^2}} \int_\Omega e^{-\frac{x^2}{2\sigma^2}} \left[-\frac{x^2}{2\sigma^2} - \ln\sqrt{2\pi\sigma^2} \right] dx \sim \frac{1}{2} + \ln\left(\sqrt{2\pi\sigma^2}\right) \to -\infty,$$

as $\sigma^2 \to 0$.

While entropy is associated with probability space and random variables, *entropy production* is associated with Markov process $\mathbf{X}(t)$. Take the discrete-time, discrete-state Markov process as an example. In this case, the relative probability of the joint distribution of $[\mathbf{X}(t), \mathbf{X}(t+1)]$ with respect to $[\mathbf{X}(t+1), \mathbf{X}(t)]$, or to the stationary probability of the Markov process, gives two different types of entropy production:

$$e_{tot}(t) = \sum_{i,j \in \Omega} p_i(t) P_{ij} \ln \left(\frac{p_i(t) P_{ij}}{p_j(t+1) P_{ji}} \right) \geq 0, \tag{3.42}$$

and

$$e_{free}(t) = \sum_{i,j\in\Omega} p_i(t)P_{ij}\ln\left(\frac{p_i(t)\pi_j}{p_j(t+1)\pi_i}\right) \geq 0, \tag{3.43}$$

in which $e_{tot}(t)$ is known as the instantaneous total entropy production rate and $e_{free}(t)$ is known as the free energy dissipation rate.

It is a triumph of the mathematical theory of probability that this highly abstract definition of entropy production turns out to be precisely in agreement with the original idea of physicists who developed the theory of *thermodynamics*. In particular:

(i) If a Markov chain is detail balanced, e.g., for each i and j, $\pi_i P_{ij} = \pi_j P_{ji}$, and a uniform stationary distribution $\pi_i \equiv c$, that is constant, then

$$e_{tot}(t) = e_{free}(t) = S(t+1) - S(t), \tag{3.44}$$

where

$$S(t) = -\sum_{i\in\Omega} p_i(t)\ln p_i(t)$$

can be identified as entropy in physics. In (3.44), the nonnegative entropy production rate implies that *entropy never decreases*.

(ii) If the stationary distribution π_i of a detail-balanced Markov chain is nonuniform, then

$$e_{tot}(t) = e_{free}(t) = \sum_{i,j\in\Omega} p_i(t)P_{ij}\ln\left(\frac{p_i(t)\pi_j}{p_j(t+1)\pi_i}\right) = F(t) - F(t+1), \tag{3.45}$$

where $F(t) = \overline{E}(t) - S(t)$ should be identified as free energy in thermal physics, with

$$\overline{E}(t) = -\sum_{i\in\Omega} p_i(t)\ln\pi_i$$

being the mean internal energy. (3.45) implies that *free energy never increases*.

(iii) For a Markov chain without detailed balance,

$$S(t+1) - S(t) = \sum_{i,j\in\Omega} p_i(t)P_{ij}\ln\left(\frac{p_i(t)}{p_j(t+1)}\right)$$
$$= e_p(t) - h_{ex}(t), \tag{3.46}$$

where

$$h_{ex}(t) = \sum_{i,j\in\Omega} p_i(t)P_{ij}\ln\left(\frac{P_{ij}}{P_{ji}}\right)$$

is interpreted as heat exchange (with temperature being 1), and Eqn. (3.46) is known as the ***entropy balance equation*** in nonequilibrium thermodynamics. For a system with detailed balance, $h_{ex}(t) = \overline{E}(t) - \overline{E}(t+1)$: the internal energy change is the heat.

For a diffusion process with diffusion $D(x)$ and drift $V(x)$, the instantaneous entropy production rate is

$$e_p(t) = \int_{\mathbb{R}^n} \Big(D(x)\nabla f(x,t) - V(x)f(x,t) \Big) \Big(\nabla \ln f(x,t) - D^{-1}(x)V(x) \Big) dx. \quad (3.47)$$

The corresponding heat exchange is

$$h_{ex}(t) = -\int_{\mathbb{R}^n} \Big(D(x)\nabla f(x,t) - V(x)f(x,t) \Big) D^{-1}(x)V(x)dx. \quad (3.48)$$

It is easy to verify that

$$-\frac{d}{dt} \int_{\mathbb{R}^n} f(x,t)\ln f(x,t)dx = e_p(t) - h_{ex}(t), \quad (3.49)$$

the entropy balance equation for a diffusion process.

A more comprehensive study of "stochastic thermodynamics" will be presented in Chapter 6.

Chapter 4
Large Deviations and Kramers' rate formula

The materials in Chapter 3 are widely available in several textbooks, for example, in [262] and [85]. They have been a part of the toolboxes of statistical physicists for the past 100 years, beginning from M. von Smoluchowski's and A. Einstein's diffusion theory and P. Langevin's stochastic differential equation for Brownian motions. However, in recent years, it has become very clear that the theory of large deviations provides the mathematical foundation of statistical thermodynamics. In this chapter, we shall provide the readers with a glimpse of the most basics of this important subject. For stochastic dynamics, the large deviations theory (LDT) is intimately related to a method widely used in quantum mechanics and applied mathematics called the Wentzel–Kramers–Brillouin (WKB) approximation or WKB ansatz in connection to singular perturbation [22]; the ansatz is termed the *large deviations principle* in the mathematical theory of probability, in which the goal is to prove the ansatz as limit theorems under appropriate mathematical conditions. The materials presented in this chapter are based on computations of the limit. Currently, there is only a handful of writings by physicists introducing this theory, to which we refer the readers for further readings [197, 259, 254, 93]. The chapter culminates in Section 4.6, where Kramers' celebrated diffusion theory of reaction rates and rate formula are presented. Laplace's method for integrals is the common thread between the two subjects.

4.1 The Law of Large Numbers and the Central Limit Theorem

4.1.1 Statistical Estimation of Expectation and Variance

Consider a random variable $\mathbf{X}$ with expected value $\mathbb{E}[\mathbf{X}]$ and variance $\mathrm{Var}[\mathbf{X}]$. Let $\mathbf{X}_1$, $\mathbf{X}_2$, ..., $\mathbf{X}_n$ be the values obtained from independently repeated n trails. In statistics, an estimator is a numerical construction from the measured samples that estimates the statistical properties of an RV. The simplest example is

$$\overline{\mathbf{X}}(n) = \frac{\mathbf{X}_1 + \mathbf{X}_2 + \cdots + \mathbf{X}_n}{n}, \tag{4.1}$$

which is an estimator of $\mathbb{E}[\mathbf{X}]$. It is important to realize that $\overline{\mathbf{X}}$ is itself an RV, a function of the independent and identically distributed (i.i.d.) $\{X_k; 1 \leq k \leq n\}$, with an expectation and variance of its own! Computing the expectation and variance of an estimator is one of the main themes in statistics. We encourage the reader to show that $\mathbb{E}\left[\overline{\mathbf{X}}\right] = \mathbb{E}[\mathbf{X}]$ and $\mathrm{Var}\left[\overline{\mathbf{X}}\right] = \frac{1}{n}\mathrm{Var}[\mathbf{X}]$.

How do we estimate the variance of an RV? Let us use

$$\mu_2 = \frac{(\mathbf{X}_1 - \overline{\mathbf{X}})^2 + (\mathbf{X}_2 - \overline{\mathbf{X}})^2 + \cdots + (\mathbf{X}_n - \overline{\mathbf{X}})^2}{n}$$
$$= \frac{\mathbf{X}_1^2 + \mathbf{X}_2^2 + \cdots + \mathbf{X}_n^2}{n} - \overline{\mathbf{X}}^2.$$

Again, μ_2 is an RV itself! Then, we have,

$$\mathbb{E}[\mu_2] = \frac{n\mathbb{E}[\mathbf{X}]}{n} - \mathbb{E}\left[\overline{\mathbf{X}}^2\right] = \mathbb{E}\left[\mathbf{X}^2\right] - \mathbb{E}\left[\overline{\mathbf{X}}^2\right],$$

in which

$$\mathbb{E}\left[\overline{\mathbf{X}}^2\right] = \mathbb{E}\left[\frac{\sum_{i,j} X_i X_j}{n^2}\right] = \frac{n\mathbb{E}[\mathbf{X}^2] + n(n-1)\mathbb{E}[\mathbf{X}]^2}{n^2}.$$

Hence,

$$\mathbb{E}[\mu_2] = \mathbb{E}[\mathbf{X}^2] - \frac{1}{n}\mathbb{E}[\mathbf{X}^2] + \frac{n-1}{n}\mathbb{E}[\mathbf{X}]^2 = \frac{n-1}{n}\left(\mathbb{E}[\mathbf{X}^2] - \mathbb{E}[\mathbf{X}]^2\right) = \frac{n-1}{n}\mathrm{Var}[\mathbf{X}].$$

That is, μ_2 is biased: $\mathbb{E}[\mu_2] \neq \mathrm{Var}[\mathbf{X}]$! Rather, $\frac{n}{n-1}\mathbb{E}[\mu_2] = \mathrm{Var}[\mathbf{X}]$. Your calculator will show:

$$\sigma_{n-1}^2 = \frac{n}{n-1}\mu_2 = \frac{\mathbf{X}_1^2 + \mathbf{X}_2^2 + \cdots + \mathbf{X}_n^2}{n-1} - \frac{n}{n-1}\overline{\mathbf{X}}^2$$

to be the most frequently used unbiased estimator for $\mathrm{Var}[\mathbf{X}]$.

4.1.2 Central Limit Theorem For Independent and Identically Distributed RVs

Eqn. (4.1) estimates the expectation of an RV with multiple independent observations. The law of large numbers tells us that

$$\Pr\left\{ \left| \overline{\mathbf{X}} - \mathbb{E}[\mathbf{X}] \right| \leq \varepsilon \right\} \longrightarrow 1 \quad \text{as} \quad n \longrightarrow \infty,$$

for any small $\varepsilon > 0$.

If this random variable is merely the indicator function of a random event A, then its expectation is just its probability, p, which can be estimated from the frequency of observing A. As n goes to infinity, we have

$$\Pr\left\{ \left| \frac{k}{n} - p \right| \leq \varepsilon \right\} \longrightarrow 1 \quad \text{as} \quad n \longrightarrow \infty,$$

for any small $\varepsilon > 0$, in which k is the number of occurrences of event A in the first n trials. In other words, the statistical frequency converges to the probability. This result is a weaker version of Borel's law of large numbers, which states that

$$\Pr\left\{ \frac{k}{n} \longrightarrow p, \text{ as } n \longrightarrow \infty \right\} = 1.$$

Moreover, in the nineteenth century, De Moivre and Laplace proved that the probability distribution of $(k - np)$, which is *binomial distributed*, asymptotically approaches a Gaussian distribution after proper scaling. It is the very first version of the central limit theorems.

In the early 20th century, the central limit theorem was generalized to any distribution with finite mean and variance. Let $\mathbf{X}_1, \mathbf{X}_2, \cdots, \mathbf{X}_n$ be independent and identically distributed (i.i.d.) copies of a random variable $\mathbf{X}$ with finite mean and variance. The central limit theorem states that the probability distribution of

$$\sqrt{\frac{n}{\text{Var}[\mathbf{X}]}} \left[\overline{\mathbf{X}} - \mathbb{E}[\mathbf{X}] \right] \tag{4.2}$$

will converge to the standard normal distribution as n goes to infinity.

Therefore, combining the law of large numbers and the central limit theorem, one can heuristically obtain

$$\overline{\mathbf{X}}(n) \simeq \mathbb{E}[\mathbf{X}] + \sqrt{\frac{\text{Var}[\mathbf{X}]}{n}} \mathbf{Y} + o\left(n^{-\frac{1}{2}}\right), \tag{4.3}$$

in which $\mathbf{Y}$ is a Gaussian random variable with zero mean and unit variance. Such a series can be proven, term by term, by taking the limits appropriately, step by step:

$$\lim_{n\to\infty} \overline{\mathbf{X}}(n) = \mathbb{E}[\mathbf{X}],$$

$$\lim_{n\to\infty} \sqrt{n}\left(\overline{\mathbf{X}}(n) - \mathbb{E}[\mathbf{X}]\right) = \sqrt{\mathrm{Var}[\mathbf{X}]}\,\mathbf{Y},$$

$$\lim_{n\to\infty} n^{\alpha+\frac{1}{2}}\left(\overline{\mathbf{X}}(n) - \mathbb{E}[\mathbf{X}] - \sqrt{\frac{\mathrm{Var}[\mathbf{X}]}{n}}\,\mathbf{Y}\right) = ...,$$

which is the fundamental idea of *asymptotic series* [22, 184], originally proposed by Henri Poincaré (1854–1912).

4.2 Laplace's Method for Integrals

The above $\overline{\mathbf{X}}(n)$ is an example of a sequence of probability distributions. In a nutshell, the theory of large deviations describes the rate of a sequence of convergent probability measures. For instance, assume these probability measures have density functions $\phi_n(x)$ and that the law of large numbers holds, which implies $\phi_n(x) \to \delta(x - x_0)$, the Dirac δ function located at $x = x_0$, when n tends to infinity. One of the key steps in the theory of large deviations is the computation of the asymptotic behavior of

$$\int_{|x-x_0|\geq\varepsilon} \phi_n(x)\,\mathrm{d}x, \tag{4.4}$$

and the general conclusion is that there exists a function $I(x)$ such that

$$\int_{|x-x_0|\geq\varepsilon} \phi_n(x)\,\mathrm{d}x \sim \exp\left\{-n \inf_{|x-x_0|\geq\varepsilon} I(x)\right\}, \quad I(x) \geq 0, \ I(x_0) = 0. \tag{4.5}$$

Intuitively, one can regard the so-called large deviation rate function $I(x)$ as the limit of $-(1/n)\ln\phi_n(x)$. If $I(x)$ is sufficiently smooth near x_0, then $I(x) \simeq I(x_0) + \frac{1}{2}I''(x_0)(x-x_0)^2$. The location x_0 signifies the limit of the law of large numbers, and the quadratic term represents the central limit theorem.

Theoretically, the large deviation principle is closely related to the Laplace method in functional spaces. Here, we introduce the traditional Laplace's method, which will be used later in several examples.

Assume $I(x)$ is sufficiently smooth and has a single global minimum at $x = x_0$. Then

$$\int_{\mathbb{R}} e^{-nI(x)} dx = \int_{|x-x_0|<\varepsilon} e^{-nI(x)} dx + \int_{|x-x_0|\geq\varepsilon} e^{-nI(x)} dx, \qquad (4.6)$$

in which the second term is exponentially smaller than the first term. With this observation, we have

$$\int_{\mathbb{R}} e^{-nI(x)} dx = \int_{\mathbb{R}} e^{-n\left[I(x_0)+\frac{1}{2}I''(x_0)(x-x_0)^2\right]}$$

$$\times \exp\left[-\frac{nI'''(x_0)}{6}(x-x_0)^3 - \frac{nI''''(x_0)}{24}(x-x_0)^4 + \cdots\right] dx$$

$$= e^{-nI(x_0)}\int_{\mathbb{R}} e^{-\frac{n}{2}I''(x_0)(x-x_0)^2}\left[1 - \frac{nI'''(x_0)}{6}(x-x_0)^3 + O\left(n(x-x_0)^4\right)\right] dx$$

$$= e^{-nI(x_0)}\sqrt{\frac{2\pi}{nI''(x_0)}}\left\{1 + 0 + O(n^{-1})\right\}. \qquad (4.7)$$

That is, the integral is dominated by the global minimum of $I(x)$, where one can use a Gaussian integral to evaluate the integral asymptotically.

Now consider the second integral in (4.6); the global minimum of $I(x)$, on the interval $|x - x_0| \geq \varepsilon$, is assumed to be at the boundary, such as when $I(x)$ is convex. Then, denoting $x_0 + \varepsilon = x_1$ and $x_0 - \varepsilon = x_2$, we have

$$\int_{|x-x_0|\geq\varepsilon} e^{-nI(x)} dx = \int_{x_1}^{\infty} e^{-nI(x)} dx + \int_{-\infty}^{x_2} e^{-nI(x)} dx$$

$$= \int_{x_1}^{\infty} e^{-n[I(x_1)+I'(x_1)(x-x_1)]}\exp\left[-\frac{nI''(x_1)}{2}(x-x_1)^2 + O\left(n(x-x_1)^3\right)\right] dx$$

$$+ \int_{-\infty}^{x_2} e^{-n[I(x_2)+I'(x_2)(x-x_2)]}\exp\left[-\frac{nI''(x_2)}{2}(x-x_2)^2 + O\left(n(x-x_2)^3\right)\right] dx$$

$$= e^{-nI(x_1)}\int_0^{\infty} e^{-nI'(x_1)z}\left[1 - \frac{nI''(x_1)}{2}z^2 + O(nz^3)\right] dz$$

$$+ e^{-nI(x_2)}\int_0^{\infty} e^{-n|I'(x_2)|z}\left[1 - \frac{nI''(x_2)}{2}z^2 + O(nz^3)\right] dz$$

$$= \left(\frac{e^{-nI(x_1)}}{nI'(x_1)} + \frac{e^{-nI(x_2)}}{n|I'(x_2)|}\right)\left\{1 + o(n^{-1})\right\}. \qquad (4.8)$$

Therefore, when the global minimum of $I(x)$ is at a boundary, the integral can also be asymptotically evaluated in terms of exponential integrals.

4.3 Legendre–Fenchel Transformation and LDT of sample mean

The central limit theorem states that the distribution of the partial sum $\overline{\mathbf{X}}$ can be approximated by a normal distribution with mean $\mathbb{E}[\mathbf{X}]$ and variance $\frac{1}{n}\mathrm{Var}[\mathbf{X}]$. In this section, we will determine the small probability of $\Pr\{|\overline{\mathbf{X}} - \mathbb{E}(\mathbf{X})| \geq \delta\}$ when $n \to \infty$. It turns out that this probability is exponentially small, and we are interested in the exponent, i.e., the limit

$$\lim_{n\to\infty} \frac{1}{n}\ln\Pr\left\{|\overline{\mathbf{X}} - \mathbb{E}(\mathbf{X})| \geq \delta\right\}. \tag{4.9}$$

First, a few words about the notion of an "exponentially small, infinitesimal" quantity $e^{-\alpha/\varepsilon}$, with $\varepsilon, \alpha > 0$ and as $\varepsilon \to 0^+$. It is also known as asymptotically small beyond all order [196]: Order here refers to the exponent in ε^k ($k \geq 0$).

We know that for larger k, ε^k goes to zero faster. By applying l'Hôspital's rule,

$$\lim_{\varepsilon\to 0^+} \frac{e^{-\alpha/\varepsilon}}{\varepsilon^k} = \lim_{\eta\to+\infty} \frac{\eta^k}{e^{\alpha\eta}} = \lim_{\eta\to+\infty} \left(\frac{k}{\alpha}\right)\frac{\eta^{k-1}}{e^{\alpha\eta}} = \cdots = 0.$$

This indicates that $e^{-\alpha/\varepsilon}$ goes to zero faster than ε^k for any k.

We now show, as a simple example, that if $\mathbf{X}$ is normally distributed, then the limit in (4.9) is $-\delta^2/(2\sigma^2)$, in which $\sigma^2 = \mathrm{Var}[\mathbf{X}]$. Note that

$$\Pr\{|\overline{\mathbf{X}} - \mathbb{E}(\mathbf{X})| \geq \delta\}$$

$$= \frac{\sqrt{n}}{\sqrt{2\pi}\sigma}\int_{|y|\geq\delta} e^{-ny^2/2\sigma^2}\,dy = \frac{2\sqrt{n}}{\sqrt{2\pi}\sigma}\int_{\delta}^{\infty} e^{-ny^2/2\sigma^2}\,dy$$

$$= \frac{\sigma}{\delta}\sqrt{\frac{2}{n\pi}}e^{-(\delta^2/2\sigma^2)n}\left(1 - \frac{\sigma^2}{n\delta^2} + \frac{3\sigma^4}{n^2\delta^4} + \cdots\right),$$

where the integration is computed approximately with Laplace's method for integrals.

This result is only true for a strictly Gaussian $\mathbf{X}$. For an i.i.d. $\{\mathbf{X}_i\}$ whose distribution satisfies certain growth conditions and $\overline{\mathbf{X}}$, even though the latter has an asymptotic normal distribution, the limit in Eqn. (4.9) contains more information than the variance σ^2 of $\mathbf{X}$: while the limit in Eqn. (4.9) always exists, its value depends on the details of the distribution of $\mathbf{X}_i$.

More specifically, for any $A \subset \mathbb{R}$,

$$\lim_{n\to\infty} \frac{1}{n}\ln\Pr\{\mathbf{Y}_n = \overline{X} - \mathbb{E}[\mathbf{X}] \in A\} = -\inf_{x\in A} \Lambda^*(x) \tag{4.10}$$

where $\Lambda^*(x)$ is related to the logarithmic moment generating function (LMGF) via the Legendre–Fenchel transform:

$$\Lambda^*(x) \triangleq \sup_{\lambda\in\mathbb{R}} \left\{ \lambda x - \ln\mathbb{E}\left[e^{\lambda\mathbf{X}}\right] \right\}. \tag{4.11}$$

Eqn. (4.10) is called the large deviation principle of the sample mean $\overline{X}$ of i.i.d. random variables, and $\Lambda^*(x)$ is exactly the large deviation rate function. Both $\Lambda^*(x)$ and the LMGF are convex functions. In this case, the Legendre–Fenchel transform of Eqn. (4.11) yields

$$\Lambda^*(x) = \left[\lambda x - \ln\mathbb{E}\left[e^{\lambda\mathbf{X}}\right] \right]_{\lambda=g^{-1}(x)},$$

in which the monotonic $g(x)$ represents a dual relationship between λ and x:

$$x = \frac{d}{d\lambda}\ln\mathbb{E}\left[e^{\lambda\mathbf{X}}\right] \equiv g(\lambda), \quad \lambda = \frac{d}{dx}\Lambda^*(x) = g^{-1}(x). \tag{4.12}$$

The rate function $\Lambda^*(x)$, therefore, can also be expressed as:

$$\Lambda^*(x) = \int_{\mathbb{E}[\mathbf{X}]}^{x} g^{-1}(y)dy.$$

One can also derive another relation

$$\ln\mathbb{E}\left[e^{\lambda\mathbf{X}}\right] = -\frac{d}{d(1/\lambda)}\left(\frac{\Lambda^*(g(\lambda))}{\lambda}\right). \tag{4.13}$$

We can check for the case of a Gaussian-distributed $\mathbf{X}$:

$$\mathbb{E}\left[e^{\lambda\mathbf{X}}\right] = e^{\sigma^2\lambda^2/2}, \quad \frac{d}{d\lambda}\ln\mathbb{E}\left[e^{\lambda\mathbf{X}}\right] = \sigma^2\lambda, \quad \Lambda^*(x) = \frac{x^2}{2\sigma^2}.$$

Substituting this into Eqn. (4.10) and choosing $A = [\delta,\infty)$, we obtain Eqn. (4.9). We observe that $\ln\mathbb{E}[e^{\lambda\mathbf{X}}] \propto \sigma^2$, $\Lambda^*(x) \propto \sigma^{-2}$.

The theory of large deviations is the mathematical foundation for statistical thermodynamics. The LMGF $\ln\mathbb{E}[e^{\lambda\mathbf{X}}]$ is also called free entropy, or the Massieu–Planck function in the old thermodynamic literature, if one identifies $-\lambda^{-1}$ as temperature. Applying the formalism to a polymer chain, $g(\lambda)$ is the extension-force curve characterizing the elastic string. $\Lambda^*(x)$ is the work done by the external force λ. It is important to note a key difference, especially in research practice, between

"statistical mechanics" and "statistical thermodynamics": in the former, all attention is usually given to *fluctuations*, *correlations*, and other statistical properties of a
system, while *energy*, *entropy*, etc., are rarely discussed since these thermodynamic
quantities are in principle completely understood and are independent of the details of the statistical mechanical model. Statistical thermodynamics, however, turns
out to be very powerful in the studies of nonmacroscopic systems, such as proteins
[239, 14], particularly in conjunction with laboratory measurements of heat and observables as functions of temperature.

4.4 LDT of Empirical Distribution and Contraction

Let us now consider an RV, $\mathbf{X}$, with a finite number of possible outcomes $1, 2, ..., r$
and corresponding probabilities p_1, p_2, $...$, p_r. A second type of large deviation
calculation concern the frequency of each outcome for n trials:

$$v_i(n) = \frac{1}{n} \sum_{k=1}^{n} \delta_{\mathbf{X}_k, i}, \tag{4.14}$$

in which $\{\mathbf{X}_k\}$ are independent identically distributed random variables with the
same distribution as $\mathbf{X}$.

Again, the large deviation is exponentially small when $n \to \infty$; then, one obtains

$$\lim_{n \to \infty} \frac{1}{n} \ln \Pr \left\{ \max_{1 \le i \le r} |v_i(n) - p_i| \ge \delta \right\} = - \inf_{\max_{1 \le k \le r} |v_k - p_k| \ge \delta} \sum_{k=1}^{r} v_k \ln \left(\frac{v_k}{p_k} \right), \tag{4.15}$$

in which the relative entropy $\sum_{k=1}^{r} v_k \ln(v_k/p_k)$ is the corresponding rate function.

From this LDT for an empirical distribution, one can obtain the large deviation
rate for any random variable $g(i)$, $i = 1, 2, \cdots, r$ through an optimization process
known as *contraction* as follows. With a large number n of samples, the corresponding sample mean value for the random variable is simply

$$\bar{g} = \frac{1}{n} \sum_{i=1}^{n} g(\mathbf{X}_i) = \sum_{k=1}^{r} v_k(n) g(k), \tag{4.16}$$

whose limit as $n \to \infty$ is $\mathbb{E}[g]$. Then, the corresponding large deviation rate function
for the law of large numbers, $\psi(x)$, can be obtained from

$$\psi(x) = \begin{cases} \underset{v_k}{\inf} \sum_{k=1}^{r} v_k \ln\left(\dfrac{v_k}{p_k}\right), \\[2mm] s.t. \sum_{k=1}^{r} v_k g(k) = x, \\[2mm] \sum_{k=1}^{r} v_k = 1. \end{cases} \qquad (4.17)$$

Employing the method of the Lagrange multiplier for constrained optimization, the result is

$$\psi(x) = \left[\lambda^* x - \Lambda(\lambda^*)\right]_{\lambda^*:\partial\Lambda(\lambda^*)/\partial\lambda=x}, \qquad (4.18a)$$

$$\Lambda(\lambda) = \ln \sum_{k=1}^{r} p_k e^{\lambda g(k)} = \ln\mathbb{E}\left[e^{\lambda g}\right], \qquad (4.18b)$$

where $\Lambda(\lambda)$ is the LMGF of random variable g and Eqn. (4.18a) is equivalent to the Legendre–Fenchel transformation

$$\psi(x) = \sup_{\lambda \in \mathbb{R}} \left\{\lambda x - \Lambda(\lambda)\right\}.$$

This proves the result in Eqn. (4.11) for the simple case of finite sample space $\Omega = \{1, 2, \cdots, r\}$.

4.5 Large Deviations for Stationary Diffusion on a Circle and $\vec{\lambda}$-Surgery

In addition to the large deviation result for a sequence of converging random variables, stochastic dynamics also exhibit a large deviation principle as the randomness in stochastic dynamics vanishes. Large deviation rate functions are not always smooth; one may also wonder how the rate function varies when a bifurcation occurs in stochastic dynamics. In this section, we illustrate these concepts through a simple solvable example. Let us now consider the LDT for a sequence of diffusion processes on a circle in the limit of diffusion coefficient $D = \varepsilon \to 0$. We are particularly interested in the limit of the stationary distribution $\pi_\varepsilon(x)$, which satisfies a singularly perturbed ordinary differential equation

$$\varepsilon\left(\frac{d^2 \pi_\varepsilon(x)}{dx^2}\right) - \frac{d}{dx}\left(V(x)\pi_\varepsilon(x)\right) = 0, \qquad (4.19)$$

with $x \in \mathbb{S}[-1,1]$ under a periodic boundary condition. The solution $\pi_\varepsilon(x)$ has to satisfy the normalization condition

$$\int_{-1}^{1} \pi_\varepsilon(x)dx = 1.$$

We are particularly interested in the problem with a $V(x)$ that has an unstable fixed point at $x = 0$ and a stable fixed point in $[-1,0)$. We shall define $U(x) = -\int_{-1}^{x} V(y)dy$; hence, $V(x) = -U'(x)$. $U(x)$ has its global maximum at $x = 0$, and since there is no zero on $(0,1]$ for $V(x)$, $U(x)$ is monotonically decreasing on $(0,1]$. We shall further assume $U(-1) = 0 > U(1)$. These functions are shown in Fig. 4.1.

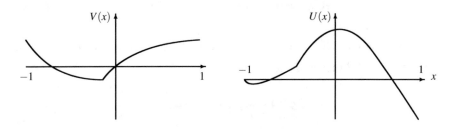

Fig. 4.1 The roots of function $V(x) = 0$ correspond to the extrema of function $U(x)$. Since function $V(x)$ is defined on a circle, it is necessarily true that $V(-1) = V(1)$. However, the function $U(x)$ is not periodic in general. $U(-1) = U(1)$ if and only if $\int_{-1}^{1} V(y)dy = 0$. Otherwise, J_ε in Eqn. (4.23) is not zero; there is a stationary circular flux on the circle.

One notices that if

$$U(1) = U(-1), \tag{4.20}$$

then

$$\pi_\varepsilon(x) = \hat{C}_\varepsilon e^{-\frac{U(x)}{\varepsilon}} \quad \text{where} \quad \hat{C}_\varepsilon = \int_{-1}^{1} e^{-\frac{U(y)}{\varepsilon}} dy. \tag{4.21}$$

The equation satisfies both the periodic and normalization conditions.

However, if Eqn. (4.20) is not true, then one has

$$\pi_\varepsilon(x) = C_\varepsilon \cdot e^{-\frac{U(x)}{\varepsilon}} \left[e^{\frac{U(1)}{\varepsilon}} + J_\varepsilon \int_x^1 e^{\frac{U(z)}{\varepsilon}} dz \right], \tag{4.22}$$

in which

$$J_\varepsilon = \left\{ e^{\frac{U(-1)}{\varepsilon}} - e^{\frac{U(1)}{\varepsilon}} \right\} \left(\int_{-1}^{1} e^{\frac{U(z)}{\varepsilon}} dz \right)^{-1}. \tag{4.23}$$

and C_ε is the normalization constant.

We see that $J_\varepsilon = 0$ when (4.20) holds true. Additionally, since $U(-1) > U(1)$, $J_\varepsilon > 0$. J_ε has a clear physical meaning; it represents the stationary *flux* in the positive direction of x under a *nondetailed balanced* boundary conditions $U(-1) \neq U(1)$. In physics, the term $\{\cdots\}$ in Eqn. (4.23) represents the "driving force" that is responsible for the nonzero flux, with the term inside $(\cdots)$ being the "resistance".

With the unique global maximum of $U(x)$ at $x = 0$ and following the method in Section 4.2, in the limit of $\varepsilon \to 0$ we have the following leading-order asymptotics,

$$
\begin{aligned}
J_\varepsilon &= \left\{ e^{\frac{U(-1)}{\varepsilon}} - e^{\frac{U(1)}{\varepsilon}} \right\} \left(\int_{-1}^{1} e^{\frac{U(z)}{\varepsilon}} dz \right)^{-1} \\
&\simeq \left\{ e^{\frac{U(-1)}{\varepsilon}} - e^{\frac{U(1)}{\varepsilon}} \right\} \left(e^{\frac{U(0)}{\varepsilon}} \int_{-1}^{1} e^{-\frac{|U''(0)|z^2}{\varepsilon}} dz \right)^{-1} \\
&\simeq \left\{ e^{\frac{U(-1)-U(0)}{\varepsilon}} - e^{\frac{U(1)-U(0)}{\varepsilon}} \right\} \sqrt{\frac{|U''(0)|}{2\pi\varepsilon}},
\end{aligned}
\tag{4.24}
$$

and

$$
\int_{x}^{1} e^{\frac{U(z)}{\varepsilon}} dz \simeq
\begin{cases}
e^{\frac{U(0)}{\varepsilon}} \sqrt{\frac{2\pi\varepsilon}{|U''(0)|}} & -1 \leq x < 0 \\
\frac{1}{2} e^{\frac{U(0)}{\varepsilon}} \sqrt{\frac{2\pi\varepsilon}{|U''(0)|}} & x = 0 \\
\frac{1}{|U'(x)|} e^{\frac{U(x)}{\varepsilon}} & 1 \geq x > 0
\end{cases}
\tag{4.25}
$$

Therefore

$$
\begin{aligned}
y_\varepsilon &\overset{def}{=} e^{-\frac{U(x)}{\varepsilon}} \left[e^{\frac{U(1)}{\varepsilon}} + J_\varepsilon \int_{x}^{1} e^{\frac{U(z)}{\varepsilon}} dz \right] \\
&\simeq e^{\frac{U(1)-U(x)}{\varepsilon}} + e^{\frac{U(1)-U(0)}{\varepsilon}} \left(e^{\frac{\Delta\mu}{\varepsilon}} - 1 \right)
\begin{cases}
e^{-\frac{U(x)-U(0)}{\varepsilon}} & -1 \leq x < 0 \\
\frac{1}{2} & x = 0 \\
\frac{1}{|U'(x)|} \sqrt{\frac{|U''(0)|}{2\pi\varepsilon}} & 1 \geq x > 0
\end{cases}
\end{aligned}
\tag{4.26}
$$

in which $\Delta\mu \overset{\Delta}{=} U(-1) - U(1) > 0$.

In the limit as ε tends to zero, we have

$$
-\lim_{\varepsilon \to 0} \varepsilon \ln y_\varepsilon(x) =
\begin{cases}
U(x) - U(1) - \Delta\mu & -1 \leq x \leq 0 \\
U(0) - U(1) - \Delta\mu & 0 \leq x \leq \hat{x} \\
U(x) - U(1) & 1 \geq x \geq \hat{x}
\end{cases}
\tag{4.27}
$$

where at $\hat{x}$, $U(\hat{x}) = U(0) - \Delta\mu$. Since we have assumed $U(0) > U(-1)$, $0 < \hat{x} < 1$ exists.

Note that the minimum of $-\lim_{\varepsilon\to 0}\varepsilon \ln y_\varepsilon(x)$ is 0 at 1 and -1, and $C_\varepsilon = \int_{-1}^{1} y_\varepsilon(x)dx$ satisfies

$$-\lim_{\varepsilon\to 0}\varepsilon \ln C_\varepsilon = 0.$$

Hence, the large deviation rate function for $\pi_\varepsilon(x)$ is

$$-\lim_{\varepsilon\to 0}\varepsilon \ln \pi_\varepsilon(x) = \begin{cases} U(x)-U(1)-\Delta\mu & -1\leq x\leq 0 \\ U(0)-U(1)-\Delta\mu & 0\leq x\leq \hat{x} \\ U(x)-U(1) & 1\geq x\geq \hat{x} \end{cases} \qquad (4.28)$$

The function in (4.28) has the shape of a λ, shown in Fig. 4.2 below, with the nonzero flux J

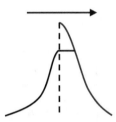

Fig. 4.2 A schematic diagram illustrating the large deviation rate function given in (4.28). The left- and right-halves relative to the maximum are shifted by $\Delta\mu$. There, the function is differentiable with zero derivative but is nonsmooth.

4.5.1 Caricature of a Turning Point: An Example

Point $x = 0$ in the above discussion, at which the drift $V(x) = 0$ leads to the "surgical incision" in Fig. 4.2, is known as a turning point in the theory of singular perturbation [196]. From Eqn. (4.19), near $x = 0$,

$$\varepsilon\left(\frac{d^2\pi_\varepsilon(x)}{dx^2}\right) + \frac{d}{dx}\left(V(x)\pi_\varepsilon(x)\right) \simeq \varepsilon\left(\frac{d^2\pi_\varepsilon(x)}{dx^2}\right) + V'(0)\pi_\varepsilon(x) = 0,$$

which implies that π_ε is highly oscillatory when $V'(0) > 0$.

The differential equation is highly problematic nearby. The "surgical incision" for the large deviation rate function at the local minimum of $U(x)$ is actually a generic behavior in nonsmooth large deviation rate functions, transitioning from a nonconstant function to a constant function continuously with zero derivative at

$x = 0$. The following example, in terms of the error function and with a simple calculation, helps us to understand the nature of the nonsmoothness, due to a nonuniform convergence, in the middle of $\vec{\lambda}$:

$$e^{\frac{x^2}{\varepsilon}} \int_x^\infty e^{-\frac{z^2}{\varepsilon}}\, dz = \frac{\sqrt{\pi\varepsilon}}{2}\, e^{\frac{x^2}{\varepsilon}} \left\{ 1 - \mathrm{erf}\left(\frac{x}{\sqrt{\varepsilon}}\right) \right\}. \tag{4.29}$$

According to Laplace's method for integrals:

$$e^{\frac{x^2}{\varepsilon}} \int_x^\infty e^{-\frac{z^2}{\varepsilon}}\, dz \simeq \begin{cases} \sqrt{\pi\varepsilon}\, e^{\frac{x^2}{\varepsilon}} & x < 0 \\[2mm] \dfrac{1}{2}\sqrt{\pi\varepsilon} & x = 0 \\[2mm] \dfrac{\varepsilon}{|U'(x)|} & x > 0 \end{cases} \tag{4.30}$$

The limit as $\varepsilon \to 0$ is taken for

$$\varepsilon \ln \left\{ e^{\frac{x^2}{\varepsilon}} \int_x^\infty e^{-\frac{z^2}{\varepsilon}}\, dz \right\}$$

at each fixed x. Fig. 4.3 illustrates the function near $x = 0$ as ε tends to zero.

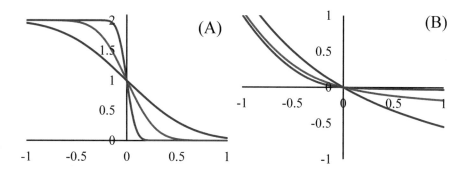

Fig. 4.3 (A) Function $\left[1 - \mathrm{erf}(x/\sqrt{\varepsilon})\right]$, and (B) Function $\varepsilon \ln\{e^{x^2/\varepsilon}\left[1 - \mathrm{erf}(x/\sqrt{\varepsilon})\right]\}$, with $\varepsilon = 0.5, 0.1$ and 0.01 in blue, orange, and red, respectively. The red curve in (B), placed upside-down, constitutes the left half of λ in Fig. 4.2 and becomes flat for $x \geq 0$.

In the theory of phase transition in physics, a nonsmooth, large deviation rate function is indicative of nonconvexity, with the possibility of a phase transition. Fig. 4.3 shows such a behavior at $x = 0$, the turning point, as $\varepsilon \to 0$.

4.5.2 Saddle-node Bifurcation of Diffusion On a Circle

For bifurcation problems, we are now interested in $U(x, \eta)$, with parameter η, which is monotonically decreasing and x when $\eta > 0$, e.g., no fixed point. For $\eta < 0$, the function has a local minimum at $x_1^* < 0$ and a maximum at $x = 0$. In other words, the periodic drift function $V(x; \eta)$ undergoes a saddle-node bifurcation near $x = 0$ at $\eta = 0$. $\pi_0(x; \eta)$ is the limit of the stationary probability distribution with finite ε, and $\pi_\varepsilon(x; \eta)$ as $\varepsilon \to 0$. Fig. 4.4, adapted from [94], gives a visual example of $\pi_0(x; \eta)$ when bifurcation parameter η approaches zero from above. This distribution tells

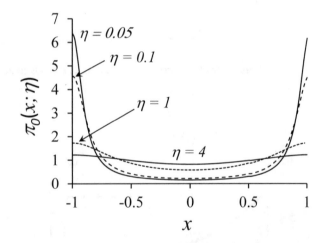

Fig. 4.4 $\pi_0(x; \eta)$ is the limit of stationary probability distribution $\pi_\varepsilon(x; \eta)$ with finite ε, as $\varepsilon \to 0$. For a deterministic dynamic system $V(x; \eta) = \eta + 1 - \cos[\pi(x - 1)]$ with fixed points, this distribution is a Dirac δ function centered at the stable fixed point. For the deterministic dynamic system with limit cycle behavior, $\pi_0(x; \eta)$ is still fully supported on $\mathbb{S}[-1, 1]$ and represents the local dwell time of the periodic, cyclic motion. As $\eta \to 0^+$ approaching the saddle-node bifurcation point, the periodic motion spends most of its time in the region where $|V(x; \eta)|$ is very near zero, e.g., $x = \pm 1$.

When $\eta > 0$, the global maximum of $U(x)$, $x \in \mathbb{S}[-1, 1]$, is located at $U(-1)$. Therefore, for $\eta < 0$ but sufficiently near the bifurcation point, the global maximum remains at $x = -1$, and $x_1^* < 0$. Again applying Laplace's method, we have

$$\pi_\varepsilon(x;\eta) = e^{\frac{U(1)}{\varepsilon}} e^{-\frac{U(x)}{\varepsilon}} + e^{\frac{U(1)}{\varepsilon}} e^{-\frac{U(x)}{\varepsilon}} \left(e^{\frac{\Delta\mu}{\varepsilon}} - 1\right) \left(\int_{-1}^{1} e^{\frac{U(z)}{\varepsilon}} dz\right)^{-1} \int_{x}^{1} e^{\frac{U(z)}{\varepsilon}} dz$$

$$\simeq e^{\frac{U(1)}{\varepsilon}} e^{-\frac{U(x)}{\varepsilon}} + e^{\frac{U(1)}{\varepsilon}} \left(e^{\frac{\Delta\mu}{\varepsilon}} - 1\right) |U'(-1)| e^{-\frac{U(-1)}{\varepsilon}}$$

$$\times \begin{cases} \dfrac{1}{|U'(x)|} & x < x_2^* \text{ or } x > 0 \\[2mm] \sqrt{\dfrac{\pi}{\varepsilon|U''(0)|}} e^{\frac{U(0)-U(x)}{\varepsilon}} & x_2^* < x < 0 \end{cases} \tag{4.31}$$

in which $x_2^*(\eta)$ is determined from $U(x_2^*;\eta) = U(0;\eta)$. The local minimum x_1^* of $U(x;\eta)$ is located within the interval $(x_2^*,0)$, which disappears when $\eta > 0$.

Noting that $U(-1) - U(1) = \Delta\mu$, therefore, we have the leading-order asymptotics as $\varepsilon \to 0$, for η slightly less than zero,

$$-\varepsilon \ln \pi_\varepsilon(x;\eta) \simeq \begin{cases} \varepsilon \ln|U'(x;\eta)| & x \le x_2^*(\eta) \text{ or } x \ge 0 \\ U(x;\eta) - U(0;\eta) + o(\varepsilon \ln\varepsilon) & x_2^*(\eta) < x < 0 \end{cases} \tag{4.32}$$

Since we did not enforce the normalization of $\pi_\varepsilon(x;\eta)$, the function in Eqn. (4.32) can include an arbitrary additive constant. The continuous function in (4.32) is once differentiable at $x = 0$ but not at $x = x_2^*$. These points correspond to the $x = 0$ and $x = \hat{x}$, respectively, in $\vec{\lambda}$ of Fig. 4.2, respectively.

The term $\varepsilon \ln|U'(x;\eta)|$ in (4.32) is very significant: when $\eta > 0$, there is no fixed point on $\mathbb{S}[0,1]$ with $x_2^* = 0$. Eqn. (4.32), however, still holds and indicates that the deterministic rotational motion on the circle has a stationary probability measure that is inversely proportional to the local velocity $|V(x;\eta)| = |U'(x;\eta)|$. This is a consequence of ergodic motion. The large deviation rate function, i.e., the limit of (4.32) as ε vanishes, on the other hand, is flat, e.g., the $O(1)$ term is zero.

4.6 Kramers' Rate Formula

From a classical mechanics standpoint, a molecule is a collection of atoms with very complex movements. For a protein with N atoms, a Newtonian mechanical description of its dynamics has $6N$ degrees of freedom, without even considering the atoms in the solvent, which is at least an order of magnitude more. This is what one observes from a molecular dynamics (MD) simulation. It is very complicated.

However, any such mechanical system has a potential energy function (its gradient is called a force field in MD simulations). Treating the solvent as a viscous medium with frictional coefficient η, the oscillatory dynamics of a protein are over-damped and spends most of the time at the bottom of an "energy well", as illustrated in Fig. 4.5. However, since the solvent is not truly continuous, but rather corpuscular, the collisions with the solvent molecules constitute a random force. Therefore, the dynamics can be described by a stochastic differential equation such as

$$d\mathbf{Y}(t) = b(\mathbf{Y})dt + AdW(t), \tag{4.33}$$

in which $b(y) = -\eta^{-1}\nabla_y U(y)$ and $A = \sqrt{2\eta^{-1}k_BT}$. $W(t)$ is standard Brownian motion.

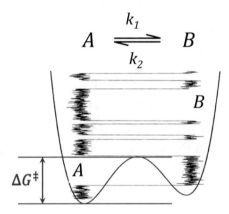

Fig. 4.5 The mathematical description of the chemical reaction of a single molecule. The is the emergent statistical law of a large number of discrete, stochastic reactions. $k_1 \propto e^{-\Delta G^{\ddagger}/k_BT}$, where $\Delta G^{\ddagger}$ is called the *activation energy*. Similarly, k_2 has its own activation barrier height. According to this description, the ratio k_1/k_2 is independent of the barrier.

With the presence of the random forcing term $W(t)$, $\mathbf{Y}(t)$ will occasionally move against the deterministic force field and even cross the barrier (a saddle point in multidimensional space). However, this is a rare event. This randomly perturbed nonlinear dynamical system thus behaves, on a very long time scale, as $A \rightleftharpoons B$, with only two parameters k_1 and k_2. The rate constants are related to the height of the barrier. H. A. Kramers first determined the mathematical theory for this type of problem in 1940 [152]. The idea is not limited to chemical reactions; it is applicable to any nonlinear dynamic system with random perturbations.

Kramers' rate theory can be formulated by the same equation as $\pi_{\varepsilon}(x)$ from the previous section under different boundary conditions, i.e.,

$$\varepsilon\left(\frac{d^2 y_\varepsilon(x)}{dx^2}\right) - \frac{d}{dx}\left(V(x)y_\varepsilon(x)\right) = 0,$$

with $y_\varepsilon(1) = 0$ and $y_\varepsilon(-1) = 1$.

To discretize the continuous interval $[-1, 1]$, let us identify $[-1, 0)$ and $(0, 1]$ as two states A and B, respectively. Let $U(x)$ have a local maximum at $x = 0$ and a local minimum at $x = -1$. The boundary condition $y_\varepsilon(1) = 0$ corresponds to absorbing state B. Then, in the steady state, the flux for reaction $A \to B$ is kc_A, where k is the rate constant of the reaction, and

$$c_A = 2\int_{-1}^{0} e^{-\frac{U(z)-U(-1)}{\varepsilon}} dz \simeq \int_{-\infty}^{\infty} e^{-\frac{U''(-1)(z+1)^2}{2\varepsilon}} dz = \sqrt{\frac{2\pi\varepsilon}{U''(-1)}}$$

is the total concentration of species A.

Then, with a similar derivation as in Eqn. (4.24), we have

$$k = \frac{J_\varepsilon}{c_A} \simeq \varepsilon e^{\frac{U(-1)-U(0)}{\varepsilon}} \sqrt{\frac{|U''(0)|}{2\pi\varepsilon}} \sqrt{\frac{U''(-1)}{2\pi\varepsilon}}$$
$$= \frac{\sqrt{|U''(0)|U''(-1)}}{2\pi} e^{-\frac{U(0)-U(-1)}{\varepsilon}}. \tag{4.34}$$

Substituting ε with $k_B T$, Eqn. (4.34) has the form of the Arrhenius law, in which $U(0) - U(-1)$ is the activation energy, commonly called the "barrier height".

Chapter 5
The Probabilistic Basis of Chemical Kinetics

In Chapter 2, we discussed the law of mass action (LMA). One might have noted that for a large number of mathematical models in ecology and population dynamics, the form of the differential equations is remarkably similar to the chemical kinetic equations using the LMA. In fact, one can see a correspondence between synthesis and degradation in chemical kinetics to birth and death in ecology. In this Chapter, we will show that the fundamentals of all dynamics of population of individuals, be they molecules, cells, or biological species in ecology, have a unified mathematical representation.

5.1 Probability and Stochastic Processes — A New Language for Population Dynamics

Let us first revisit the key concept in deterministic population kinetics in term of a real-valued $P \in \mathbb{R}^+$ and continuous time t. The instantaneous *growth rate* of a time-dependent population $P(t)$ is

$$r(t) = \frac{dP(t)}{dt}. \tag{5.1}$$

In calculus, quantities like $r(t)$ in Eqn. (5.1) are rigorously understood through the limiting procedure: if a population increases by ten people every 100 days, then it is "equivalent" to an increase of one person every 10 days and of 0.1 persons per day. The instantaneous growth rate at time t is then

$$r(t) = \lim_{\Delta t \to 0} \frac{P(t + \Delta t) - P(t)}{\Delta t}.$$

© The Author(s), under exclusive license to Springer Nature Switzerland AG 2021
H. Qian, H. Ge, *Stochastic Chemical Reaction Systems in Biology*,
Lecture Notes on Mathematical Modelling in the Life Sciences,
https://doi.org/10.1007/978-3-030-86252-7_5

This is Newton's *fluxion*, the most important concept of differential calculus. In Newtonian mechanics, the continuous nature of space and time is seriously justified. But is the same mathematical concept valid for quantifying population growth? What does a tenth of a person or one percent of a person mean? Clearly such a theory cannot be rigorously justified when Δt is too small, as *population change* can not have non-integer numbers; $P(t)$ is not even a continuous function of time, let alone differentiable. The instantaneous population growth rate articulated naively above does not exist under tighter scrutiny.

There is a second, equally serious problem in the formulation in (5.1): when counting one by one, the integer-valued function $P(t)$ is not reproducible, no matter how carefully a "cell clonal expansion" experiment is controlled. Additionally, there has never been a regular population growth of two individuals in the first 100 hours, and another two individuals exactly in the next 100 hours. Certainly, most of you would say that "that is just an average".

These two issues, *discreteness* and *probability*, are two fundamental aspects of any example of population dynamics. Both issues have been ignored in the differential equations-based description of population dynamics, including chemical kinetics. We shall, however, treat them in mathematical terms more carefully. What is the justification, if any, for the entire concentration-based kinetics formulation presented in Chapter 2? In the present Chapter, we will again discuss population kinetics, but with the newfound insights from the theory of probability and stochastic processes.

5.2 Radioactive Decay and Exponential Time

Let us revisit the simplest differential equation

$$\frac{dy}{dt} = -\lambda y, \tag{5.2}$$

in which the constant $\lambda > 0$. Eqn. (5.2) is universally known as the mathematical description of the phenomenon of radioactive decay. It gives the remaining fraction of a radioactive material at time t

$$\frac{y(t)}{y(0)} = e^{-\lambda t}. \tag{5.3}$$

With modern advances in nuclear engineering, the individual events of the emission of radioactive particles can be measured [192], for example, using a Geiger counter. What one discovers is that the times of occurrence of the events are not regular but random. If all the atomic nuclei are statistically *identical and independent*, then

$$\Pr\{\text{a nucleus remaining radioactive at time } t\} = \frac{y(t)}{y(0)} = e^{-\lambda t}. \qquad (5.4)$$

If we use $\mathbf{T}$ to denote the random time at which a radioactive decay event occurs for an individual nucleus, then

$$\Pr\{\text{the nucleus remaining radioactive at time } t\} = \Pr\{\mathbf{T} \geq t\}. \qquad (5.5)$$

$\mathbf{T}$ is a real-valued, nonnegative random variable with cumulative probability distribution function $F_T(t) = \Pr\{\mathbf{T} \leq t\} = 1 - e^{-\lambda t}$ and probability density function $f_T(t) = dF(t)/dt = \lambda e^{-\lambda t}$.

What is the possible "mechanism", e.g., explanation, for this random behavior? In other words, what are the problems or scenarios in which this type of exponentially distributed waiting time will appear? Why is it so universal? Solid answers to these questions will provide the reader with a much deeper understanding of the **mathematical foundation of population dynamics** as an emergent statistical law, in terms of the seemingly random behavior of complex individuals, from macromolecules made of large number of atoms to biological entities such as cells.

5.2.1 Basic Formulation

Let $\mathbf{T}$ be the random time at which a certain event occurs. If the occurrence of such an event is independent between any nonoverlapping time intervals, then

$$\text{Prob. of no event occurs in } [0, t + \Delta t] = \qquad (5.6)$$
$$\text{Prob. of no event occurs in } [0, t] \times \text{Prob. of no event occurs in } [t, t + \Delta t].$$

This equation simply follows the basic laws of probabilities. In mathematical notation,

$$\Pr\{\mathbf{T} > t + \Delta t\} = \Pr\{\mathbf{T} > t\} \times \text{Prob. of no event occurs in } [t, t + \Delta t].$$

Now, if the probability of one such event occurring in the time interval $[t, t + \Delta t]$ is proportional to Δt, with coefficient λ not dependent on t, and the probability of more than one events occurring is $o(\Delta t)$, then

$$\Pr\{\mathbf{T} > t + \Delta t\} = \Pr\{\mathbf{T} > t\} \times \left(1 - \lambda \Delta t + o(\Delta t)\right). \tag{5.7}$$

Then,

$$\frac{d}{dt}\Pr\{\mathbf{T} > t\} = -\lambda \Pr\{\mathbf{T} > t\}, \implies F_T(t) = 1 - e^{-\lambda t}. \tag{5.8}$$

Example. The waiting time for the first shopper to come into a store in the morning on a regular day.

5.2.2 Memoryless Property

One of the most important, in fact defining, properties of the exponentially distributed waiting time is

$$\frac{\Pr\{\mathbf{T} \geq t + \tau\}}{\Pr\{\mathbf{T} \geq t\}} = \frac{e^{-\lambda(t+\tau)}}{e^{-\lambda t}} = e^{-\lambda \tau}, \tag{5.9}$$

which is independent of t.

Example. You and your lazy brother are carrying out experiments to observe the mean time of an exponentially distributed event. Even though your brother starts counting time a whole hour later than you, his results will be exact the same as yours! It does not matter when one starts recording the time.

5.2.3 From Binomial Distribution to Poisson Distribution

In Section 3.1.5 we introduced Poisson random variables and binomial random variables, both integer valued. Consider a total of N_0 identical, independent atomic nuclei, none of which have decayed at $t = 0$. At time t, the probability of one nucleus having not yet decayed is $e^{-\lambda t}$, and the probability that it has decayed is $(1 - e^{-\lambda t})$. Therefore, the number of decay events up to time t among the N_0 total nuclei is a binomial random variable N_t^*:

$$\Pr\{N_t^* = n\} = \frac{N_0!}{n!(N_0-n)!}e^{-(N_0-n)\lambda t}\left(1-e^{-\lambda t}\right)^n.$$

On the other hand, it is well known empirically that the arrival of X-ray particles due to radioactive decay is a Poisson process. This means the number of decay events up to time t is

$$\Pr\{N_t^* = n\} = \frac{(\mu t)^n}{n!}e^{-\mu t},$$

for a certain constant μ.

The relationship between these two results is that the latter can be considered bulk material with $N_0 \approx \infty$, but $N_0\lambda \approx \mu$ is what is being observed. Actually, it can be mathematically shown that in the limit of $N_0 \to \infty$,

$$\frac{N_0!}{n!(N_0-n)!}e^{-(N_0-n)\lambda t}\left(1-e^{-\lambda t}\right)^n \longrightarrow \frac{(\mu t)^n}{n!}e^{-\mu t}, \; if \; N_0\lambda \to \mu.$$

This computation reveals that the intensity μ for the Poisson process is indeed proportional to the total amount of radioactive material, an assumption widely made but not critically evaluated in standard textbooks on radioactive decay.

5.2.4 Minimum of Independent Exponential Distributed Random Variables

The more individuals there are in a population, the faster the next event will occur. In mathematical terms, if all $T_k \sim \lambda_k e^{-\lambda_k t}$ and they are independently distributed, then $T^* = \min(T_1, T_2, \cdots, T_n)$ can also be characterized by an exponential distribution

$$\Pr\{T^* > t\} = \Pr\{T_1 > t, \cdots, T_n > t\}$$
$$= \Pr\{T_1 > t\} \times \Pr\{T_2 > t\} \times \cdots \times \Pr\{T_n > t\} = e^{-\mu t}, \qquad (5.10)$$

where $\mu = \lambda_1 + \lambda_2 + \cdots + \lambda_n$. Thus, $f_{T^*}(t) = \mu e^{-\mu t}$.

Indeed, this is the hidden mathematical result behind the Gillespie algorithm. We leave this exercise to the readers as homework.

5.3 Known Mechanisms that Yield Memoryless Time

In the previous section, we derived the exponentially distributed waiting time based on some very elementary assumptions: (i) time homogeneous and (ii) independence. Furthermore, in Section 5.3.1, we will show that for nonexponential **T**, as long as $f_T(0) \neq 0$, the minimum of a large collection of *independent and identically distributed* (i.i.d.) **T**'s will be exponential. This is a strong argument for why one can use, on an appropriate time scale, equations such as (5.2) to model population dynamics.

5.3.1 Minimal Time of a Set of Nonexponential I.I.D. Random Times

Now consider a set of random times $\{T_k\}$. These times are i.i.d. with pdf $f_T(t)$ and cumulative probability distribution $F_T(t)$. They are *arbitrarily* distributed, with the sole assumption that $f_T(0) \neq 0$. Then, $T^* = \min(T_1, T_2, \cdots, T_n)$ has distribution

$$\Pr\{T^* > t\} = \left(1 - F_T(t)\right)^n \tag{5.11}$$

Now we introduce scaled $\hat{T}^* = nT^*$. If we consider n to be very large, then its distribution is

$$\Pr\{\hat{T}^* > t\} = \left(1 - F_T\left(\frac{t}{n}\right)\right)^n \simeq \left(1 - \frac{F_T'(0)}{n}t + O\left(n^{-2}\right)\right)^n \to e^{-F_T'(0)t}. \tag{5.12}$$

Therefore, if $F_T'(0) = f_T(0) \neq 0$, one obtains exponentially distributed time.

In applications, we note that the mathematical condition $f_T(0) > 0$ implies that the time scale involved in the mechanism for the occurrence of an event is many orders of magnitude faster than the time scale in question. It can happen at any moment, in the next minute or in the next microsecond.

5.3.2 Khinchin's Theorem

Let us consider another variation on the theme: a theater that has n ceiling light bulbs. One buys a large number of new bulbs, and let us assume that all of them have an independent and identically distributed lifetime X with pdf $f_X(x)$.

Whenever a light bulb goes out, it is replaced by a new one. We assume no deterioration in the unused light bulbs. The time sequence $0, \mathbf{T}_1, \mathbf{T}_2, \cdots, \mathbf{T}_k, \cdots$ is called a *renewal process*, in which $\mathbf{T}_k = \sum_{\ell=1}^{k} \mathbf{X}^{(\ell)}$, where values of $\mathbf{X}^{(\ell)}$ with different ℓ are i.i.d. random variables drawn from the distribution $f_X(x)$. Now, for the entire theater, there are n identical, independent renewal processes. The time sequence of all bulb changes forms a *superposition* of the n renewal processes, as illustrated in Fig. 5.1.

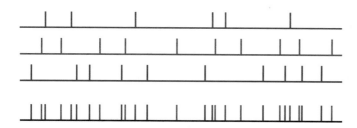

Fig. 5.1 If the red, orange, and blue point processes represent the renewal events of changing light bulbs for three ($n = 3$) different sockets, then the fourth row is the combined point process for all bulb changes. That is, it is the superposition of the three individual processes. With an increasing number of sockets, a statistical law with memoryless waiting time emerges.

Khinchin proved the following theorem in 1960 [147].

Theorem 5.1. *If* $\mathbf{T}_k^{(1)}, \mathbf{T}_k^{(2)} \cdots, \mathbf{T}_k^{(n)}$ *are n i.i.d. renewal processes with waiting time distribution* $f_X(x)$*, then the superposition of the n renewal processes has an exponential waiting time for the next event in the limit of* $n \to \infty$*, with rate parameter* $n\mathbb{E}^{-1}[\mathbf{X}]$*.*

Khinchin's theorem allows us to model the occurrence of bimolecular chemical reactions, which are the consequences of independent molecular collisions, as a Poisson process, i.e., the waiting time between adjacent occurrences follows an exponential distribution.

5.3.3 *Kramers' Theory and Saddle-crossing As a Rare Event*

We have discussed the minimal time of a large collection of i.i.d. waiting times and
the superposition of parallel renewal processes. While these two results justify the
use of exponential waiting time in the birth and death events of a large population
of individuals, it does not explain why the radioactive decay of a *single* nucleus
exhibits an exponentially distributed waiting time. We now turn to a completely
different mechanism: the emergence of discrete molecular transformation events
from a description of the continuously moving atoms within a molecule situated in
a thermal environment.

In the last section of Chapter 4, we derived Kramers' rate formula, $k \propto e^{-\Delta G^{\ddagger}/k_B T}$
(see Fig. 4.5). All the detailed atomic motions are deemed irrelevant; only two parameters, called the forward and backward rate constants, are useful to a biochemist
who is interested in cell biology. Furthermore, the theory shows that the transition
from $A \to B$ spends most of its time waiting; the actual transition event is instantaneous! Indeed, one can mathematically prove that in the limit of $\Delta G^{\ddagger}/k_B T \to \infty$, the
waiting time distribution asymptotically approaches an exponential curve [44, 195].
This provides the physical foundation for the master-equation model for unimolecular, first-order reactions. From a molecular biological function perspective, the notion of discrete conformational states and the events of transitions among them are
fundamental.

5.4 Stochastic Population Growth

We have shown that the law of radioactive decay fundamentally originates from exponentially distributed, memoryless waiting time. The same logic can be naturally
applied to population dynamics involving only death. On the other hand, traditionally, the proportionality of the population growth rate to the current population has
always been presented as a consequence of geometric growth. Just like death, birth
is a random event. Khinchin's theorem explains that for the birth process of an asynchronized population of biological organisms, the waiting time for the next birth is
also expected to be exponentially distributed. The birth rate is expected to be proportional to the number of individuals currently in the population, say $X(t)$, not

because of geometric growth but because of the exponential distribution of the time of a birth event. *On average, birth produces* 1 *additional person in* $\left(r\mathbb{E}[X]\right)^{-1}$ *time*:

$$\frac{\mathrm{d}}{\mathrm{d}t}\mathbb{E}\big[\mathbf{X}(t)\big] = r\mathbb{E}\big[\mathbf{X}(t)\big]. \tag{5.13}$$

Death is an event, birth is an event, and state transition is an event. Most studies on biological dynamics concern population counting and biological events that lead to changes in the population. The stochasticity lies in the timings of the various events. This is why J. D. Murray stated in [185] that continuous growth models for a species at time t share a universal conservation equation:

$$\frac{\mathrm{d}Y}{\mathrm{d}t} = \text{births} - \text{deaths} + \text{migration}, \tag{5.14}$$

where $Y(t)$ is the population density.

Note several important differences between Eqs. (5.14) and (3.17). The former is used for population *density* $Y(t)$, while the latter is used for the probability of population size $p_k(t) \equiv \Pr\{N(t) = k\}$; the right-hand side of the former is usually a nonlinear function of Y, while the latter is necessarily linear. The dimensionality of the latter ODE system, however, is much higher than that of the former.

5.5 Theory of Chemical and Biochemical Reaction Systems

Let us recall the general representation for complex chemical reaction systems in Chapter 2, Eqn. (2.19). We shall now consider *irreversible* reactions. Therefore, for the reversible reactions in (2.19), we separate the forward and backward reactions. Then, we have

$$v_{j1}^+X_1 + v_{j2}^+X_2 + \cdots + v_{jN}^+X_N \xrightarrow{k_j} v_{j1}^-X_1 + v_{j2}^-X_2 + \cdots + v_{jN}^-X_N, \tag{5.15}$$

$1 \leq j \leq M$. There are N species and M reactions. Again, recall that $(v_{ji}^+ - v_{ji}^-)$ are called stoichiometric coefficients, which relate species i to reaction j. A "reaction" is a random "transforming event".

5.5.1 Delbrück–Gillespie Process (DGP)

Let us now probabilistically consider the discrete, individual events of the M possible reactions in Eqn. (5.15), one at a time. We shall abuse the notions a little, using X_i to represent the ith chemical species, as in Eqn. (5.15), and $\mathbf{X}(t) = (X_1, X_2, \cdots, X_N)$ as the integer-valued random variables that count the number of molecules of each species in a reaction vessel of size V at time t.

The Delbrück–Gillespie process (DGP) assumes that the jth reaction occurs following an exponentially distributed waiting time, with rate parameter $\varphi_j(\mathbf{X})$. Note the $\varphi_j(\mathbf{X})$ has a dimension of $[\text{time}]^{-1}$. $\varphi_j(\mathbf{X})$ is called the *propensity* of the jth reaction.

The first reaction that occurs also follows an exponential time, with the rate being the sum of the rates of the m reactions

$$\sum_{j=1}^{M} \varphi_j(\mathbf{X}), \tag{5.16}$$

and independently, the probability that the first reaction that occurs is the $k - th$ reaction is

$$\frac{\varphi_k(\mathbf{X})}{\sum_{j=1}^{M} \varphi_j(\mathbf{X})}. \tag{5.17}$$

It is based on a simple fact in probability theory: among M independent, random times $\mathbf{T}_1, \mathbf{T}_2, \cdots, \mathbf{T}_M$, all exponentially distributed with respective rate parameters $\lambda_1, \lambda_2, \cdots, \lambda_M$, the probability of the smallest one being $\mathbf{T}_k$ is

$$\Pr\{\mathbf{T}^* = T_k\} = \Pr\Big\{\mathbf{T}_k \leq \min\big(\mathbf{T}_1, \cdots, \mathbf{T}_{k-1}, \mathbf{T}_{k+1}, \cdots, \mathbf{T}_M\big)\Big\}$$
$$= \frac{\lambda_k}{\lambda_1 + \cdots + \lambda_M}, \tag{5.18}$$

and more importantly,

$$\Pr\{\mathbf{T}^* = T_k, \mathbf{T}^* \geq t\}$$
$$= \Pr\Big\{\mathbf{T}_1 \geq \mathbf{T}_k, \cdots, \mathbf{T}_{k-1} \geq \mathbf{T}_k, \mathbf{T}_k \geq t, \mathbf{T}_{k+1} \geq \mathbf{T}_k, \mathbf{T}_M \geq \mathbf{T}_k,\Big\}$$
$$= \int_t^\infty \lambda_k e^{-\lambda_k t_k} \prod_{\ell=1,\ell \neq k}^{M} \left(\int_{t_k}^\infty \lambda_\ell e^{-\lambda_\ell t_\ell} \mathrm{d}t_\ell \right)$$
$$= \int_t^\infty \lambda_k e^{-\lambda_k t_k} \prod_{\ell=1,\ell \neq k}^{M} \left(\int_{t_k}^\infty \lambda_\ell e^{-\lambda_\ell t_\ell} \mathrm{d}t_\ell \right)$$

$$= \left(\frac{\lambda_k}{\lambda_1 + \cdots + \lambda_M} \right) e^{-(\lambda_1 + \cdots + \lambda_M)t}. \tag{5.19}$$

This means the smallest time among $\{\mathbf{T}_k\}$ gives two random variables: $\mathbf{T}^* \equiv \min_k\{\mathbf{T}_k\}$ and $k^* \equiv \arg\min_k\{\mathbf{T}_k\}$; the smallest time $\mathbf{T}^*$ and the identity k^* are statistically independent.

All the results so far apply to general rate laws. In a Delbrück–Gillespie process, one has specifically

$$\varphi_j(\mathbf{X}) = k_j V \prod_{\ell=1}^{N} \left(\frac{X_\ell!}{(X_\ell - v_{j\ell}^+)! V^{v_{j\ell}^+}} \right), \tag{5.20}$$

when the molecular number of the ith chemical species is X_i. This expression is the stochastic counterpart of the law of mass action.

5.5.2 Integral Representations With Random Time Change

5.5.2.1 Poisson Process

Recall that, from Section 3.3.1, a standard Poisson process $\mathbf{Y}(t)$ is an integer-valued, continuous-time Markov process with distribution

$$\Pr\left\{ \mathbf{Y}(t) = k \right\} = \frac{t^k}{k!} e^{-t}. \tag{5.21}$$

A Poisson process has both a *point process* representation, $\mathbf{T}_1, \mathbf{T}_2, \cdots, \mathbf{T}_n$, and a *counting process* representation $\mathbf{Y}(t)$. The latter is a positive, real-valued, discrete-time Markov process with independent increments, and $\mathbf{T}_{i+1} - \mathbf{T}_i$ is exponentially distributed with rate 1.

5.5.2.2 Random Time-Changed Poisson Representation

In Chapter 3, we introduced the random time-changed Poisson representation for the stochastic trajectory of a discrete-state, continuous-time Markov process. Similarly, one can use Poisson processes to represent the stochastic trajectory of a DGP for the integer number of molecule X_i at time t,

$$\mathbf{X}_i(t) = \mathbf{X}_i(0) + \sum_{j=1}^{M} \left(v_{ji}^- - v_{ji}^+ \right) \mathbf{Y}_j \left(\int_0^t \varphi_j\left(\mathbf{X}(t)\right) dt \right) \tag{5.22}$$

in which $\varphi_j(\mathbf{X})$ is given in (5.20).

We see that in the limit of $\mathbf{X} \to \infty$ and $V \to \infty$, while keeping $\mathbf{x} = \frac{\mathbf{X}}{V}$, the classic propensity

$$\varphi_j(\mathbf{X}) \to k_j V \prod_{\ell=1}^{N} \left(\frac{X_\ell}{V} \right)^{v_{j\ell}^+} = k_j V \prod_{\ell=1}^{N} x_\ell^{v_{j\ell}^+} = V \hat{\varphi}_j(\mathbf{x}), \tag{5.23}$$

where

$$\hat{\varphi}_j(\mathbf{x}) = k_j \prod_{\ell=1}^{N} x_\ell^{v_{j\ell}^+} \tag{5.24}$$

is exactly the flux of the jth reaction according to the law of mass action.

5.5.3 Single-population System

Consider the stochastic population kinetics of a single species. Let $p_n(t)$ be the probability of having n individuals in the population at time t. Then, $p_n(t)$ satisfies the master equation

$$\frac{dp_0(t)}{dt} = -u_0 p_0 + w_1 p_1,$$

$$\frac{dp_n(t)}{dt} = p_{n-1} u_{n-1} - p_n \left(u_n + w_n \right) + p_{n+1} w_{n+1}, \tag{5.25}$$

in which u_k and w_k are the birth (or immigration) rate and death (or emigration) rate of the population with exactly k individuals.

The system in Eqn. (5.25) has been traditionally known for a general birth-death process; it has had been studied by others long before Delbrück or Gillespie, for example, by G. U. Yule (1871–1951) [279]. Regardless, we used the term "Delbrück–Gillespie process" throughout the book to emphasize the nature of the stochastic-process description of chemical kinetics. Eqn. (5.25) is one of the simplest cases of the DGP.

The stationary distribution of Eqn. (5.25) can be obtained by setting the rhs of Eqn. (5.25) to zero:

$$\frac{p_n^{ss}}{p_{n-1}^{ss}} = \frac{u_{n-1}}{w_n}. \tag{5.26}$$

Therefore,

$$p_n^{ss} = p_0^{ss} \prod_{k=1}^{n} \left(\frac{u_{k-1}}{w_k} \right),$$ (5.27)

in which p_0^{ss} is to be determined by normalization.

Eqn. (5.25) is the DGP corresponding to the nonlinear population dynamics of a single species with birth and death rates $\hat{u}(x)$ and $\hat{w}(x)$, with $x(t) \equiv \frac{X(t)}{V}$,

$$\frac{dx}{dt} = \hat{u}(x) - \hat{w}(x),$$ (5.28)

where,

$$\hat{u}(x) = \lim_{V \to \infty} \frac{u_{xV}}{V}, \quad \hat{w}(x) = \lim_{V \to \infty} \frac{w_{xV}}{V}.$$ (5.29)

It is easy to verify that the peaks and troughs of stationary probability distribution p_n^{ss} correspond nicely to the stable and unstable fixed points of Eqn. (5.29). Eqn. (5.27) gives the probability distribution for the number of individuals in a stationary, fluctuating population. Such a distribution can be either unimodal or multimodal. The location of a modal value n^* occurs when the product in (5.27) changes from increasing with n to decreasing with n. This implies

$$\frac{u_{n^*-1}}{w_{n^*}} \geq 1$$

and

$$\frac{u_{n^*}}{w_{n^*+1}} \leq 1.$$

Noting the relations in (5.29), this corresponds exactly to $\hat{u}(x) = \hat{w}(x)$ for $x^* = n^*/V$, $\hat{u}(x) > \hat{w}(x)$ for $x < x^*$, and $\hat{u}(x) < \hat{w}(x)$ for $x > x^*$, for V sufficiently large. In other words, x^* is a stable fixed point of Eqn. (5.28). Similarly, it is easy to show that a trough of the probability distribution p_n^{ss} corresponds to an unstable fixed point of (5.28). This result is quite general: It is true also for multi-species DGPs and their corresponding macroscopic ODEs. We shall return to this problem in Section 7.4.1, after we build our mathematical arsenals.

One highly novel aspect arises in the stochastic representation of single-species population kinetics. Let us consider the case in which all the individuals are i.i.d. and thus $u_n = nu_1$ and $w_n = nw_1$, where u_1 and w_1 are the per capita birth and death rates, respectively, and $u_1 - w_1 \equiv g$ is the per capita growth rate. Then, for the expected value of the population, we have:

$$\frac{d}{dt}\mathbb{E}[\mathbf{X}(t)] = \frac{d}{dt}\left(\sum_{k=0}^{\infty} kp_k(t)\right) = \sum_{k=0}^{\infty} k\left(\frac{dp_k(t)}{dt}\right)$$

$$= \sum_{k=0}^{\infty} k\Big(p_{k-1}(k-1)u_1 - kp_k(u_1 + w_1) + (k+1)p_{k+1}w_1\Big)$$

$$= (u_1 - w_1)\mathbb{E}[\mathbf{X}(t)] = g\mathbb{E}[\mathbf{X}(t)]. \tag{5.30}$$

This is precisely what one expects from the stochastic model. One notices, however, that the per capita birth and death rates can not be separated in this equation: the only parameter is the per capita growth rate g. This is a key feature of the deterministic model. It is impossible to observe an increased instantaneous birth rate balanced by an increased instantaneous death rate.

However, for the stochastic population kinetics, one can measure the variance of the population:

$$\mathrm{Var}[\mathbf{X}(t)] = \mathbb{E}[\mathbf{X}^2(t)] - \Big(\mathbb{E}[\mathbf{X}(t)]\Big)^2. \tag{5.31}$$

which satisfies:

$$\frac{d}{dt}\mathrm{Var}[\mathbf{X}(t)] = \frac{d}{dt}\left(\sum_{k=0}^{\infty} k^2 p_k(t)\right) - 2\mathbb{E}[\mathbf{X}(t)]\frac{d}{dt}\mathbb{E}[\mathbf{X}(t)]$$

$$= \sum_{k=0}^{\infty} k^2\Big(p_{k-1}(k-1)u_1 - kp_k(u_1 + w_1) + (k+1)p_{k+1}w_1\Big) - 2g\Big(\mathbb{E}[\mathbf{X}(t)]\Big)^2$$

$$= 2g\mathrm{Var}[\mathbf{X}(t)] + (u_1 + w_1)\mathbb{E}[\mathbf{X}(t)]. \tag{5.32}$$

We see that the change in variance is dependent upon $(u_1 + w_1)$. There is additional information on the birth and death rates separately in the variance of stochastic population kinetics.

We also notice that for nonlinear population kinetics, the expectation of a nonlinear function $\mathbf{X}(t)$ is not equal to the nonlinear function $\mathbb{E}[\mathbf{X}(t)]$. In this case, one will not be able to derive a closed-form differential equation for the time-dependent moments of $\mathbf{X}(t)$ in general. A deterministic kinetic equation arises only in the limit of the law of large numbers. This will be discussed in Section 7.4.1.

Part II
From Chemical Kinematics to Mesoscopic and Macroscopic Thermodynamics

We are articulating in the present book a changing perspective that is based on a mechanical world made of featureless point masses moving continuously in space and time to a biological world made of complex individual molecules and cells that change their states intermittently. It is instructive, therefore, to recall that there are two parts to the theory of classical mechanics: *kinematics* and *dynamics*. The former deals with mechanical motion in terms of instantaneous velocities and accelerations under various constraints; it is essentially a subject of applied calculus. The core physics of classical mechanics, however, is embodied in dynamics, where the notions of mechanical force and mechanical energy are first defined, and the idea of conservation of energy, linear and angular momenta plays central roles.

The chemical kinetics we have thus far studied, either deterministic or stochastic, correspond mainly to kinematics: it is a mathematical description of the number of molecules in a system of chemical species, and how fast and how slow the reactions occur among the species, as events. This description is completely devoid of the notion of energy and force. On the other hand, chemical energy and force, which we have touched upon at certain points in Part I, are a part of Gibbsian *chemical thermodynamics*, a theory that until now contains no time and can only be applied to dead substances as equilibrium matters.

In the following three chapters, we shall show a surprising turn of events: that one can actually mathematically derive Gibbsian chemical thermodynamics for an isothermal reaction system, whether or not it is in equilibrium. In terms of the stochastic kinematics, the notion of chemical energy and force naturally emerge from the concept of *entropy* and the related *entropic force*. Kinematics dictates thermodynamics.

The situation has a deep mathematical root: the stochastic kinetic formulation implies that deterministic chemical kinetics truly is an idealization, a limit, of a stochastic system. The thermodynamic potential function, it turns out, is a part of the mathematics that characterize the limiting process in terms of the theory of large deviations introduced in Chapter 4.

However, nonlinear stochastic dynamics with multiple attractors implies an emergent Markov description of *state transitions* with an exponential, memoryless waiting time. Part II concludes with Chapter 8 in which the fundamental notions and concepts from the theory of phase transitions are captured in terms of simple mathematics. Extensive discussions are given on the nature of *symmetry breaking* and a hierarchical organization of complex "individuals" as well as complex systems with nonlinear interacting complex individuals. Even though this book focuses on stochastic chemical reaction systems and cellular biochemistry, the theoretical principles contained in Part I and Part II are applicable to higher level biological systems and organisms and thus the entirety of *mathematical biology*.

Chapter 6
Mesoscopic Thermodynamics of Markov Processes

By merely introducing the mathematical function identified as entropy, $S(E,V) = k_B \ln \Omega(E,V)$, into Hamiltonian dynamics, L. Boltzmann was able to "derive" a relation, the Gibbs equation, $T\mathrm{d}S = \mathrm{d}E + p\mathrm{d}V$, which represents the fundamental thermodynamic relation, identifying $p\mathrm{d}V$ as work and $T\mathrm{d}S$ as heat, where $T = (\partial S/\partial E)_V^{-1}$ and $p/T = (\partial S/\partial V)_E$. Here we note the significance of k_B, Boltzmann's constant: to a mathematician, it is of no consequence; but to a physicist, it connects mathematical concepts to reality. Given the definition of Boltzmann's constant, $k_B = 1.3807 \times 10^{-23} J \cdot K^{-1}$, $k_B T$ is a unit for energy, where T is the absolute temperature.

In the present chapter, we will follow Boltzmann's spirit and mathematically introduce the notion of entropy into stochastic dynamics as represented by Markov processes. The result is spectacular: one is able to "derive" several equations that echo the fundamental relations in classic thermodynamics. In fact, an entire mesoscopic thermodynamic structure emerges based on the stochastic dynamics description without the need for the mechanical theory of motion. Here, "mesoscopic" refers to any stochastic system between the microscopic and macroscopic scales whose evolution can be described by Markov processes. Microscopic few-body systems are described by Newton's equation; macroscopic systems again follow deterministic dynamics due to the law of large numbers of one form or another. A macroscopic system can be understood as the limit of a sequence of mesoscopic systems.

© The Author(s), under exclusive license to Springer Nature Switzerland AG 2021
H. Qian, H. Ge, *Stochastic Chemical Reaction Systems in Biology*,
Lecture Notes on Mathematical Modelling in the Life Sciences,
https://doi.org/10.1007/978-3-030-86252-7_6

6.1 Entropy Balance Equation

We consider a Markov process $\mathbf{X}(t)$ with discrete state space $\mathscr{S}$ and transition probability rate q_{ij}, $i, j \in \mathscr{S}$, from state i to state j. We further assume that $\mathbf{X}(t)$ is irreducible and $q_{ij} \neq 0$ if and only if $q_{ji} \neq 0$. Here, "irreducibility" means that each state in the state space can be reached from any other state by multiple transitions. The first assumption then implies the existence of a unique, positive stationary probability $\pi = \{\pi_i | i \in \mathscr{S}\}$, and the second assumption means one can introduce terms such as (q_{ij}/q_{ji}), as we shall do below. Under these assumptions and assume the process is stationary, for a consecutive pair we have $P\{X(t) = i, X(t+dt) = j\} = \pi_i q_{ij} dt$. One then immediately obtains an interesting equality for random variable $Z[X(t), X(t+dt)]$, as a function of the pair defined on $\mathscr{S} \times \mathscr{S}$, $Z[i, j] \triangleq \ln(\pi_i q_{ij}/\pi_j q_{ji})$:

$$\frac{P\{Z = z\}}{P\{Z = -z\}} = e^z. \tag{6.1}$$

The ratio is meaningful since the probabilities of $Z = z$ and $Z = -z$ are simultaneously nonzero. Corresponding to $(i, j) \in \mathscr{S} \times \mathscr{S}$, the numerator is equal to $\pi_i p_{ij}$ if and only if the denominator is equal to $\pi_j q_{ji}$. Therefore, their ratio is e^z. Furthermore, the characteristic function of Z, $\mathbb{E}\left[e^{\lambda Z}\right] = \mathbb{E}\left[e^{(1-\lambda)Z}\right]$. We leave proof of this symmetry of the characteristic function to the readers as an exercise.

The instantaneous Shannon entropy of the Markov process at time t is defined as

$$S[p(t)] = -\sum_{i \in \mathscr{S}} p_i(t) \ln p_i(t). \tag{6.2}$$

Since probability distribution $p(t)$ satisfies the master equation (3.17),

$$\frac{dp_i(t)}{dt} = \sum_{j \in \mathscr{S}} \left(p_j q_{ji} - p_i q_{ij}\right), \tag{6.3}$$

we have

$$\frac{d}{dt}S[p(t)] = -\sum_{i \in \mathscr{S}} \frac{dp_i(t)}{dt} \ln p_i(t) = \sum_{i,j \in \mathscr{S}} \left(p_i q_{ij} - p_j q_{ji} \right) \ln p_i$$

$$= \frac{1}{2} \sum_{i,j \in \mathscr{S}} \left(p_i q_{ij} - p_j q_{ji} \right) \ln \left(\frac{p_i}{p_j} \right)$$

$$= \frac{1}{2} \sum_{i,j \in \mathscr{S}} \left[p_i q_{ij} - p_j q_{ji} \right] \ln \left[\frac{p_i q_{ij}}{p_j q_{ji}} \right] - \frac{1}{2} \sum_{i,j \in \mathscr{S}} \left[p_i q_{ij} - p_j q_{ji} \right] \ln \left[\frac{q_{ij}}{q_{ji}} \right].$$

$$(6.4)$$

We note that the first term in (6.4) is nonnegative. A reader with some knowledge of elementary chemistry will recognize that the logarithmic part in the first term has a remarkable resemblance to $\Delta \mu$ of a unimolecular chemical reaction, while the logarithmic part in the second term remarkably resembles the $\Delta \mu^o$ term of the reaction. We will call the first term in (6.4) the instantaneous *entropy production rate*, e_p. The second term then is related to the *entropy exchange* with the environment. In the theory of nonequilibrium thermodynamics, the latter is often related to the heat exchange rate divided by temperature, h_{ex}/T:

$$\frac{d}{dt}S[p(t)] = e_p - \frac{h_{ex}}{T}, \tag{6.5a}$$

$$e_p[p] = \frac{1}{2} \sum_{i,j \in \mathscr{S}} \left(p_i q_{ij} - p_j q_{ji} \right) \ln \left(\frac{p_i q_{ij}}{p_j q_{ji}} \right) \geq 0, \tag{6.5b}$$

$$\frac{h_{ex}[p]}{T} = \frac{1}{2} \sum_{i,j \in \mathscr{S}} \left(p_i q_{ij} - p_j q_{ji} \right) \ln \left(\frac{q_{ij}}{q_{ji}} \right). \tag{6.5c}$$

Equation (6.5a) is known as the entropy balance equation, and herein we will use this terminology. In physics, this equation originated from the Second Law of Thermodynamics and Clausius' inequality. Stochastic dynamics provides this fundamental equation with two explicit expressions, given in (6.5b) and (6.5c). Since the notion of temperature does not exist in Markov theory, we shall let $T = 1$. This implies that Markov dynamics is a theory of *isothermal* systems and processes.

In L. Onsager's theory of irreversible thermodynamics, the entropy production rate is the sum of many terms, each representing an irreversible process, in the form of "thermodynamic flux $\times$ thermodynamic force". This suggests that we identify $(p_i q_{ij} - p_j q_{ji})$ as a flux and $\ln(p_i q_{ij}/p_j q_{ji})$ as the corresponding force, or affinity. One example of Onsager's theory is "electrical current $\times$ voltage = power ", which can never be negative.

Eqn. (6.5c) tells us that if all $q_{ij} = q_{ji}$ for any $i, j \in \mathscr{S}$, then $h_{ex} = 0$. For such an isolated system, Eqn. (6.5a) becomes $dS/dt = e_p \geq 0$. Without exchanging heat with its environment, an isolated system's instantaneous entropy continuously increases until it reaches its "thermodynamic equilibrium" when the entropy attains its maximum. Entropy does not monotonically increase in systems with nonzero h_{ex}, however.

Note that all this discussion is just a "narrative", or "interpretation", of the mathematical results concerning general Markov processes.

6.2 Detailed Balance and Reversibility

The above discussion suggests that one could identify $\ln(q_{ij}/q_{ji})$ as the amount of heat exchanged between a stochastic system and its environment. If $q_{ij} = q_{ji}$, then there is no heat exchange. Such a system is called isolated or *adiabatic*. Furthermore, for a system in thermal equilibrium with its environment with regard to heat exchange, if it starts in state i_0, passes through $i_1, i_2, \cdots, i_n$ and eventually returns to initial state i_0, the total amount of heat exchange has to be zero:

$$\ln \left(\frac{q_{i_0 i_1}}{q_{i_1 i_0}} \right) + \ln \left(\frac{q_{i_1 i_2}}{q_{i_2 i_1}} \right) + \cdots + \ln \left(\frac{q_{i_{n-1} i_n}}{q_{i_n i_{n-1}}} \right) + \ln \left(\frac{q_{i_n i_0}}{q_{i_0 i_n}} \right) = 0.$$

Motivated by this observation, we now introduce the very important notion of *detailed balance* in Markov processes, which is a subclass of general Markov processes: its stationary probability distribution satisfies $\pi_i q_{ij} = \pi_j q_{ji}, \forall i, j \in \mathscr{S}$. Detailed balance is a property of the transition Q matrix.

Theorem 6.1. *Assume an irreducible Markov process with matrix Q has a stationary distribution π. π is detailed balanced if and only if for every sequence of distinct states $i_0, i_1, \cdots, i_{n-1}, i_n \in \mathscr{S}$:*

$$q_{i_0 i_1} q_{i_1 i_2} \cdots q_{i_{n-1} i_n} q_{i_n i_0} = q_{i_1 i_0} q_{i_2 i_1} \cdots q_{i_n i_{n-1}} q_{i_0 i_n}. \tag{6.6}$$

The equation in (6.6) is called the Kolmogorov cycle condition. Theorem 6.1 implies that detailed balance is actually a possible property of a Q matrix.

Proof. Necessity: From detailed balance, we have

$$1 = \left(\prod_{k=0}^{n-1} \frac{\pi_{i_k} q_{i_k i_{k+1}}}{\pi_{i_{k+1}} q_{i_{k+1} i_k}} \right) \frac{\pi_{i_n} q_{i_n i_0}}{\pi_{i_0} q_{i_0 i_n}} = \left(\prod_{k=0}^{n-1} \frac{q_{i_k i_{k+1}}}{q_{i_{k+1} i_k}} \right) \frac{q_{i_n i_0}}{q_{i_0 i_n}},$$

which yields relation (6.6).

Sufficiency: From (6.6), introduce a set of positive, normalized values for each and every state $i \in \mathscr{S}$, ξ_i as:

$$\frac{\xi_i}{\xi_j} = \frac{q_{j,i_1} q_{i_1 i_2} \cdots q_{i_{n-1},i}}{q_{i_1,j} q_{i_2 i_1} \cdots q_{i,i_{n-1}}} \left(= \frac{q_{ji}}{q_{ij}}, \; if \; q_{ij} > 0 \right).$$

Then

$$\sum_{i,j \in \mathscr{S}} \left(\xi_j q_{ji} - \xi_i q_{ij} \right) = \sum_{i,j \in \mathscr{S}} \xi_i q_{ij} \left(\frac{\xi_j q_{ji}}{\xi_i q_{ij}} - 1 \right) = 0.$$

Therefore, $\{\xi_i\}$ is a stationary solution to the master equation. Since the normalized stationary solution is unique, $\xi_i = \pi_i$; thus, $\pi_i q_{ij} = \pi_j q_{ji} \; \forall i, j \in \mathscr{S}$. $\qquad \square$

Theorem 6.2. *The following six statements regarding an irreducible stationary Markov process with matrix Q and stationary distribution π are equivalent [137].*

(i) Its stationary distribution satisfies detailed balance: $\pi_i q_{ij} = \pi_j q_{ji}$, $\forall i, j \in \mathscr{S}$.

(ii) Any path connecting states i and j, $i \equiv i_0, i_1, i_2, \cdots, i_n \equiv j$, has a path independent

$$\ln \left(\frac{q_{i_0 i_1}}{q_{i_1 i_0}} \right) + \ln \left(\frac{q_{i_1 i_2}}{q_{i_2 i_1}} \right) + \cdots + \ln \left(\frac{q_{i_{n-1} i_n}}{q_{i_n i_{n-1}}} \right) = \ln \pi_{i_n} - \ln \pi_{i_0}. \tag{6.7}$$

(iii) It is a time-reversible, stationary Markov process.

(iv) Its matrix Q satisfies the Kolmogorov cycle condition for every sequence of states.

(v) There exists a positive diagonal matrix Π such that matrix $Q\Pi$ is symmetric.

(vi) Its stationary process has a zero entropy production rate.

Proof. $(i) \Rightarrow (ii)$:

Using (i), we have

$$\ln \left(\frac{q_{i_0 i_1}}{q_{i_1 i_0}} \right) + \ln \left(\frac{q_{i_1 i_2}}{q_{i_2 i_1}} \right) + \cdots + \ln \left(\frac{q_{i_{n-1} i_n}}{q_{i_n i_{n-1}}} \right) \tag{6.8a}$$

$$= \ln \left(\frac{\pi_{i_0} q_{i_0 i_1}}{\pi_{i_1} q_{i_1 i_0}} \right) + \ln \left(\frac{\pi_{i_1} q_{i_1 i_2}}{\pi_{i_2} q_{i_2 i_1}} \right) + \cdots + \ln \left(\frac{\pi_{i_{n-1}} q_{i_{n-1} i_n}}{\pi_n q_{i_n i_{n-1}}} \right) + \ln \pi_{i_n} - \ln \pi_{i_0}$$

$$= \ln \pi_{i_n} - \ln \pi_{i_0}. \tag{6.8b}$$

This means the term in (6.8a) is independent of the path; it is completely determined by initial state $i \equiv i_0$ and final state $j \equiv i_n$.

$(ii) \Rightarrow (i)$: It is straightforward.

$(i) \Rightarrow (iii)$:

It is easy to show that the time-reversed process of the original stationary process $\mathbf{X}(t)$ is also a continuous-time Markov process with matrix $\tilde{Q}$, with $\tilde{q}_{ij} = \pi_j q_{ji}/\pi_i$ and the same stationary distribution π.

Detailed balance implies $Q = \tilde{Q}$, which means that the original process $\mathbf{X}(t)$ and its time-reversed process share the same statistical laws. That is, process $\mathbf{X}(t)$ is time reversible: its statistical behaviors are identical when observed forward or backward in time.

$(iii) \Rightarrow (i)$:

A reversible Markov process has joint distribution $\pi_i p_{ij}(t) = \pi_j p_{ji}(t)$ for any t. Therefore,

$$\pi_i \left(e^{Qt}\right)_{ij} = \pi_j \left(e^{Qt}\right)_{ji},$$

$$\pi_i \left(\delta_{ij} + q_{ij}t + \frac{t^2}{2}(Q^2)_{ij} \cdots \right) = \pi_j \left(\delta_{ji} + q_{jit} + \frac{t^2}{2}(Q^2)_{ji} \cdots \right),$$

$$\pi_i \left(q_{ij} + \frac{t}{2}(Q^2)_{ij} \cdots \right) = \pi_j \left(q_{ji} + \frac{t}{2}(Q^2)_{ji} \cdots \right), \; if \; i \neq j.$$

Let $t = 0$. We then have $\pi_i q_{ij} = \pi_j q_{ji}$.

$(iv) \Longleftrightarrow (i)$:

This is contained in Theorem 6.1.

$(i) \Rightarrow (v)$:

Simply choose the positive diagonal matrix $\Pi = \mathrm{diag}(\pi_1, \pi_2, \cdots)$.

$(v) \Rightarrow (i)$:

If there exists a set of positive numbers $\{v_i\}$ such that $v_i q_{ij} = v_j q_{ji}$, then it is an eigenvector of Q with eigenvalue 0. Therefore, since Q has a unique eigenvector with the eigenvalue 0, the normalized values are

$$\frac{v_i}{\sum_{k \in \mathscr{S}} v_k} = \pi_i,$$

satisfying the detailed balance condition.

(i) $\iff$ (vi):

The entropy production rate is defined in (6.5b):

$$e_p\left[\pi\right] = \frac{1}{2}\sum_{i,j\in\mathscr{S}}\left(\pi_i q_{ij} - \pi_j q_{ji}\right)\ln\left(\frac{\pi_i q_{ij}}{\pi_j q_{ji}}\right),$$

in which every term inside the summation $(\pi_i q_{ji} - \pi_j q_{ji})\ln(\pi_i q_{ji}/\pi_j q_{ji}) \geq 0$. Therefore, the entire sum is zero if and only if $\pi_i q_{ji} - \pi_j q_{ji} = 0$. This is detailed balance.
$\square$

Statement (ii) is similar to the path-independent work in a conservative force
field, which implies the existence of a potential function. In this discrete case, (6.8)
indicates that the potential function is the negative logarithm of the stationary distribution. Equilibrium probability distribution is the exponential function of a potential
energy, known as Boltzmann's law in statistical mechanics.

In thermodynamics, a stationary system with fluctuations is said to be at equilibrium if it satisfies the detailed balance condition. If not, then it is called a nonequilibrium steady -state (NESS) system [281, 100]. An equilibrium steady state has no
sense of direction in time; it is time reversible. Theorem 6.2 suggests that in such a
system, any sequence of events that occurs will have equal probability to "deoccur":
nothing can be truly accomplished in a system at equilibrium.

From statement (ii), one naturally thinks of $-\ln\pi_i$ as an "internal energy function" of state i. Then, Eqn. (6.7) becomes a kind of conservation of internal energy
and heat: the left-hand side of (6.7) is the amount of heat released when the system
moves along path $i_0, i_1, \cdots, i_n$, and the right-hand side of (6.7) is the internal energy
difference between states i_0 and i_n.

Because of statement (v), matrix Q is similar to a symmetric matrix:

$$\left(\Pi^{-\frac{1}{2}}Q\Pi^{\frac{1}{2}}\right) = \Pi^{\frac{1}{2}}Q^T\Pi^{-\frac{1}{2}} = \left(\Pi^{-\frac{1}{2}}Q\Pi^{\frac{1}{2}}\right)^T;$$

all eigenvalues of Q are, therefore, real. There is no kinetic oscillation in a system
with detailed balance.

Equilibrium systems are well understood through the theory of equilibrium statistical mechanics and thermodynamics. There is currently no universally accepted
theory of nonequilibrium statistical mechanics or thermodynamics.

A living cell, even when it is considered a stationary process, is not an equilibrium system; its entropy production is strictly positive. As we shall see in detail in Part III, most biochemical reactions in a cell are not balanced; cycle kinetics are a hallmark of the chemistry of living cells.

6.2.1 Free Energy

Markov systems with detailed balanced stationarity π have another important property. With detailed balance, let us now revisit the entropy exchange h_{ex}/T in (6.5c):

$$\frac{h_{ex}}{T} = \frac{1}{2} \sum_{i,j \in \mathscr{S}} \left(p_i q_{ij} - p_j q_{ji} \right) \ln \left(\frac{q_{ij}}{q_{ji}} \right) = \frac{d}{dt} \sum_{i \in \mathscr{S}} p_i(t) \ln \pi_i. \tag{6.9}$$

If we identify $-\ln \pi_i$ as the internal energy of state i, then

$$\bar{E}(t) = \sum_{i \in \mathscr{S}} \left(-\ln \pi_i \right) p_i(t) \tag{6.10}$$

is the mean internal energy of the entire system at time t. In this case, the entropy balance equation (6.5a) can be rewritten as

$$\frac{d}{dt} \left(\bar{E}(t) - TS(t) \right) = -Te_p[p] \leq 0. \tag{6.11}$$

The term inside $(\cdots)$ is known in thermodynamics as free energy. For an isothermal system with temperature T, the Second Law of Thermodynamics states that its free energy never increases, and its rate of decrease is the entropy production rate times temperature.

Only for systems with detailed balance and a uniform $\pi_i = C$, one has $h_{ex} = 0$ and $dS/dt \geq 0$: the Second Law of Thermodynamics, as popularly known among laypersons.

6.3 Free Energy Balance Equation

Actually, an important conclusion from classical thermodynamics is that while entropy is the appropriate *thermodynamic potential* for an isolated system, free energy is the appropriate thermodynamic potential for an isothermal system. Taking this key insight one step further, as suggested by the Eqns. (6.10) and (6.11), we now introduce, in the context of a general irreducible Markov process with or without detailed balance, a mathematical definition of instantaneous *generalized free energy*

$$F[p(t)] = \sum_{i \in \mathscr{S}} p_i(t) \ln \left(\frac{p_i(t)}{\pi_i} \right). \tag{6.12}$$

In information theory, the quantity in (6.12) is called relative entropy or Kullback–Leibler divergence. As we shall show below, it has "much nicer" properties than the entropy in (6.2) for the thermodynamic theory of Markov processes.

Concerning the $F[p]$ in (6.12), since $\ln x \leq x - 1$, we first have

$$F[p] = \sum_{i \in \mathscr{S}} p_i \ln \left(\frac{p_i}{\pi_i} \right) = -\sum_{i \in \mathscr{S}} p_i \ln \left(\frac{\pi_i}{p_i} \right)$$

$$\geq -\sum_{i \in \mathscr{S}} p_i \left(\frac{\pi_i}{p_i} - 1 \right) = -\sum_{i \in \mathscr{S}} (\pi_i - p_i) = 0. \tag{6.13}$$

This equality holds true if and only if $p = \pi$. Furthermore, we have

$$\frac{d}{dt} F[p(t)] = \sum_{i \in \mathscr{S}} \frac{dp_i(t)}{dt} \ln \left(\frac{p_i(t)}{\pi_i} \right) = \sum_{i,j \in \mathscr{S}} \left(p_j q_{ji} - p_i q_{ij} \right) \ln \left(\frac{p_i}{\pi_i} \right)$$

$$= \frac{1}{2} \sum_{i,j \in \mathscr{S}} \left(p_j q_{ji} - p_i q_{ij} \right) \ln \left(\frac{p_i \pi_j}{\pi_i p_j} \right)$$

$$= \frac{1}{2} \sum_{i,j \in \mathscr{S}} \left(p_j q_{ji} - p_i q_{ij} \right) \ln \left(\frac{\pi_j q_{ji}}{\pi_i q_{ij}} \right) - e_p[p]. \tag{6.14}$$

The first term in (6.14) is clearly zero if the detailed balance equation holds true;. In general, however, it is not zero. More importantly,

$$\frac{1}{2} \sum_{i,j \in \mathscr{S}} \left(p_j q_{ji} - p_i q_{ij} \right) \ln \left(\frac{\pi_j q_{ji}}{\pi_i q_{ij}} \right) = -\sum_{i,j \in \mathscr{S}, i \neq j} p_i q_{ij} \ln \left(\frac{\pi_j q_{ji}}{\pi_i q_{ij}} \right)$$

$$\geq \sum_{i,j \in \mathscr{S}, i \neq j} p_i q_{ij} \left(\frac{\pi_j q_{ji}}{\pi_i q_{ij}} - 1 \right) = \sum_{i,j \in \mathscr{S}} \left(\frac{\pi_j q_{ji} p_i}{\pi_i} - p_i q_{ij} \right)$$

$$= \sum_{i \in \mathscr{S}} \frac{p_i}{\pi_i} \sum_{j \in \mathscr{S}} \pi_j q_{ji} - \sum_{i,j \in \mathscr{S}} p_i q_{ij} = 0 - 0 = 0, \tag{6.15}$$

and the equality holds if and only if the detailed balance condition holds for π, regardless of $\{p_i\}$.

Therefore, the generalized free energy $F[p]$ satisfies a balance equation of its own [90, 64, 97]:

$$\frac{d}{dt}F[p(t)] = E_{in} - e_p, \tag{6.16a}$$

$$E_{in}[p] = \frac{1}{2}\sum_{i,j\in\mathscr{S}}\left(p_i q_{ij} - p_j q_{ji}\right)\ln\left(\frac{\pi_i q_{ij}}{\pi_j q_{ji}}\right) \geq 0, \tag{6.16b}$$

$$e_p[p] = \frac{1}{2}\sum_{i,j\in\mathscr{S}}\left(p_i q_{ij} - p_j q_{ji}\right)\ln\left(\frac{p_i q_{ij}}{p_j q_{ji}}\right) \geq 0. \tag{6.16c}$$

Eqn. (6.16a) is "nicer" than Eqn. (6.5a): we say this because both E_{in} and e_p in (6.16a) are always nonnegative, but h_{ex} in (6.5a) is not. Therefore, Eqn. (6.16a) has a more legitimate interpretation: *a Markov system's instantaneous free energy changes with time, with an energy input rate (source term) E_{in} and an energy dissipation rate (sink term) e_p. The energy left the system is considered as "wasted", thus implying entropy production.*

In addition to the generalized free energy balance equation (6.16a), one further has:

$$\frac{d}{dt}F[p(t)] = \frac{1}{2}\sum_{i,j\in\mathscr{S}}\left(p_j q_{ji} - p_i q_{ij}\right)\ln\left(\frac{p_i \pi_j}{\pi_i p_j}\right) = \sum_{i,j\in\mathscr{S}} p_j q_{ji}\ln\left(\frac{p_i \pi_j}{\pi_i p_j}\right)$$

$$\leq \sum_{i,j\in\mathscr{S}} p_j q_{ji}\left(\frac{p_i \pi_j}{\pi_i p_j} - 1\right)$$

$$= \sum_{i\in\mathscr{S}}\frac{p_i}{\pi_i}\sum_{j\in\mathscr{S}}\pi_j q_{ji} - \sum_{i,j\in\mathscr{S}} p_j q_{ji} = 0. \tag{6.17}$$

So the free energy is always nonincreasing for any Markov system, isolated or isothermal, with or without detailed balance.

6.4 First and Second Mesoscopic Thermodynamic Laws of Markov Processes

The generalized free energy balance equation (6.16a) bears a remarkable resemblance to the First Law of Thermodynamics, and the inequalities in Eqns. (6.17, 6.16b and 6.16c) have been widely considered as different faces of the Second Law of Thermodynamics[63, 90]. Together, the mathematics of Markov processes have provided a rigorous representation for the theory of thermodynamics for equilibrium and nonequilibrium systems.

The three nonnegative quantities in (6.16a), $-dF/dt$, E_{in}, and e_p, can be rearranged into

$$e_p = \left(-\frac{dF}{dt} \right) + E_{in}. \tag{6.18}$$

This equation can be interpreted as total entropy production having two contributions: from transient relaxation toward stationarity and from dissipation driven by sustained environmental energy. These contributions fittingly reflect Boltzmann's and Prigogine's theses on irreversibility, respectively.

6.4.1 Application to Population Kinetic Processes

Let us consider the continuous-time, integer-valued Markov process for counting a population system that consists of N kinetic species $X_1, X_2, \cdots, X_N$ and M reversible reactions with stoichiometric coefficients $\nu_\ell = \nu_\ell^- - \nu_\ell^+$, and rate laws $r_{+\ell}(\mathbf{n}, V)$ and $r_{-\ell}(\mathbf{n}, V)$, $1 \le \ell \le M$ (see Chapter 5). $\mathbf{n} = (n_1, n_2, \cdots, n_N)$ represents the number of individuals within each species, and V stands for the volume of a continuous, stirred-tank reactor (CSTR). Then, the master equation for the probability distribution $p_V(\mathbf{n}, t)$ takes the general form

$$\frac{dp_V(\mathbf{n}, t)}{dt} = \sum_{\ell=1}^{M} \Big(p_V(\mathbf{n} - \nu_\ell) r_{+\ell}(\mathbf{n} - \nu_\ell) - p_V(\mathbf{n}) r_{-\ell}(\mathbf{n}) - p_V(\mathbf{n}) r_{+\ell}(\mathbf{n})$$
$$+ p_V(\mathbf{n} + \nu_\ell) r_{-\ell}(\mathbf{n} + \nu_\ell) \Big), \tag{6.19}$$

with the thermodynamic quantities given as follows:

$$e_p[p_V(\mathbf{n})] = \frac{1}{2} \sum_{\text{all } \mathbf{n}} \sum_{\ell=1}^{M} \Big(p_V(\mathbf{n}) r_{+\ell}(\mathbf{n}) - p_V(\mathbf{n} + \nu_\ell) r_{-\ell}(\mathbf{n} + \nu_\ell) \Big)$$
$$\times \ln \left(\frac{p_V(\mathbf{n}) r_{+\ell}(\mathbf{n})}{p_V(\mathbf{n} + \nu_\ell) r_{-\ell}(\mathbf{n} + \nu_\ell)} \right), \tag{6.20a}$$

$$h_{ex}[p_V(\mathbf{n})] = \frac{1}{2} \sum_{\text{all } \mathbf{n}} \sum_{\ell=1}^{M} \Big(p_V(\mathbf{n}) r_{+\ell}(\mathbf{n}) - p_V(\mathbf{n} + \nu_\ell) r_{-\ell}(\mathbf{n} + \nu_\ell) \Big)$$
$$\times \ln \left(\frac{r_{+\ell}(\mathbf{n})}{r_{-\ell}(\mathbf{n} + \nu_\ell)} \right), \tag{6.20b}$$

$$E_{in}\left[p_V(\mathbf{n})\right] = \frac{1}{2}\sum_{\text{all }\mathbf{n}}\sum_{\ell=1}^{M}\Big(p_V(\mathbf{n})r_{+\ell}(\mathbf{n}) - p_V(\mathbf{n}+v_\ell)r_{-\ell}(\mathbf{n}+v_\ell)\Big)$$

$$\times \ln\left(\frac{\pi_V(\mathbf{n})r_{+\ell}(\mathbf{n})}{\pi_V(\mathbf{n}+v_\ell)r_{-\ell}(\mathbf{n}+v_\ell)}\right),\tag{6.20c}$$

$$F[p_V(\mathbf{n})] = \sum_{\text{all }\mathbf{n}} p_V(\mathbf{n})\ln\left(\frac{p_V(\mathbf{n})}{\pi_V(\mathbf{n})}\right),\tag{6.20d}$$

$$\frac{dF[p_V(\mathbf{n},t)]}{dt} = \frac{1}{2}\sum_{\text{all }\mathbf{n}}\sum_{\ell=1}^{M}\Big(p_V(\mathbf{n})r_{+\ell}(\mathbf{n}) - p_V(\mathbf{n}+v_\ell)r_{-\ell}(\mathbf{n}+v_\ell)\Big)$$

$$\times \ln\left(\frac{\pi_V(\mathbf{n})p_V(\mathbf{n}+v_\ell)}{\pi_V(\mathbf{n}+v_\ell)p_V(\mathbf{n})}\right).\tag{6.20e}$$

In the context of stochastic dynamics, the quantities in Eqns. (6.20a) to (6.20e) represent, respectively, the physicochemical concepts of entropy production rate, heat exchange rate, energy input rate (or power supply), and the generalized free energy of a system and its rate of dissipation. The mathematical formulae are quite foreign to a physical chemist. However, in the next chapter we shall show that in the macroscopic limit, as $V \to \infty$, they will take on not only more familiar expressions but also novel expressions.

Chapter 7
Emergent Macroscopic Chemical Thermodynamics

Compared with the macroscopic chemical thermodynamics we have touched upon in Sections 2.2 and 2.5, J. W. Gibbs' equilibrium statistical mechanics is a much more widely known contribution to the theoretical science, in which he first formulated the *free energy functional* and discussed a variational principle for equilibrium. In recent years, the stochastic, mesoscopic thermodynamics presented in the previous chapter has slowly emerged as the dynamic counterpart of Gibbs' static, statistical theory of complex systems [219, 228]. Even though we only presented the theory in terms of discrete-state Markov processes in Chapter 6, a counterpart for stochastic diffusion processes in continuous state space also exists [87, 278].

In deterministic chemical kinetics, Waage–Guldberg's law of mass action (LMA) was proven to be closely related to the variational principle with respect to Gibbs' functional for ideal solutions, connecting equilibrium thermodynamics with kinetics [249] (see Shear's theorem in Section 2.5.2).

The present chapter concerns the macroscopic limit of mesoscopic nonequilibrium thermodynamics for nonlinear chemical reaction systems with reversible reactions,

$$v_{\ell 1}^{+}X_1 + v_{\ell 2}^{+}X_2 + \cdots + v_{\ell N}^{+}X_N \underset{r_{-\ell}}{\overset{r_{+\ell}}{\rightleftharpoons}} v_{\ell 1}^{-}X_1 + v_{\ell 2}^{-}X_2 + \cdots + v_{\ell N}^{-}X_N, \qquad (7.1)$$

in which $1 \le \ell \le M$: again, there are N species and M reactions. $v_{ij} = (v_{ij}^{-} - v_{ij}^{+})$ are the *stoichiometric coefficients*, which relate species to reactions. As we already established in Chapter 5, in a reaction tank with rapidly stirred chemical solutions, the concentrations of the species at time t, $x_i(t)$ for X_i, satisfy the system of ordinary differential equations that describes the mass balance

© The Author(s), under exclusive license to Springer Nature Switzerland AG 2021
H. Qian, H. Ge, *Stochastic Chemical Reaction Systems in Biology*,
Lecture Notes on Mathematical Modelling in the Life Sciences,
https://doi.org/10.1007/978-3-030-86252-7_7

$$\frac{dx_i(t)}{dt} = \sum_{\ell=1}^{M} v_{\ell i}\Big(R_{+\ell}(\mathbf{x}) - R_{-\ell}(\mathbf{x})\Big),\tag{7.2}$$

in which $\mathbf{x} = (x_1, x_2, \cdots, x_N)$ denote the concentrations of these chemical species, and $R_{+\ell}(\mathbf{x})$ and $R_{-\ell}(\mathbf{x})$ are the general forms of the forward and backward fluxes for the ℓ–th reaction, respectively. For a meaningful thermodynamic analysis, we further assume that both $R_{\pm\ell}(\mathbf{x}) \geq 0$ when all the components of $\mathbf{x}$ are positive.

Chemical reactions at the individual molecule level, with a time resolution of one reaction at a time, are stochastic. This can be represented by the mathematical theory of the Chemical Master Equation (CME) [224, 169, 62], which first appeared in the work of Leontovich [162] and Delbrück [47] and whose fluctuating trajectories can be exactly computed using the stochastic simulation method discussed by Doob, Boltz–Kalos–Lebowitz, Gillespie, and others [50, 28, 103]. The deterministic rate equations in (7.2), it turns out, are the macroscopic infinite-volume limit of this stochastic chemical reaction system. This has been mathematically proven by T. G. Kurtz in 1970s [155, 17, 252].

Recall that the fundamental postulate in stochastic theory is the notion of the *stochastic elementary reaction*, , e.g., each and every one of the chemical reactions in (7.1) is a random event that occurs with an exponentially distributed waiting time.

The functions $r_{+\ell}(\mathbf{n}; V)$ and $r_{-\ell}(\mathbf{n}; V)$ in (7.1) are the rates of the ℓth reaction in the forward and backward directions, respectively, represented as two stochastic events in a finite volume V.

The stochastic trajectory of the molecular counting numbers $\mathbf{n} = (n_1, n_2, \cdots, n_N)$ can be expressed in terms of the random time-changed Poisson representation [5]:

$$n_j(t) = n_j(0) + \sum_{\ell=1}^{M} v_{\ell j}\left\{Y_{+\ell}\left(\int_0^t r_{+\ell}\big(\mathbf{n}(s)\big)ds\right) - Y_{-\ell}\left(\int_0^t r_{-\ell}\big(\mathbf{n}(s)\big)ds\right)\right\},\tag{7.3}$$

in which $Y_{+\ell}(s)$ and $Y_{-\ell}(s)$ are independent standard Poisson processes with mean $\mathbf{E}[Y_{\pm\ell}(s)] = s$. Note that the stochastic process $\mathbf{n}(t)$ in (7.3) is a function of parameter V since both $r_{\pm\ell}(\mathbf{n}, V)$ are functions of V; when necessary, it is expressed as $\mathbf{n}_V(t)$. As $V \to \infty$, Kurtz assumed the rate functions satisfy the scaling condition as [5]

$$\lim_{V\to\infty} \frac{r_{+\ell}(V\mathbf{x}; V)}{V} = R_{+\ell}(\mathbf{x}), \quad \lim_{V\to\infty} \frac{r_{-\ell}(V\mathbf{x}; V)}{V} = R_{-\ell}(\mathbf{x}).\tag{7.4}$$

As in Eqn. (6.19), the corresponding CME for the stochastic reactions, which is the Kolmogorov forward equation for the continuous-time, integer-value Markov process, is

$$\frac{\mathrm{d}p_V(\mathbf{n},t)}{\mathrm{d}t} = \sum_{\ell=1}^{M} \Big[p_V(\mathbf{n} - \boldsymbol{v}_\ell,t) r_{+\ell}(\mathbf{n} - \boldsymbol{v}_\ell;V) \tag{7.5}$$
$$- p_V(\mathbf{n},t)\Big(r_{+\ell}(\mathbf{n};V) + r_{-\ell}(\mathbf{n};V) \Big) + p_V(\mathbf{n} + \boldsymbol{v}_\ell,t) r_{-\ell}(\mathbf{n} + \boldsymbol{v}_\ell;V) \Big].$$

For the class of kinetics with complex balance under the law of mass action [125, 124, 67, 68], it was recently shown [6] that the steady-state, large-deviation rate function $\varphi^{ss}(\mathbf{x})$ for the stochastic chemical reaction model is actually Shear's Lyapunov function for the chemical kinetics discussed in Section 2.5.2:

$$A[\mathbf{x}] = \sum_{i=1}^{N} x_i \ln\left(\frac{x_i}{x_i^{ss}}\right) - x_i + x_i^{ss}, \tag{7.6}$$

where $\mathbf{x}^{ss}$ is a positive stable fixed point of (7.2). In [96], it was further shown that $A[\mathbf{x}]$ for a kinetic system with complex balance is also an emergent quantity of the large-volume limit of the mesoscopic free energy function $F[p_V(\mathbf{n},t)]$ in (6.20d). Hence, macroscopic chemical thermodynamics for complex-balance kinetics, which include detail balance and many driven chemical kinetics, naturally emerges.

We give an outline of the steps in this chapter below. We do not assume detailed balance nor complex balance. Starting from the CME, using a WKB ansatz, and performing a Taylor expansion in the order of V^{-1}, in Section 7.1 we derive a nonlinear partial differential equation for the large-deviation rate function $\varphi_t(\mathbf{x})$. In Section 7.2, we seek the limits for the thermodynamic quantities introduced in Chapter 6 as $V \to \infty$, the Kurtz limit. The mathematical theory shows how, in a multiscale system, a macroscopic dynamic law emerges from the mesoscopic kinetics one level below. Based on these general results, Section 7.3 provides a rather thorough study of a kinetic system that has detailed balance, representing *closed, nondriven* reaction systems that spontaneously approach chemical equilibrium. This is the domain of classic Gibbsian theory. In Section 7.4 and Section 7.5, mathematically more rigorous materials with proofs are presented. These Sections can be skipped for readers who are satisfied with the heuristic nature of the proofs in Sections 7.1 to 7.4.

7.1 Kurtz's Law of Large Numbers, WKB Ansatz and Hamilton–Jacobi Equation

The solution to the CME in (7.5), $p_V(\mathbf{n}, t)$, provides the probability distribution for the Markov process $\mathbf{n}_V(t)$ in (7.3), in which the standard Poisson process has distribution

$$P\left\{Y(\lambda(s)) = k\right\} = \frac{[\lambda(s)]^k}{k!} e^{-\lambda(s)}, \tag{7.7}$$

where $\lambda(s) \geq 0$. It is easy to verify that $\mathbb{E}[Y(\lambda)] = \lambda$ and $\mathrm{Var}[Y(\lambda)] = \lambda$. Therefore, as $\lambda \to \infty$, $Y(\lambda) \simeq \lambda + O(\sqrt{\lambda})$. Applying this asymptotic approximation to Eqn. (7.3) and noting the relations in (7.4), we have, as $V \to \infty$ and $\mathbf{x}(t) = \frac{\mathbf{n}_V(t)}{V}$,

$$\begin{aligned}
x_j(t) &\simeq x_j(0) + \sum_{\ell=1}^M v_{\ell j} \left\{ \frac{1}{V} \int_0^t r_{+\ell}\big(\mathbf{n}(s)\big) ds - \frac{1}{V} \int_0^t r_{-\ell}\big(\mathbf{n}(s)\big) ds \right\} \\
&\to x_j(0) + \sum_{\ell=1}^M v_{\ell j} \left\{ \int_0^t R_{+\ell}\big(\mathbf{x}(s)\big) ds - \int_0^t R_{-\ell}\big(\mathbf{x}(s)\big) ds \right\}.
\end{aligned} \tag{7.8}$$

One immediately recognizes that this is the integral form of the ordinary differential equations in (7.2) with initial value $\mathbf{x}(0)$.

This result is known as Kurtz's theorem (see Sec. 7.4 for a rigorous statement). It indicates that as $V \to \infty$, one recovers the deterministic kinetic equation (7.2) from the Markov stochastic description: in other words, the probability distribution

$$p_V(V\mathbf{z}, t) \to \delta\big(\mathbf{z} - \mathbf{x}(t)\big). \tag{7.9}$$

This is a form of the law of large numbers. Associated with the law of large numbers in Eqn. (7.9), the large deviation principle says that $p_V(V\mathbf{z}, t)$ has an asymptotic expression

$$p_V(V\mathbf{z}, t) \simeq e^{-V\varphi_t(\mathbf{z})}, \tag{7.10}$$

in which $\varphi_t(\mathbf{z})$ is a large deviation rate function, whose global minimum is located precisely at $\mathbf{x}(t)$ in (7.9).

Taking this expression as given, called the WKB ansatz, we can substitute the expression in (7.10) into the CME (7.5), keep the leading order term, and obtain

$$\frac{\partial \varphi_t(\mathbf{x})}{\partial t} \simeq -\frac{1}{V} e^{V \varphi_t(\mathbf{x})} \left(\frac{dp_V(V\mathbf{x},t)}{dt} \right)$$

$$\simeq -\sum_{\ell=1}^{M} \left[\exp\left\{ V\left[\varphi_t(\mathbf{x}) - \varphi_t\left(\mathbf{x} - V^{-1}\mathbf{v}_\ell\right) \right] \right\} R_{+\ell}(\mathbf{x}) - \left(R_{+\ell}(\mathbf{x}) + R_{-\ell}(\mathbf{x}) \right) \right.$$

$$\left. + \exp\left[V\left[\varphi_t(\mathbf{x}) - \varphi_t\left(\mathbf{x} + V^{-1}\mathbf{v}_\ell\right) \right] \right] \right\} R_{-\ell}(\mathbf{x}) \right]$$

$$\simeq -\sum_{\ell=1}^{M} \left\{ \exp\left[\mathbf{v}_\ell \cdot \nabla_{\mathbf{x}} \varphi_t(\mathbf{x}) \right] R_{+\ell}(\mathbf{x}) - \left(R_{+\ell}(\mathbf{x}) + R_{-\ell}(\mathbf{x}) \right) \right.$$

$$\left. + \exp\left[-\mathbf{v}_\ell \cdot \nabla_{\mathbf{x}} \varphi_t(\mathbf{x}) \right] R_{-\ell}(\mathbf{x}) \right\}. \tag{7.11}$$

This yields a nonlinear partial differential equation with first order in $(\partial/\partial t)$ and first order in $(\partial/\partial x)$ but without explicit dependence on φ_t:

$$\frac{\partial \varphi_t(\mathbf{x})}{\partial t} + H\left(\nabla_{\mathbf{x}} \varphi_t, \mathbf{x} \right) = 0, \tag{7.12a}$$

$$H(\mathbf{y},\mathbf{x}) = \sum_{\ell=1}^{M} \left\{ R_{+\ell}(\mathbf{x}) e^{\mathbf{v}_\ell \cdot \mathbf{y}} - R_{+\ell}(\mathbf{x}) - R_{-\ell}(\mathbf{x}) + R_{-\ell}(\mathbf{x}) e^{-\mathbf{v}_\ell \cdot \mathbf{y}} \right\}. \tag{7.12b}$$

Eqn. (7.12) was first derived by Gang Hu in 1986 [126]. It is known as the Hamilton–Jacobi equation (HJE) as well as the Eikonal equation in stochastic dynamics.

The existence and uniqueness of the solutions to the HJE, such as Eqn. (7.12), a nonlinear first-order partial differential equation, are not well defined until more regularity conditions are provided [69].

7.1.1 Stationary Solution to the HJE As a Lyapunov Function

A function $\varphi^{ss}(\mathbf{x})$ that satisfies $H(\nabla_{\mathbf{x}}\varphi^{ss}(\mathbf{x}),\mathbf{x}) = 0$ is called a stationary solution to the HJE (7.12). We notice that one trivial solution is $\varphi^{ss}(\mathbf{x}) = \text{const}$. The large-deviation rate function of the sequence of stationary distribution $\pi_V(\mathbf{n})$, as $V \to \infty$, is also a solution. However, all stationary solutions have the following important property:

$$0 = \sum_{\ell=1}^{M} \left\{ R_{+\ell}(\mathbf{x}) \left(e^{\nu_\ell \cdot \nabla_\mathbf{x} \varphi^{ss}} - 1 \right) + R_{-\ell}(\mathbf{x}) \left(e^{-\nu_\ell \cdot \nabla_\mathbf{x} \varphi^{ss}} - 1 \right) \right\}$$

$$\geq \sum_{\ell=1}^{M} \left\{ R_{+\ell}(\mathbf{x}) \nu_\ell \cdot \nabla_\mathbf{x} \varphi^{ss} - R_{-\ell}(\mathbf{x}) \nu_\ell \cdot \nabla_\mathbf{x} \varphi^{ss} \right\}$$

$$= \frac{d\mathbf{x}(t)}{dt} \cdot \nabla_\mathbf{x} \varphi^{ss} = \frac{d}{dt} \varphi^{ss}(\mathbf{x}(t)). \tag{7.13}$$

Therefore, $\varphi^{ss}(\mathbf{x})$ is a Lyapunov function in J. P. LaSalle's sense for the nonlinear differential equation (7.2). The inequality in (7.13) is the mathematical foundation for the use of $\varphi^{ss}(\mathbf{x})$ as a thermodynamic potential function.

If the $R_{\pm\ell}(\mathbf{x})$ for system (7.1) both satisfy, for each ℓ, $1 \leq \ell \leq M$,

$$R_{+\ell}(\mathbf{x}) e^{\nu_\ell \cdot \nabla_\mathbf{x} \varphi^{ss}} - R_{-\ell}(\mathbf{x}) = 0, \tag{7.14}$$

which yields

$$\nu_\ell \cdot \nabla_\mathbf{x} \varphi^{ss}(\mathbf{x}) = -\ln \left(\frac{R_{+\ell}(\mathbf{x})}{R_{-\ell}(\mathbf{x})} \right). \tag{7.15}$$

This is the weak detailed balance in deterministic chemical kinetics, which is equivalent to the Wegscheider–Lewis cycle condition, a concept discovered by chemists independent of the theory of Markov processes (see Section 7.3). Furthermore, if both $R_{\pm\ell}(\mathbf{x})$ follow the law of mass action,

$$R_{+\ell}(\mathbf{x}) = k_{+\ell} \prod_{j=1}^{N} x_j^{\nu_{\ell j}^+}, \quad R_{-\ell}(\mathbf{x}) = k_{-\ell} \prod_{j=1}^{N} x_j^{\nu_{\ell j}^-}. \tag{7.16}$$

Then, it is easy to verify that $A[\mathbf{x}]$ in (7.6) satisfies Eqn. (7.15):

$$\sum_{i=1}^{N} \nu_{\ell i} \frac{\partial A[\mathbf{x}]}{\partial x_i} = \sum_{i=1}^{N} \nu_{\ell i} \ln \left(\frac{x_i}{x_i^{ss}} \right) = -\ln \left[\frac{k_{+\ell} \prod_{i=1}^{n} x_i^{\nu_{\ell i}^+}}{k_{-\ell} \prod_{i=1}^{n} x_i^{\nu_{\ell i}^-}} \right]. \tag{7.17}$$

7.1.2 Mass-action Kinetics With Complex Balance

The stationary state in reaction systems with detailed balance (7.15) is an equilibrium steady state. However, we know that not every stationary state in a mesoscopic stochastic reaction system (7.1) with stochastic trajectory (7.3) and CME (7.5) is an equilibrium state. We now consider reaction systems with complex balance, introduced in Section 2.5.3. We note that stationary HJE (7.12) can be rewritten as

$$\sum_{\ell=1}^{M} \left[R_{+\ell}(\mathbf{x}) - R_{-\ell}(\mathbf{x}) e^{-\boldsymbol{\nu}_\ell \cdot \nabla_{\mathbf{x}} \varphi^{ss}(\mathbf{x})} \right] \left[1 - e^{\boldsymbol{\nu}_\ell \cdot \nabla_{\mathbf{x}} \varphi^{ss}(\mathbf{x})} \right] = 0. \qquad (7.18)$$

Hence, in general, the validity of this equation at NESS is not due to each and every ℓ term being equal to zero as in (7.14) but due to a balance between positive and negative terms with different ℓs.

Again, if $R_{\pm\ell}(\mathbf{x})$ follow the law of mass action (7.16), then the gradient of $A[\mathbf{x}]$ in (7.6) $\nabla_{\mathbf{x}} A[\mathbf{x}] = \ln(\mathbf{x}/\mathbf{x}^{ss})$, and

$$\begin{aligned} \boldsymbol{\nu}_\ell \cdot \nabla_{\mathbf{x}} A[\mathbf{x}] &= \sum_{j=1}^{N} \nu_{\ell j} \ln \left(\frac{x_j}{x_j^{ss}} \right) = \ln \prod_{j=1}^{N} \left(\frac{x_j}{x_j^{ss}} \right)^{\nu_{\ell j}} \\ &= \ln \left(\frac{R_{-\ell}(\mathbf{x}) R_{+\ell}(\mathbf{x}^{ss})}{R_{+\ell}(\mathbf{x}) R_{-\ell}(\mathbf{x}^{ss})} \right). \end{aligned} \qquad (7.19)$$

Substituting this into (7.18), we have

$$\begin{aligned} &\sum_{\ell=1}^{M} \left(R_{+\ell}(\mathbf{x}^{ss}) - R_{-\ell}(\mathbf{x}^{ss}) \right) \left(\frac{R_{+\ell}(\mathbf{x})}{R_{+\ell}(\mathbf{x}^{ss})} - \frac{R_{-\ell}(\mathbf{x})}{R_{-\ell}(\mathbf{x}^{ss})} \right) \\ &= \sum_{\ell=1}^{M} \left(R_{+\ell}(\mathbf{x}^{ss}) - R_{-\ell}(\mathbf{x}^{ss}) \right) \left[\prod_{j=1}^{N} \left(\frac{x_j}{x_j^{ss}} \right)^{\nu_{\ell j}^+} - \prod_{j=1}^{N} \left(\frac{x_j}{x_j^{ss}} \right)^{\nu_{\ell j}^-} \right]. \qquad (7.20) \end{aligned}$$

Recall that complex balanced mass-action kinetics guarantee that the right-hand side of (7.20) is zero for all $\mathbf{x}$ (see Section 2.5.3). In other words, for a kinetic system with mass-action rate laws and assuming $\mathbf{x}^{ss}$ is one of its steady state, the $A[\mathbf{x}]$ defined in (7.6) is a solution to Eqn. (7.12) for kinetics with complex balance. We further note that $A[\mathbf{x}]$ is convex, and the macroscopic kinetics (7.2) always decrease according to $A[\mathbf{x}]$. Therefore, $\mathbf{x}^{ss}$ is the unique steady state. Actually, it can be mathematically shown that the large-deviation rate function is $\varphi^{ss}(\mathbf{x}) = A[\mathbf{x}]$ for kinetic systems with mass-action rate laws and complex balance.

7.2 Macroscopic Nonequilibrium Thermodynamic Formalism

In the Kurtz limit as $V \to \infty$, the mesoscopic thermodynamic quantities in Eqns. (6.20a) to (6.20e) take on expressions familiar to physical chemists [98]:

$$\lim_{V \to \infty} \frac{F^{meso}[p_V(\mathbf{n})]}{V} = \lim_{V \to \infty} \frac{1}{V} \sum_{\text{all } \mathbf{n}} p_V(\mathbf{n}) \ln \left(\frac{p_V(\mathbf{n})}{\pi_V(\mathbf{n})} \right)$$

$$\simeq \lim_{V \to \infty} \sum_{\text{all } \tilde{\mathbf{x}}} p_V(V\tilde{\mathbf{x}}) \left(\varphi^{ss}(\tilde{\mathbf{x}}) - \varphi_t(\tilde{\mathbf{x}}) \right)$$

$$= \int_{\mathbb{R}^N} d\mathbf{x} \, \delta(\tilde{\mathbf{x}} - \mathbf{x}(t)) \left(\varphi^{ss}(\tilde{\mathbf{x}}) - \varphi_t(\tilde{\mathbf{x}}) \right), \qquad (7.21)$$

in which $\mathbf{x}(t)$ is a solution to the nonlinear ODE (7.2). Note that $\mathbf{x}(t)$ is located at the global minimum of $\varphi_t(\tilde{\mathbf{x}})$, where $\varphi_t(\tilde{\mathbf{x}}) = 0$. Therefore,

$$\lim_{V \to \infty} \frac{F^{meso}[p_V(\mathbf{n},t)]}{V} = \varphi^{ss}[\mathbf{x}(t)]. \qquad (7.22)$$

The mesoscopic generalized free energy $F^{meso}[p_V(\mathbf{n},t)]$ becomes the large-deviation rate function $\varphi^{ss}(\mathbf{x}(t))$, also called landscape. The monotonicity of the latter, proven in (7.13), is actually the consequence of the monotonicity of the former, shown in (6.17) for every V.

Eqn. (7.22) is a very significant result: it has a very clear macroscopic thermodynamic meaning. $\varphi^{ss}(\mathbf{x})$ is not just one of the Lyapunov functions for the nonlinear dynamics in (7.2), it is the macroscopic limit of the mesoscopic generalized free energy.

Similarly, the total entropy production in (6.20a) is

$$\sigma^{tot}[\mathbf{x}] = \lim_{V \to \infty} \frac{e_p[p_V(V\mathbf{x},t)]}{V} = \sum_{\ell=1}^{M} \left(R_{+\ell}(\mathbf{x}) - R_{-\ell}(\mathbf{x}) \right) \ln \left(\frac{R_{+\ell}(\mathbf{x})}{R_{-\ell}(\mathbf{x})} \right), \qquad (7.23)$$

which is the same as the macroscopic entropy exchange

$$h_{ex}^{macro}[\mathbf{x}] = \lim_{V \to \infty} \frac{h_{ex}^{meso}[p_V(V\mathbf{x},t)]}{V} = \sigma^{tot}[\mathbf{x}]. \qquad (7.24)$$

The macroscopic entropy change per unit volume is negligible.

In terms of the total entropy production $\sigma^{tot}[\mathbf{x}]$, the "law" of mesoscopic generalized free energy balance equation in Chapter 6 becomes the following, the macroscopic free energy dissipation in transient kinetics

$$f_d^{macro}[\mathbf{x}] \equiv -\frac{d\varphi^{ss}[\mathbf{x}(t)]}{dt} = -\sum_{i=1}^{N} \left(\frac{dx_i}{dt} \right) \left(\frac{\partial \varphi^{ss}(\mathbf{x})}{\partial x_i} \right)$$

$$= \sum_{\ell=1}^{M} \sum_{i=1}^{N} v_{\ell i} \left(R_{-\ell}[\mathbf{x}] - R_{+\ell}[\mathbf{x}] \right) \left(\frac{\partial \varphi^{ss}(\mathbf{x})}{\partial x_i} \right)$$

$$= \sum_{\ell=1}^{M} \left(R_{-\ell}[\mathbf{x}] - R_{+\ell}[\mathbf{x}] \right) \left\{ \mathbf{v}_{\ell} \cdot \nabla_{\mathbf{x}} \varphi^{ss}(\mathbf{x}) \right\}, \tag{7.25}$$

which can be decomposed into

$$f_d^{\text{macro}}[\mathbf{x}] = \sigma^{\text{tot}}[\mathbf{x}] + E_{in}^{\text{macro}}[\mathbf{x}], \tag{7.26}$$

where

$$
\begin{aligned}
E_{in}^{\text{macro}}[\mathbf{x}] &= \lim_{V \to \infty} \frac{E_{in}\left[p_V(V\mathbf{x},t)\right]}{V} \\
&= \sum_{\ell=1}^{M} \left(R_{-\ell}[\mathbf{x}] - R_{+\ell}[\mathbf{x}] \right) \left\{ \mathbf{v}_{\ell} \cdot \nabla_{\mathbf{x}} \varphi^{ss}(\mathbf{x}) + \ln\left(\frac{R_{+\ell}(\mathbf{x})}{R_{-\ell}(\mathbf{x})} \right) \right\} \\
&= \sum_{\ell=1}^{M} \left(R_{-\ell}[\mathbf{x}] - R_{+\ell}[\mathbf{x}] \right) \ln\left\{ \frac{R_{+\ell}(\mathbf{x})}{R_{-\ell}(\mathbf{x})} e^{\mathbf{v}_{\ell} \cdot \nabla_{\mathbf{x}} \varphi^{ss}(\mathbf{x})} \right\}.
\end{aligned} \tag{7.27}
$$

We prove in Section 7.4 that E_{in} is nonnegative and $E_{in} = 0$ is equivalent to the weak detailed balance condition (7.15).

Once the weak detailed balance condition is violated, the local kinetics, in terms of $R_{\pm\ell}(\mathbf{x})$, are in disagreement with the global potential $\varphi^{ss}(\mathbf{x})$, i.e., they can not determine each other. Such a disagreement constitutes the chemostatic driving force of a macroscopic open system.

7.3 Kinetics with Detailed Balance and Chemical Equilibrium

We have shown that the driving force of a reaction system under a chemostat is represented by E_{in}. Historically, R. Wegscheider's condition for kinetic cycles and G. N. Lewis' law of entire equilibrium (e.g., the principle of detailed balance in chemistry) have been cornerstone concepts for the consistency between chemical kinetics and equilibrium thermodynamics [272, 164, 245].

We now provide a comprehensive discussion of the subject given meoscopic and macroscopic nonequilibrium thermodynamics (NET).

We again consider the general reaction kinetic system in (7.1). Let there be M reversible reactions, and let $\xi = (\xi_1, \cdots, \xi_M)$ be a M-dimensional vector that satisfies

$$\sum_{\ell=1}^{M} \xi_{\ell} \nu_{\ell i} = 0, \ \forall i = 1, 2, \cdots, N.$$

This ξ is in the left null space of the $M \times N$ stoichiometric matrix v_{ij}. If, for every vector ξ in the left null space of v_{ij}, the rate laws $R_{\pm\ell}(\mathbf{x})$, $1 \leq \ell \leq M$, satisfy

$$\sum_{\ell=1}^{M} \xi_\ell \ln \left(\frac{R_{+\ell}(\mathbf{x})}{R_{-\ell}(\mathbf{x})} \right) = 0, \ \forall \mathbf{x}, \tag{7.28}$$

then we can say the kinetic system satisfies the Wegscheider–Lewis cycle condition, or the loop law.

This cycle condition turns out to be fundamental to nondriven chemical reaction systems that necessarily approach a chemical equilibrium in the long time limit. Sometime this type of system is said to be *closed*. There are many important properties that serve as the defining characteristics of this type of system. We shall now demonstrate a number of such properties.

In fact, the following four statements are equivalent [98]:

(*i*) A **weak** detailed balance condition (7.15) is satisfied for every $\mathbf{x}$ and all ℓ;

(*ii*) The chemical reaction model satisfies the Wegscheider–Lewis cycle condition (7.28);

(*iii*) The macroscopic energy input rate $e_{in}^{(macro)}[\mathbf{x}] = 0$ for all $\mathbf{x}$.

(*iv*) The time-evolution of the macroscopic general free energy function

$$\frac{d\varphi^{ss}[\mathbf{x}(t)]}{dt} = -\sigma^{tot}[\mathbf{x}(t)]. \tag{7.29}$$

Furthermore, (*iv*) also implies that any stable steady state of the macroscopic kinetics (7.2), $\mathbf{x}^{ss}$, satisfies the **strong** detailed balance condition: $R_{+\ell}(\mathbf{x}^{ss}) = R_{-\ell}(\mathbf{x}^{ss})$.

The proof of (*i*) $\rightarrow$ (*ii*) straightforward. Now, we demonstrate (*ii*) $\rightarrow$ (*i*); due to the Wegscheider–Lewis cycle condition, we can choose the rates $r_{+\ell}(\mathbf{n};V)$ for the forward reaction and $r_{-\ell}(\mathbf{n};V)$ for the backward reaction such that for sufficiently large V, the stochastic chemical reaction system with finite V satisfies the Kolmogorov cycle condition, i.e., for each possible cycle in the state space, the product of the reaction rates for the forward cycle and the product of the reaction rates for the backward cycle are equal to each other. The strategy can be first set by letting $r_{\pm\ell}(\mathbf{n};V) = V R_{\pm\ell}(\frac{\mathbf{n}}{V})$ and then fine-tuning the subsequent elementary cycles with the smallest numbers of states.

The Kolmogorov cycle condition is equivalent to the stationary distribution that satisfies the detailed balance condition

$$\frac{p_V^{ss}(\mathbf{n}+\mathbf{v}_\ell)}{p_V^{ss}(\mathbf{n})} = \frac{r_{+\ell}(\mathbf{n};V)}{r_{-\ell}(\mathbf{n}+\mathbf{v}_\ell;V)}. \tag{7.30}$$

Therefore, in the macroscopic limit, it yields (see Section 7.4.2, Lemma 2)

$$-\mathbf{v}_\ell \cdot \nabla_\mathbf{x} \varphi^{ss}(\mathbf{x}) = \ln\left(\frac{R_{+\ell}(\mathbf{x})}{R_{-\ell}(\mathbf{x})}\right). \tag{7.31}$$

Note that the definition of $\varphi^{ss}(\mathbf{x})$ is only dependent on $R_{\pm\ell}(\mathbf{x})$; as the limit of $r_{\pm\ell}(V\mathbf{x};V)/V$, the weak detailed balance condition holds independent of the choices of $r_{\pm\ell}(\mathbf{n};V)$.

The equivalence between $(i) \Leftrightarrow (iii)$ will be found in Proposition 3 in Section 7.4.4. To show $(iii) \Leftrightarrow (iv)$, we note that $\sigma^{tot}[\mathbf{x}] = f_d^{macro}[\mathbf{x}]$ if and only if $E_{in}^{macro}[\mathbf{x}] = 0$.

7.3.1 Law of Mass Action and Ideal Solution

For the specific mass-action kinetics given in (7.16), the weak detailed balance condition for every $\mathbf{x}$ means $\mathbf{v}_\ell \cdot \nabla_\mathbf{x}\varphi^{ss}(\mathbf{x}) = \ln(k_{-\ell}/k_{+\ell}) + \mathbf{v}_\ell \cdot \ln\mathbf{x} = \Delta\mu_\ell$, the chemical potential *difference* of the ℓth reaction. Then, we see that $\varphi^{ss}(\mathbf{x})$ is precisely the macroscopic Gibbs function for the chemical reaction system in an ideal solution, and $\mu_k(\mathbf{x}) = \partial\varphi^{ss}(\mathbf{x})/\partial x_k$ is the chemical potential of the kth species. It is now worth going back to take a fresh look at how Gibbs was able to connect his two major accomplishments, statistical mechanics and macroscopic chemical thermodynamics, using the idea of "variational method for virtual change" to demonstrate the Lyapunov function. This is *a fundamental theorem of nonequilibrium chemical thermodynamics*.

For the general rate law, which represents a nonideal solution, the weak detailed balance condition satisfied by each $\mathbf{x}$ might not always guarantee a unique steady state x^{ss} for (7.2). However, the following is a necessary condition for having a convex $\varphi(\mathbf{x})$:

$$\mathbf{v}_\ell \cdot \nabla_\mathbf{x}\ln\left(\frac{R_{+\ell}(\mathbf{x})}{R_{-\ell}(\mathbf{x})}\right) < 0 \tag{7.32}$$

for all $\mathbf{x}$ and ℓ. Actually, from Eqn. (7.31) and after performing differentiation, we have

$$-\sum_{k,j=1}^{N} v_{\ell j}\left(\frac{\partial^2 \varphi^{ss}(\mathbf{x})}{\partial x_j \partial x_k}\right) v_{\ell k} = \mathbf{v}_\ell \cdot \nabla_\mathbf{x}\ln\left(\frac{R_{+\ell}(\mathbf{x})}{R_{-\ell}(\mathbf{x})}\right). \tag{7.33}$$

Therefore, once the weak detailed balance condition is satisfied, if $\varphi^{ss}(\mathbf{x})$ is convex, then the left-hand side of (7.33) is negative for all $\boldsymbol{v}_\ell$ and $\mathbf{x} > 0$. Convex $\varphi(\mathbf{x})$ is a sufficient condition for the uniqueness of $\mathbf{x}^{ss}$.

It is easy to verify that Eqn. (7.32) is always satisfied for mass-action kinetics:

$$\boldsymbol{v}_\ell \cdot \nabla_{\mathbf{x}} \ln\left(\frac{R_{+\ell}(\mathbf{x})}{R_{-\ell}(\mathbf{x})}\right) = \sum_{j=1}^{N} v_{\ell j} \frac{\partial}{\partial x_j} \sum_{k=1}^{N} \ln x_k^{-v_{\ell k}} = -\sum_{j=1}^{N} \frac{v_{\ell j}^2}{x_j} < 0. \qquad (7.34)$$

Hence, in this case, a weak detailed balance can guarantee the uniqueness of $\mathbf{x}^{ss}$.

7.4 Mathematical Theory of the Macroscopic NET

In this section, we establish a mathematical basis for the macroscopic chemical thermodynamics we presented heuristically in the previous two Sections [98]. First, in Sections 7.4.1 and 7.4.2, Kurtz's law of large numbers, some results on the large-deviations principle, and a key lemma are presented. Then, in Section 7.4.3 and Section 7.4.4 the set of macroscopic NET quantities as the Kurtz limits of the mesoscopic NET quantities introduced in Chapter 6 are approached.

7.4.1 Kurtz's Theorem

Denote the right-hand side of the rate equation (7.2) as $\mathbf{B}(\mathbf{x}) = (B_1(\mathbf{x}), B_2(\mathbf{x}), \cdots, B_N(\mathbf{x}))$, with

$$B_i(\mathbf{x}) = \sum_{\ell=1}^{M} v_{\ell i}\left(R_{+\ell}(\mathbf{x}) - R_{-\ell}(\mathbf{x})\right). \qquad (7.35)$$

The following theorem regarding the law of large numbers for stochastic chemical reaction models was proven by T. G. Kurtz in 1978 [155, 252].

Theorem 7.1. *Assume that there exist constants ε_ℓ and Γ such that*

$$\left|\frac{r_{+\ell}(V\mathbf{x};V)}{V}\right| \leq \varepsilon_\ell\left(1+|\mathbf{x}|\right), \quad \left|\frac{r_{-\ell}(V\mathbf{x};V)}{V}\right| \leq \varepsilon_\ell\left(1+|\mathbf{x}|\right),$$

$$\left|\frac{r_{+\ell}(V\mathbf{x};V)}{V} - R_{+\ell}(\mathbf{x})\right| \leq \frac{\Gamma\varepsilon_\ell}{V}\left(1+|\mathbf{x}|\right),$$

$$\left|\frac{r_{-\ell}(V\mathbf{x};V)}{V} - R_{-\ell}(\mathbf{x})\right| \leq \frac{\Gamma\varepsilon_\ell}{V}\left(1+|\mathbf{x}|\right). \qquad (7.36)$$

Denote $\mathbf{v}_\ell = (v_{\ell 1}, v_{\ell 1}, \cdots, v_{\ell N})$. *Further assume* $\sum_\ell |\mathbf{v}_\ell| \varepsilon_\ell < \infty$ *and that there exists a positive constant M such that*

$$|\mathbf{B}(\mathbf{x}) - \mathbf{B}(\mathbf{y})| \leq M|\mathbf{x} - \mathbf{y}|. \tag{7.37}$$

Then, if $\mathbf{n}_V(0) = V\mathbf{x}(0)$, *for any* $T > 0$,

$$\lim_{V \to \infty} \sup_{t \leq T} \left| \frac{\mathbf{n}_V(t)}{V} - \mathbf{x}(t) \right| = 0, \quad a.s. \tag{7.38}$$

7.4.2 Large-Deviation Principle

Denote

$$g(\mathbf{x}, \boldsymbol{\theta}) = \sum_{\ell=1}^{M} \left\{ R_{+\ell}(\mathbf{x}) \left[e^{\mathbf{v}_\ell \cdot \boldsymbol{\theta}} - 1 \right] + R_{-\ell}(\mathbf{x}) \left[e^{-\mathbf{v}_\ell \cdot \boldsymbol{\theta}} - 1 \right] \right\}, \quad \boldsymbol{\theta} \in \mathbb{R}^N,$$

and

$$l(\mathbf{x}, \mathbf{y}) = \sup_{\boldsymbol{\theta}} \left\{ \boldsymbol{\theta} \cdot \mathbf{y} - g(\mathbf{x}, \boldsymbol{\theta}) \right\},$$

which is called the local rate function.

Then, we can define the Freidlin–Wentzell-type rate function [80]

$$I_0^T \left(\{ \mathbf{r}(t) : 0 \leq t \leq T \} \right) = \begin{cases} \int_0^T l(\mathbf{r}(s), \mathbf{r}'(s)) \, ds & \text{if } \mathbf{r}'(t) \text{ exists,} \\ \infty & \text{otherwise.} \end{cases} \tag{7.39}$$

The theorem below is the Freidlin–Wentzell-type large-deviation theory for stochastic chemical reaction models, following reference [252]. Recently, a more general and rigorous proof of the large-deviation principle for chemical reaction networks was presented [2], which also gives nice conditions for when the large-deviation principle holds.

$D_{[0,T]}^N$ is the space containing all the functions of parameter $t \in [0, T]$ with values in $\mathbb{R}^N$ that are right continuous with left limits. Let Λ be a collection of strictly increasing functions λ on $[0, T]$ such that $\lambda(0) = 0$ and $\lambda(T) = T$ and such that

$$\gamma(\lambda) \stackrel{\triangle}{=} \sup_{0 \leq s \leq t \leq T} \left| \ln \frac{\lambda(s) - \lambda(t)}{s - t} \right| < \infty.$$

The standard metric on $D_{[0,T]}^N$ is

$$d_d\left(\{\mathbf{x}(t):0\le t\le T\},\{\mathbf{y}(t):0\le t\le T\}\right)$$

$$=\inf_{\lambda\in\Lambda}\left\{\max\left(\gamma(\lambda),\sup_{0\le t\le T}\left|x(t)-y(\lambda(t))\right|\right)\right\}. \tag{7.40}$$

$\left(D^N_{[0,T]},d_d\right)$ is a complete, severable metric space. It is also called the Skorohod space [252].

Theorem 7.2. *Assume that for each ℓ, $\ln R_{+\ell}(\mathbf{x})$ and $\ln R_{-\ell}(\mathbf{x})$ are bounded and Lipschitz continuous. Then, I_0^T is a good rate function in $\left(D^N_{[0,T]},d_d\right)$, and*
(i) for every closed set $F\in D^N_{[0,T]}$ and every $\mathbf{x}$,

$$\limsup_{V\to\infty}\frac{1}{V}\ln P_{\mathbf{x}}\left(\left\{\frac{\mathbf{n}_V(t)}{V}:0\le t\le T\right\}\in F\right)$$

$$\le-\inf\left\{I_0^T\left(\{\mathbf{r}(t):0\le t\le T\}\right):\{\mathbf{r}(t):0\le t\le T\}\in F,\mathbf{r}(0)=\mathbf{x}\right\}, \tag{7.41}$$

(ii) for every open set $G\in D^N_{[0,T]}$, uniformly for $\mathbf{x}$ *in compact sets,*

$$\liminf_{V\to\infty}\frac{1}{V}\ln P_{\mathbf{x}}\left(\left\{\frac{\mathbf{n}_V(t)}{V}:0\le t\le T\right\}\in G\right)$$

$$\ge-\inf\left\{I_0^T\left(\{\mathbf{r}(t):0\le t\le T\}\right):\{\mathbf{r}(t):0\le t\le T\}\in G,\mathbf{r}(0)=\mathbf{x}\right\}. \tag{7.42}$$

Further following the Freidlin–Wentzell large-deviation theory for stochastic chemical reaction models and again from [252],

Theorem 7.3. *Assume*
(a) $\mathbf{q}$ is the global attractive, unique stable fixed point of rate equation (7.2);
(b) for each ℓ, $\ln R_{+\ell}(\mathbf{x})$ and $\ln R_{-\ell}(\mathbf{x})$ are bounded and Lipschitz continuous in some neighborhood of $\mathbf{q}$;
(c) for each V, the stochastic chemical reaction models are positive recurrent with steady state probability p_V^{ss} for stochastic process $\mathbf{n}_V(t)$.
Then, for any ε,

$$\lim_{V\to\infty}\pi_V\left(B_\varepsilon(\mathbf{q})\right)=1, \tag{7.43}$$

in which $B_\varepsilon(\mathbf{q})$ is the ε-neighborhood of position $\mathbf{q}$ in $\mathbb{R}^N$, and $\pi_V(B_\varepsilon)$ is defined as $\pi_V(B_\varepsilon)=\sum_{\frac{\mathbf{n}}{V}\in B_\varepsilon}p_V^{ss}(\mathbf{n})$.

Let $D\subset\mathbb{R}^N$ be a smooth, bounded open set. Define

$$S\stackrel{\triangle}{=}\left\{\{\mathbf{r}(t)\}:r(0)=\mathbf{q},r(T)\in\bar{D}\ for\ some\ T>0\right\}$$

and

$$I^* \overset{\triangle}{=} \inf \left\{ I_0^T \left(\{\mathbf{r}(t) : 0 \le t \le T\} \right) : \{\mathbf{r}(t) : 0 \le t \le T\} \in S \right\}.$$

Further assume

(d) for each $\delta > 0$, *there is a* $T < \infty$ *such that, uniformly over* $\mathbf{x}_0 \in B_\varepsilon(\mathbf{q})$, $\mathbf{x}(0) = \mathbf{x}_0$ *such that*

$$|\mathbf{x}(t) - \mathbf{q}| < \delta \forall t > T.$$

(e) S is a continuity set, and every point in S is the limit of points in the interior of S;

(f) there is some ε_0 *such that* $D \subset B_{\varepsilon_0}(\mathbf{x})$ *and* $\varphi^{ss}(\mathbf{y}) > I^*(S) + 1$ *whenever* $\mathbf{y}$ *is outside* $B_{\varepsilon_0}(\mathbf{x})$.

Then,

$$\lim_{V \to \infty} \frac{1}{V} \ln \pi_V(D) = - \inf_{\mathbf{x} \in D} \varphi^{ss}(\mathbf{x}), \tag{7.44}$$

in which

$$\varphi^{ss}(\mathbf{x}) = \inf_{t \ge 0} \inf_{r(0) = \mathbf{q}, r(t) = \mathbf{x}} \left\{ I_0^t \left(\{r(s) : 0 \le s \le t\} \right) \right\}.$$

Remark 1 *Applying the contraction principle, one can straightforwardly prove that the distribution of* $\mathbf{n}_V(t)/V$ *as a function of V satisfies the large-deviation principle with good rate function* $\varphi_t(\mathbf{x})$ *for any given t. Additionally, one can use the techniques in Chapter 6 of reference [80] to generalize (7.44) to cases with multiple attractors.*

Next, we derive the nonlinear partial differential equation satisfied by $\varphi^{ss}(\mathbf{x})$. We first need a lemma.

Lemma 2 *For each* ε, *assume*

$$k^\varepsilon(V, \mathbf{x}) = \pi_V(B_\varepsilon(\mathbf{x})) \exp \left\{ V \inf_{\mathbf{y} \in B_\varepsilon(\mathbf{x})} \varphi^{ss}(\mathbf{y}) \right\}$$

is continuous and there exists another positive continuous function $f^\varepsilon(V)$ *of V such that*

$$\lim_{V \to \infty, \varepsilon \to 0^+} \frac{k^\varepsilon(V, \mathbf{x})}{f^\varepsilon(V)} = K(\mathbf{x})$$

exists uniformly at any sufficiently small neighborhood of each $\mathbf{x}$. *We also assume that both functions* $K(\mathbf{x})$ *and* $\varphi^{ss}(\mathbf{x})$ *are at least twice continuously differentiable and that* $K(\mathbf{x})$ *is positive.*

Then, for any positive function $\tilde{\varepsilon}(V)$ *such that* $V\tilde{\varepsilon}(V) \ge \frac{1}{2}$ *and*

$$\lim_{V\to\infty} V\tilde{\varepsilon}^2(V) = 0,$$

we have $\pi_V(B_{\tilde{\varepsilon}(V)}(\mathbf{x})) > 0$ *and*

$$\lim_{V\to+\infty} \frac{\pi_V(B_{\tilde{\varepsilon}(V)}(\mathbf{x}-\mathbf{v}/V))}{\pi_V(B_{\tilde{\varepsilon}(V)}(\mathbf{x}))} = e^{\mathbf{v}\cdot\nabla_{\mathbf{x}}\varphi^{ss}(\mathbf{x})}, \tag{7.45}$$

for any N-dimensional integer vector $\mathbf{v}$. *The convergence is uniform in any suffi-ciently small neighborhood of* $\mathbf{x}$.

Furthermore, denote $\mathbf{n}(\mathbf{x},V)$ *as the nearest integer vector to the point* $\mathbf{x}V$, *then*

$$\lim_{V\to\infty} \frac{p_V^{ss}(\mathbf{n}(\mathbf{x},V)-\mathbf{v})}{p_V^{ss}(\mathbf{n}(\mathbf{x},V))} = e^{\mathbf{v}\cdot\nabla_{\mathbf{x}}\varphi^{ss}(\mathbf{x})}.$$

Proof. According to these assumptions, we know that for each $\mathbf{x}$, given any $\delta > 0$, there exists a constant V_0 and ε_0, such that when $V > V_0$ and $\varepsilon < \varepsilon_0$,

$$\left| \frac{k^{\varepsilon}(V,\mathbf{x})}{f^{\varepsilon}(V)} - K(\mathbf{x}) \right| < \frac{\delta}{3},$$

$$\left| \frac{k^{\varepsilon}(V,\mathbf{x}-\mathbf{v}/V)}{f^{\varepsilon}(V)} - K(\mathbf{x}-\mathbf{v}/V) \right| < \frac{\delta}{3}.$$

Combined with the fact that for sufficiently large V

$$\left| K(\mathbf{x}) - K(\mathbf{x}-\mathbf{v}/V) \right| < \frac{\delta}{3},$$

we therefore have

$$\lim_{V\to\infty, \varepsilon\to 0^+} \frac{k^{\varepsilon}(V,\mathbf{x}-\mathbf{v}/V)}{f^{\varepsilon}(V)} = K(\mathbf{x}),$$

followed by

$$\lim_{V\to\infty, \varepsilon\to 0^+} \frac{k^{\varepsilon}(V,\mathbf{x}-\mathbf{v}/V)}{k^{\varepsilon}(V,\mathbf{x})} = 1.$$

Noticing that

$$\pi_V(B_{\tilde{\varepsilon}(V)}(\mathbf{x}-\mathbf{v}/V)) = k^{\varepsilon}(V,\mathbf{x}-\mathbf{v}/V)\exp\left\{ -V \inf_{\mathbf{y}\in B_{\tilde{\varepsilon}(V)}(\mathbf{x}-\mathbf{v}/V)} \varphi^{ss}(\mathbf{y}) \right\},$$

$$\pi_V(B_{\tilde{\varepsilon}(V)}(\mathbf{x})) = k^{\varepsilon}(V,\mathbf{x})\exp\left\{ -V \inf_{\mathbf{y}\in B_{\tilde{\varepsilon}(V)}(\mathbf{x})} \varphi^{ss}(\mathbf{y}) \right\},$$

we only need to prove

$$\lim_{V\to\infty} V\left[\inf_{\mathbf{y}\in B_{\tilde{\varepsilon}(V)}(\mathbf{x})} \varphi^{ss}(\mathbf{y}) - \inf_{\mathbf{y}\in B_{\tilde{\varepsilon}(V)}(\mathbf{x}-\mathbf{v}/V)} \varphi^{ss}(\mathbf{y}) \right] = \mathbf{v}\cdot\nabla_{\mathbf{x}}\varphi^{ss}(\mathbf{x}).$$

Since the function $\varphi^{ss}(\mathbf{x})$ is at least twice continuously differentiable, when V is large, the first and second derivatives of $\varphi^{ss}(\mathbf{x})$ inside $B_{\tilde{\varepsilon}(V)}(\mathbf{x}) \cup B_{\tilde{\varepsilon}(V)}(\mathbf{x} - \mathbf{v}/V)$ are bounded. Hence

$$\inf_{\mathbf{y} \in B_{\tilde{\varepsilon}(V)}(\mathbf{x}-\mathbf{V}/V)} \varphi^{ss}(\mathbf{y}) - \varphi^{ss}(\mathbf{x}-\mathbf{v}/V) = -\sum_{i=1}^{N} \left| \frac{\partial \varphi^{ss}(\mathbf{x}-\mathbf{v}/V)}{\partial x_i} \right| \tilde{\varepsilon}(V) + O(\tilde{\varepsilon}^2(V)),$$

and

$$\inf_{\mathbf{y} \in B_{\tilde{\varepsilon}(V)}(\mathbf{x})} \varphi^{ss}(\mathbf{y}) - \varphi^{ss}(\mathbf{x}) = -\sum_{i=1}^{N} \left| \frac{\partial \varphi^{ss}(\mathbf{x})}{\partial x_i} \right| \tilde{\varepsilon}(V) + O(\tilde{\varepsilon}^2(V)).$$

Therefore

$$\left| \inf_{\mathbf{y} \in B_{\tilde{\varepsilon}(V)}(\mathbf{x}-\mathbf{V}/V)} \varphi^{ss}(\mathbf{y}) - \inf_{\mathbf{y} \in B_{\tilde{\varepsilon}(V)}(\mathbf{x})} \varphi^{ss}(\mathbf{y}) + \varphi^{ss}(\mathbf{x}) - \varphi^{ss}(\mathbf{x}-\mathbf{v}/V) \right|$$

$$= \left| \sum_{i=1}^{N} \left[\left| \frac{\partial \varphi^{ss}(\mathbf{x})}{\partial x_i} \right| - \left| \frac{\partial \varphi^{ss}(\mathbf{x}-\mathbf{v}/V)}{\partial x_i} \right| \right] \tilde{\varepsilon}(V) \right| + O\left(\tilde{\varepsilon}^2(V) \right)$$

$$= O\left(\frac{1}{V} \tilde{\varepsilon}(V) \right) + O\left(\tilde{\varepsilon}^2(V) \right) = o\left(\frac{1}{V} \right), \tag{7.46}$$

and the fact

$$\lim_{V \to \infty} V \left[\varphi^{ss}(\mathbf{x}) - \varphi^{ss}(\mathbf{x}-\mathbf{v}/V) \right] = \mathbf{v} \cdot \nabla_{\mathbf{x}} \varphi^{ss}(\mathbf{x})$$

concludes the proof of Eqn. (7.45).

Once $\tilde{\varepsilon}(V) = \frac{1}{2V}$, $\pi_V^{ss}(B_{\tilde{\varepsilon}(V)}(\mathbf{x})) = p_V^{ss}(\mathbf{n}(\mathbf{x},V))$ and then $\pi_V(B_{\tilde{\varepsilon}(V)}(\mathbf{x}-\mathbf{v}/V)) = p_V(\mathbf{n}(\mathbf{x},V) - \mathbf{v})$. $\square$

Denote the stoichiometric matrix $\mathscr{S}_{N \times M}$, in which $s_{ij} = v_{ji}$. Any vector in the left null space sets a conservation law to the chemical reaction system, i.e., once any vector $\eta = (\eta_1, \eta_2, \cdots, \eta_N)$ satisfies

$$\sum_{i=1}^{N} \eta_i s_{ij} = 0, \ \forall j,$$

then

$$\frac{d}{dt} \left(\sum_{i=1}^{N} \eta_i x_i \right) = 0.$$

Therefore, the surviving space of the reaction scheme (7.1), regardless of whether it is stochastic or deterministic, can be described by $\mathscr{L}^{\mathbf{x}} = \{ \mathbf{x} + \sum_{\ell=1}^{M} \xi_\ell \mathbf{v}_\ell^T, \forall \xi = (\xi_1, \xi_2, \cdots, \xi_M) \}$ (called the surviving space indicated by $\mathbf{x}$), in which $\mathbf{x}$ is any given state in $\mathbb{R}^N$.

Theorem 7.4. *Under the assumptions in Theorem 7.2 and Lemma 2 and further assuming that the limit*

$$\lim_{V\to\infty} \frac{r_{\pm\ell}(V\mathbf{x};V)}{V} = R_{\pm\ell}(\mathbf{x})$$

*is locally uniform for any sufficiently small neighborhood of each $\mathbf{x}$, the function $\varphi^{ss}(\mathbf{x})$, also called the **quasi-potential**, satisfies*

$$\sum_{\ell=1}^{M} R_{+\ell}(\mathbf{x})\left[1 - e^{\mathbf{v}_\ell \cdot \nabla_{\mathbf{x}}\varphi^{ss}(\mathbf{x})}\right] + R_{-\ell}(\mathbf{x})\left[1 - e^{-\mathbf{v}_\ell \cdot \nabla_{\mathbf{x}}\varphi^{ss}(\mathbf{x})}\right] = 0, \qquad (7.47)$$

i.e.,

$$g(\mathbf{x}, \nabla_{\mathbf{x}}\varphi^{ss}(\mathbf{x})) = 0.$$

Then, $\varphi^{ss}(\mathbf{x})$ satisfies $\frac{d}{dt}\varphi^{ss}(\mathbf{x}(t)) \leq 0$, and the equality holds if and only if the gradient of $\varphi^{ss}(\mathbf{x})$ in the surviving space $\mathscr{L}^{\mathbf{x}}$ indicated by $\mathbf{x}$ vanishes at $\mathbf{x}$, i.e., $\mathbf{v}_\ell \cdot \nabla_{\mathbf{x}}\varphi^{ss}(\mathbf{x}) = 0$ for all ℓ.

Furthermore, the steady state condition at $\mathbf{x}$, i.e.,

$$F_i(\mathbf{x}) = \sum_{\ell=1}^{M} v_{\ell i}\left(R_{+\ell}(\mathbf{x}) - R_{-\ell}(\mathbf{x})\right) = 0, \ \forall i$$

is a sufficient condition for the vanishing gradient of $\varphi^{ss}(\mathbf{x})$ in $\mathscr{L}^{\mathbf{x}}$, while if $\varphi^{ss}(\mathbf{x})$ is twice differentiable and the dimension of the Hessian matrix of $\varphi^{ss}(\mathbf{x})$ is equal to the dimension of the surviving space, i.e., the column space of the stoichiometric matrix $\mathscr{S}$, then the steady state condition is also a necessary condition for the vanishing gradient of $\varphi^{ss}(\mathbf{x})$ in $\mathscr{L}^{\mathbf{x}}$.

Proof. Given V, denote $\mathbf{n}(\mathbf{x}, V)$ as the nearest integer vector to point $\mathbf{x}V$. Then, the CME (7.5) at steady state can be rewritten as

$$\sum_{\ell=1}^{M} \frac{p_V^{ss}(\mathbf{n}(\mathbf{x}, V) - \mathbf{v}_\ell)}{p_V^{ss}(\mathbf{n}(\mathbf{x}, V))} r_{+\ell}(\mathbf{n}(\mathbf{x}, V) - \mathbf{v}_\ell; V) - r_{+\ell}(\mathbf{n}(\mathbf{x}, V); V)$$

$$+ \frac{p_V^{ss}(\mathbf{n}(\mathbf{x}, V) + \mathbf{v}_\ell)}{p_V^{ss}(\mathbf{n}(\mathbf{x}, V))} r_{-\ell}(\mathbf{n}(\mathbf{x}, V) + \mathbf{v}_\ell; V) - r_{-\ell}(\mathbf{n}(\mathbf{x}, V); V) = 0. \quad (7.48)$$

According to Lemma 2, we then have

$$\lim_{V\to\infty} \frac{p_V^{ss}(\mathbf{n}(\mathbf{x}, V) - \mathbf{v}_\ell)}{p_V^{ss}(\mathbf{n}(\mathbf{x}, V))} = e^{\mathbf{v}_\ell \cdot \nabla_{\mathbf{x}}\varphi^{ss}(\mathbf{x})};$$

$$\lim_{V\to\infty} \frac{p_V^{ss}(\mathbf{n}(\mathbf{x}, V) + \mathbf{v}_\ell)}{p_V^{ss}(\mathbf{n}(\mathbf{x}, V))} = e^{-\mathbf{v}_\ell \cdot \nabla_{\mathbf{x}}\varphi^{ss}(\mathbf{x})}. \qquad (7.49)$$

Additionally, similar to the proof of Lemma 2, we know that

$$\lim_{V \to \infty} \frac{r_{+\ell}(\mathbf{n}(\mathbf{x},V) - v_\ell; V)}{V} = \lim_{V \to \infty} \frac{r_{+\ell}(\mathbf{n}(\mathbf{x},V); V)}{V} = R_{+\ell}(\mathbf{x});$$

$$\lim_{V \to \infty} \frac{r_{-\ell}(\mathbf{n}(\mathbf{x},V) + v_\ell; V)}{V} = \lim_{V \to \infty} \frac{r_{-\ell}(\mathbf{n}(\mathbf{x},V); V)}{V} = R_{-\ell}(\mathbf{x}),$$

$$\text{(7.50)}$$

which proves Eqn. (7.47).

Finally, since $e^{\mathbf{x}} \geq \mathbf{x} + 1$,

$$\begin{aligned}
\frac{d}{dt} \varphi^{ss}(\mathbf{x}(t)) &= \frac{d\mathbf{x}(t)}{dt} \cdot \nabla_{\mathbf{x}} \varphi^{ss}(\mathbf{x}) \\
&= \sum_{\ell=1}^{M} \left(R_{+\ell}(\mathbf{x}) - R_{-\ell}(\mathbf{x}) \right) v_\ell \cdot \nabla_{\mathbf{x}} \varphi^{ss}(\mathbf{x}) \\
&\leq \sum_{\ell=1}^{M} R_{+\ell}(\mathbf{x}) \left[e^{v_\ell \cdot \nabla_{\mathbf{x}} \varphi^{ss}(\mathbf{x})} - 1 \right] + R_{-\ell}(\mathbf{x}) \left[e^{-v_\ell \cdot \nabla_{\mathbf{x}} \varphi^{ss}(\mathbf{x})} - 1 \right] \\
&= 0. \qquad\qquad\qquad\qquad\qquad\qquad\qquad\qquad\qquad\qquad \text{(7.51)}
\end{aligned}$$

The equality holds if and only if $v_\ell \cdot \nabla_{\mathbf{x}} \varphi^{ss}(\mathbf{x}) = 0$ for all ℓ.

The surviving space $\mathscr{L}^{\mathbf{x}}$ indicated by $\mathbf{x}$ is $\{\mathbf{x} + \sum_{\ell=1}^{M} \xi_\ell v_\ell^T, \forall \xi\}$. Therefore, the gradient of $\varphi^{ss}(\mathbf{x})$ in $\mathscr{L}^{\mathbf{x}}$ with respect to ξ can be depicted as

$$\left[\frac{\partial \varphi^{ss}(\mathbf{x} + \sum_{\ell=1}^{M} \xi_\ell v_\ell^T)}{\partial \xi_\ell} \right]_{\xi=0} = v_\ell \cdot \nabla_{\mathbf{x}} \varphi^{ss}(\mathbf{x}).$$

At steady state, $F(\mathbf{x}) = 0$, resulting in $\frac{d}{dt} \varphi^{ss}(\mathbf{x}(t)) = F^T(\mathbf{x}) \cdot \nabla_{\mathbf{x}} \varphi^{ss}(\mathbf{x}) = 0$, i.e., the equality in (7.51) holds.

On the other hand, taking the derivative of the left-hand side of Eqn. (7.47), we obtain

$$\sum_{\ell=1}^{M} \left\{ \frac{\partial R_{+\ell}(\mathbf{x})}{\partial \mathbf{x}_k} \left[1 - e^{v_\ell \cdot \nabla_{\mathbf{x}} \varphi^{ss}(\mathbf{x})} \right] + \frac{\partial R_{-\ell}(\mathbf{x})}{\partial \mathbf{x}_k} \left[1 - e^{-v_\ell \cdot \nabla_{\mathbf{x}} \varphi^{ss}(\mathbf{x})} \right] \right\}$$

$$+ \sum_{\ell=1}^{M} \left[R_{-\ell}(\mathbf{x}) e^{-v_\ell \cdot \nabla_{\mathbf{x}} \varphi^{ss}(\mathbf{x})} - R_{+\ell}(\mathbf{x}) e^{v_\ell \cdot \nabla_{\mathbf{x}} \varphi^{ss}(\mathbf{x})} \right] \sum_{1 \leq i \leq N} v_{\ell i} \frac{\partial^2 \varphi^{ss}(\mathbf{x})}{\partial \mathbf{x}_i \partial \mathbf{x}_k}$$

$$= 0, \qquad\qquad\qquad\qquad\qquad\qquad\qquad\qquad\qquad\qquad\qquad \text{(7.52)}$$

for each $1 \leq k \leq N$.

Once $v_\ell \cdot \nabla_{\mathbf{x}} \varphi^{ss}(\mathbf{x}) = 0$ for each ℓ, the above equation becomes

$$\sum_{1 \leq i \leq N} F_i(\mathbf{x}) \frac{\partial^2 \varphi^{ss}(\mathbf{x})}{\partial \mathbf{x}_i \partial \mathbf{x}_k} = 0, \tag{7.53}$$

for each $1 \leq k \leq N$.

Hence, if the dimension of the Hessian matrix $\left\{ \frac{\partial^2 \varphi^{ss}(\mathbf{x})}{\partial \mathbf{x}_i \partial \mathbf{x}_j} \right\}$ is the same as the dimension of the stoichiometric matrix $\mathscr{S}$ and noting that $F(\mathbf{x})$ is also in the column space of $\mathscr{S}$, we know that $F(\mathbf{x}) = 0$. $\qquad\qquad\square$

A few remarks are in order.

(i) Theorem 7.4 can also be proven through the classical mechanics approach, regarding $I_0^t(\{r(s) : 0 \leq s \leq t\})$ as the action functional of trajectory $\{r(s) : 0 \leq s \leq t\}$. The details of this approach can be found in the text by Feng and Kurtz [69]. They start from a Hamiltonian-Lagrangian convergence formalism, which considers the convergence of the semigroup

$$L(t)f(\mathbf{x}) \triangleq \frac{1}{V} \ln E \left[e^{Vf(\frac{\mathbf{n}_V(t)}{V})} | \frac{\mathbf{n}(0)}{V} = \mathbf{x} \right].$$

(ii) The solution of the partial differential equation (7.47) is not unique. To make sense of the physical solution, one should turn to the so-called weak KAM theory, which provides a representation formula of the Eikonal equations.

(iii) It is known that the solutions of (7.47) are generally only Lipschitz and semiconvex. There are points of Lebesgue measure zero that are nondifferentiable even for first-order derivatives. In this paper, we avoid involving ourselves in such a subtle and complicated situation; hence, we simply assume that the solution is at least twice differentiable.

7.4.3 Diffusion Approximation

Diffusion approximation to the chemical reaction models can be achieved either through the Kramers-Moyal expansion of the chemical master equation or Poisson representation [261, 262]. The approximated diffusion process $\mathbf{z}(t)$ is a diffusion process with drift coefficient $\mathbf{B}(\mathbf{z})$ and a very specific matrix of diffusion coefficients $\mathbf{A}(\mathbf{z})$, on the order of V^{-1}, whose entries are

$$A_{ij}(\mathbf{z}) = \frac{1}{V} \sum_{\ell=1}^{M} \left(R_{+\ell}(\mathbf{z}) + R_{-\ell}(\mathbf{z}) \right) v_{\ell i} v_{\ell j}, \quad 1 \leq i, j \leq N. \tag{7.54}$$

Kurtz proved that the difference at finite time intervals between $\mathbf{x}(t)$ and $\mathbf{z}(t)$ is on the order of $V^{-\frac{1}{2}}\ln V$ [155]. Additionally, we refer to [129, 143, 194] for more information about the mathematical theory of diffusion processes.

Boundary conditions are also crucial for diffusion approximation. It is known that the standard diffusion approximation is only well defined up to the boundary of the positive orthant, as the process can then go negative if the diffusion term does not vanish at the boundary. We will not explicitly touch upon this issue in this book.

The diffusion process close to any stable fixed point $\mathbf{q}$ of Eqn. (7.2) can be further approximated by a linear diffusion process with constant diffusion matrix $A = A(\mathbf{q})$ and linear drift coefficient $\hat{\mathbf{B}}\mathbf{z}$, in which matrix $\hat{\mathbf{B}}$ has

$$\hat{B}_{ij} = \frac{\partial B_i}{\partial \mathbf{x}_j}(\mathbf{q}). \tag{7.55}$$

Denote Ξ as the covariance matrix of the invariant measure of such a linear diffusion process; then, we have

$$\mathbf{A} = -\left(\hat{\mathbf{B}}\Xi + \Xi\hat{\mathbf{B}}\right). \tag{7.56}$$

This equation has been referred to as the fluctuation-dissipation theorem of the chemical reaction model by J. E. Keizer (1942–1999) [145]. In Section 7.5, we provide a clearer mathematical theorem for this result, especially for what Ξ represents for the original stochastic chemical reaction model, as well as a rigorous proof.

7.4.4 Macroscopic Limits

We now consider the mesoscopic thermodynamics quantities in the Kurtz limit of $V \to \infty$.

Theorem 7.5. *Assume Eqns. (7.38), (7.43), (7.44), and (7.47) as well as the assumptions of Lemma 2 hold, and assume that the limit*

$$\lim_{V \to \infty} \frac{r_{\pm\ell}(V\mathbf{x}; V)}{V} = R_{\pm\ell}(\mathbf{x})$$

is locally uniform for any sufficiently small neighborhood of each $\mathbf{x}$.

Then, denote $\mathbf{x}(t)$ as the solution to Eqn. (7.2); as $V \to \infty$, we have for each time t

(a)

$$\lim_{V\to\infty} \frac{e_p\left[p_V(\mathbf{n},t)\right]}{V} = \sigma^{(tot)}\left[\mathbf{x}(t)\right], \tag{7.57}$$

where the density of the macroscopic chemical entropy production rate

$$\sigma^{tot}[\mathbf{x}] = \sum_{\ell=1}^{M} \left(R_{+\ell}(\mathbf{x}) - R_{-\ell}(\mathbf{x})\right) \ln\left(\frac{R_{+\ell}(\mathbf{x})}{R_{-\ell}(\mathbf{x})}\right);$$

(b)

$$\lim_{V\to\infty} \frac{f_d\left[p_V(\mathbf{n},t)\right]}{V} = f_d^{macro}[\mathbf{x}(t)], \tag{7.58}$$

where the density of the macroscopic free energy dissipation rate

$$f_d^{macro}[\mathbf{x}] = \sum_{\ell=1}^{M} \left(R_{-\ell}(\mathbf{x}) - R_{+\ell}(\mathbf{x})\right) v_\ell \cdot \nabla_{\mathbf{x}} \varphi^{ss}(\mathbf{x}).$$

(c)

$$\lim_{V\to\infty} \frac{F^{(meso)}\left[p_V(\mathbf{n},t)\right]}{V} = \varphi^{ss}(\mathbf{x}(t)), \quad \frac{d\varphi^{ss}(\mathbf{x})}{dt} = -f_d^{(macro)}[\mathbf{x}], \tag{7.59}$$

and hence, $\varphi^{ss}(\mathbf{x})$ can be regarded as the density of general macroscopic free energy;

(d)

$$\lim_{V\to\infty} \frac{E_{in}\left[p_V(\mathbf{n},t)\right]}{V} = E_{in}^{(macro)}[\mathbf{x}(t)], \tag{7.60}$$

where the density of the macroscopic housekeeping heat dissipation rate [114]

$$E_{in}^{(macro)}[\mathbf{x}] = \sum_{\ell=1}^{M} \left(R_{-\ell}(\mathbf{x}) - R_{+\ell}(\mathbf{x})\right) \ln\left(\frac{R_{-\ell}(\mathbf{x})}{R_{+\ell}(\mathbf{x})} e^{-v_\ell \cdot \nabla_{\mathbf{x}} \varphi^{ss}(\mathbf{x})}\right).$$

Proof. According to the strong law of large numbers (Eqn. (7.38)), we know that given any small $\varepsilon > 0$, for sufficiently large V, the probability is concentrated on the integers satisfying $\frac{\mathbf{n}}{V} \in B_\varepsilon(\mathbf{x}(t))$.

Given any small positive number δ, for those $\mathbf{n}$, when V is sufficiently large,

$$\left| \frac{r_{+\ell}(\mathbf{n};V)}{V} - R_{+\ell}(\frac{\mathbf{n}}{V}) \right| < \delta,$$

and since $\left|R_{+\ell}(\frac{\mathbf{n}}{V}) - R_{+\ell}(\mathbf{x})\right| < C_{+\ell}(\mathbf{x})\varepsilon$ for some constant $C_{+\ell}(\mathbf{x})$, we have

$$\left| \frac{r_{+\ell}(\mathbf{n};V)}{V} - R_{+\ell}(\mathbf{x}) \right| < \delta + C_{+\ell}(\mathbf{x})\varepsilon.$$

Similarly,

$$\left| \frac{r_{-\ell}(\mathbf{n}; V)}{V} - R_{-\ell}(\mathbf{x}) \right| < \delta + C_{-\ell}(\mathbf{x}),$$

and

$$\left| \ln\left(\frac{r_{+\ell}(\mathbf{n}; V)}{r_{-\ell}(\mathbf{n} + \nu_\ell; V)} \right) - \ln\frac{R_{+\ell}(\mathbf{x})}{R_{-\ell}(\mathbf{x})} \right| < \delta + C_1(\mathbf{x})\varepsilon.$$

Then, based on the large-deviation principle (Eqn. (7.44)) and the assumption of Lemma 2, we know that for sufficiently large V, once $\frac{\mathbf{n}}{V} \in B_\varepsilon(\mathbf{x}(t))$,

$$\left| \frac{1}{V}\ln p_V^{ss}(\mathbf{n}) + \inf_{\mathbf{y} \in B_{\frac{1}{2V}}(\frac{\mathbf{n}}{V})} \varphi^{ss}(\mathbf{y}) \right| < \delta,$$

followed by

$$\left| \frac{1}{V}\ln p_V^{ss}(\mathbf{n}) + \varphi^{ss}(\mathbf{x}) \right| < \delta + C_2(\mathbf{x})\varepsilon,$$

since

$$\left| \inf_{\mathbf{y} \in B_{\frac{1}{2V}}(\frac{\mathbf{n}}{V})} \varphi^{ss}(\mathbf{y}) - \varphi^{ss}(\mathbf{x}) \right| < C_2(\mathbf{x})\varepsilon.$$

Furthermore, due to Lemma 2, for sufficiently large V, once $\frac{\mathbf{n}}{V} \in B_\varepsilon(\mathbf{x}(t))$,

$$\left| \ln\frac{p_V^{ss}(\mathbf{n})}{p_V^{ss}(\mathbf{n} + \nu_\ell)} - \nu_\ell \cdot \nabla_{\mathbf{x}}\varphi^{ss}\left(\frac{\mathbf{n}}{V}\right) \right| < \delta,$$

and since

$$\left| \nu_\ell \cdot \nabla_{\mathbf{x}}\varphi^{ss}\left(\frac{\mathbf{n}}{V}\right) - \nu_\ell \cdot \nabla_{\mathbf{x}}\varphi^{ss}(\mathbf{x}) \right| < C_3(\mathbf{x})\varepsilon,$$

it follows that

$$\left| \ln\frac{p_V^{ss}(\mathbf{n})}{p_V^{ss}(\mathbf{n} + \nu_\ell)} - \nu_\ell \cdot \nabla_{\mathbf{x}}\varphi^{ss}(\mathbf{x}) \right| < C_3(\mathbf{x})\varepsilon.$$

Finally, similar to the proof of Lemma 2 for the large-deviation principle of $\{p_V(\mathbf{n},t)\}$ with fixed t, we have

$$\left| \frac{1}{V}\ln p_V(\mathbf{n},t) + \varphi_t(\mathbf{x}(t)) \right| < \delta + C_4(\mathbf{x})\varepsilon,$$

and

$$\left| \ln \frac{p_V(\mathbf{n},t)}{p_V(\mathbf{n}+\mathbf{v}_\ell,t)} - \mathbf{v}_\ell \cdot \nabla_\mathbf{x} \varphi_t(\mathbf{x}(t)) \right| < \delta + C_5(\mathbf{x})\varepsilon.$$

Noticing the large-deviation rate function $\varphi_t(\mathbf{x})$ for transient distributions at time t, taking its global minimum at $\mathbf{x} = \mathbf{x}(t)$, i.e., $\varphi_t(\mathbf{x}(t)) = \nabla_\mathbf{x} \varphi_t(\mathbf{x}(t)) = 0$, we know

$$\left| \frac{1}{V} \ln p_V(\mathbf{n},t) \right| < \delta + C_4(\mathbf{x})\varepsilon,$$

and

$$\left| \ln \frac{p_V(\mathbf{n},t)}{p_V(\mathbf{n}+\mathbf{v}_\ell,t)} \right| < \delta + C_5(\mathbf{x})\varepsilon.$$

Taking all these together and realizing that ε and δ can be arbitrarily small, we finally prove (a)-(d). $\square$

The conditions we listed for the above Theorem 7.5 are needed for validating the law of large numbers and the large-deviation principle. Actually, one only needs to validate these two in order to prove Theorem 7.5. Hence, the results could hold for much more general settings.

Proposition 3 *The three terms $\sigma^{tot}[\mathbf{x}]$, $f_d^{macro}[\mathbf{x}]$ and $E_{in}^{macro}[\mathbf{x}]$ are all nonnegative. Furthermore,*

*(a) $\sigma^{tot}[\mathbf{x}] = 0$ if and only if the **strong** detailed balance condition is satisfied at* $\mathbf{x}$*:*

$$R_{+\ell}(\mathbf{x}) = R_{-\ell}(\mathbf{x}), \ \forall \ell;$$

*(b) $E_{in}^{macro}[\mathbf{x}] = 0$ if and only if the **weak** detailed balance condition is satisfied at* $\mathbf{x}$*:*

$$\ln\left(\frac{R_{+\ell}(\mathbf{x})}{R_{-\ell}(\mathbf{x})} \right) = -\mathbf{v}_\ell \cdot \varphi^{ss}(\mathbf{x})), \ \forall \ell; \tag{7.61}$$

(c) $f_d^{(macro)}[\mathbf{x}] = 0$ if and only if the following condition is satisfied, i.e.,

$$\mathbf{v}_\ell \cdot \nabla_\mathbf{x} \varphi^{ss}(\mathbf{x}) = 0, \ \forall \ell,$$

which is equivalent to the vanishing gradient of $\varphi^{ss}(\mathbf{x})$ in the surviving space $\mathscr{L}^\mathbf{x}$ indicated by $\mathbf{x}$. If the dimension of the Hessian matrix of $\varphi^{ss}(\mathbf{x})$ is equal to the dimension of the surviving space, i.e., the column space of the stoichiometric matrix $\mathscr{S}$, then it is also equivalent to the steady state condition

$$B_i(\mathbf{x}) = \sum_{\ell=1}^{M} v_{\ell i}\left(R_{+\ell}(\mathbf{x}) - R_{-\ell}(\mathbf{x})\right) = 0, \ \forall i.$$

Proof. (a) is straightforward.

(b) Note that $\forall x \geq 0$, $\ln x \geq 1 - \frac{1}{x}$ and $\ln x \leq x - 1$. Thus, we have

$$\ln\left(\frac{R_{-\ell}(\mathbf{x})}{R_{+\ell}(\mathbf{x})}e^{-\mathbf{v}_\ell \cdot \nabla_\mathbf{x}\varphi^{ss}(\mathbf{x})}\right) \geq 1 - \left(\frac{R_{+\ell}(\mathbf{x})}{R_{-\ell}(\mathbf{x})}\right)e^{\mathbf{v}_\ell \cdot \nabla_\mathbf{x}\varphi^{ss}(\mathbf{x})}, \qquad (7.62a)$$

and

$$\ln\left(\frac{R_{-\ell}(\mathbf{x})}{R_{+\ell}(\mathbf{x})}e^{-\mathbf{v}_\ell \cdot \nabla_\mathbf{x}\varphi^{ss}(\mathbf{x})}\right) \leq \left(\frac{R_{-\ell}(\mathbf{x})}{R_{+\ell}(\mathbf{x})}\right)e^{-\mathbf{v}_\ell \cdot \nabla_\mathbf{x}\varphi^{ss}(\mathbf{x})} - 1. \qquad (7.62b)$$

Therefore

$$e_{in}^{(macro)}[\mathbf{x}] \geq \sum_{\ell=1}^{M}\left\{R_{-\ell}(\mathbf{x})\left[1 - \left(\frac{R_{+\ell}(\mathbf{x})}{R_{-\ell}(\mathbf{x})}\right)e^{\mathbf{v}_\ell \cdot \nabla_\mathbf{x}\varphi^{ss}(\mathbf{x})}\right]\right.$$
$$\left. + R_{+\ell}(\mathbf{x})\left[1 - \left(\frac{R_{-\ell}(\mathbf{x})}{R_{+\ell}(\mathbf{x})}\right)e^{-\mathbf{v}_\ell \cdot \nabla_\mathbf{x}\varphi^{ss}(\mathbf{x})}\right]\right\} = 0, \qquad (7.63)$$

and the equality holds if and only if Eqn. (7.61) is satisfied.

(c) Note that $\forall x \in \mathbb{R}$, $\mathbf{x} \geq 1 - e^{-\mathbf{x}}$. Thus, we have

$$f_d^{(macro)} = \sum_{\ell=1}^{M}\left(R_{-\ell}(\mathbf{x}) - R_{+\ell}(\mathbf{x})\right)\mathbf{v}_\ell \cdot \nabla_\mathbf{x}\varphi^{ss}(\mathbf{x})$$
$$\geq \sum_{\ell=1}^{M} R_{-\ell}(\mathbf{x})[1 - \exp\{-\mathbf{v}_\ell \cdot \nabla_\mathbf{x}\varphi^{ss}(\mathbf{x})\}]$$
$$+ R_{+\ell}(\mathbf{x})[1 - \exp\{\mathbf{v}_\ell \cdot \nabla_\mathbf{x}\varphi^{ss}(\mathbf{x})\}] = 0, \qquad (7.64)$$

and the equality holds if and only if $\mathbf{v}_\ell \cdot \nabla_\mathbf{x}\varphi^{ss}(\mathbf{x}) = 0$ for all ℓ. Then, applying Theorem 7.4. $\qquad\square$

We thus have derived the macroscopic law of chemical (free) energy balance:

$$\frac{d}{dt}\varphi^{ss}(\mathbf{x}(t)) = e_{in}^{(macro)}[\mathbf{x}(t)] - \sigma^{(tot)}[\mathbf{x}(t)], \qquad (7.65)$$

in which all three macroscopic *densities*, including the free energy dissipation rate, $f_d^{(macro)}[\mathbf{x}(t)] \equiv -\frac{d}{dt}\varphi^{ss}(\mathbf{x}(t))$, the house-keeping heat rate, $E_{in}^{(macro)}[\mathbf{x}(t)]$ and the total entropy production rate $\sigma^{(tot)}[\mathbf{x}(t)]$ are nonnegative. $E_{in}^{(macro)}$ and $\sigma^{(tot)}$ should be identified as the "source" and the "sink", respectively, for an emergent, macroscopic chemical energy function φ^{ss}. It is clear that φ^{ss} is a consequence of one of the global, infinitely long time behaviors of the mesoscopic system.

7.5 Fluctuation-Dissipation Theorem

Here, we further provide a rigorous statement and proof for the fluctuation-dissipation theorem for general stochastic chemical reaction models, which was first discovered by Joel Keizer in the 1980s through diffusion approximation [145].

Theorem 7.6. $\mathbf{q}$ *is any stable fixed point of Eqn. (7.2). Assume* $\varphi^{ss}(\mathbf{x})$ *is at least twice differentiable. Define*

$$\Xi_{ij} = \frac{\partial^2 \varphi^{ss}(\mathbf{q})}{\partial x_i \partial x_j}, \ A_{ij} = \sum_{\ell=1}^{M} (R_{+\ell}(\mathbf{q}) + R_{-\ell}(\mathbf{q})) v_{\ell i} v_{\ell j},$$

and $\hat{B}_{ij} = \partial B_i(\mathbf{q})/\partial x_j$. *Then, we have the equality*

$$\Xi A \Xi = -\Xi \hat{B} - \hat{B} \Xi. \tag{7.66}$$

If Ξ *is invertible, then* $\mathbf{A} = -\left(\hat{B}\Xi^{-1} + \Xi^{-1}\hat{B}\right)$.

Proof. Taking the second derivative of the left-hand side of Eqn. (7.47) and noticing that $v_\ell \cdot \nabla_\mathbf{x} \varphi^{ss}(\mathbf{q}) = B_i(\mathbf{q}) = 0$ for each ℓ and i, we can obtain

$$\sum_{\ell=1}^{M} \left\{ \left[-\frac{\partial R_{+\ell}(\mathbf{q})}{\partial x_k} + \frac{\partial R_{-\ell}(\mathbf{q})}{\partial x_k} \right] \sum_{1 \le i \le N} v_{\ell i} \frac{\partial^2 \varphi^{ss}(\mathbf{q})}{\partial x_i \partial x_j} \right\}$$

$$+ \sum_{\ell=1}^{M} \left\{ \left[-\frac{\partial R_{+\ell}(\mathbf{q})}{\partial x_j} + \frac{\partial R_{-\ell}(\mathbf{q})}{\partial x_j} \right] \sum_{1 \le i \le N} v_{\ell i} \frac{\partial^2 \varphi^{ss}(\mathbf{q})}{\partial x_i \partial x_k} \right\}$$

$$- \sum_{\ell=1}^{M} (R_{+\ell}(\mathbf{q}) + R_{-\ell}(\mathbf{q})) \sum_{1 \le i \le N} v_{\ell i} \frac{\partial^2 \varphi^{ss}(\mathbf{q})}{\partial x_i \partial x_j} \sum_{1 \le i \le N} v_{\ell i} \frac{\partial^2 \varphi^{ss}(\mathbf{q})}{\partial x_i \partial x_k} = 0 \quad (7.67)$$

for each $1 \le j, k \le N$. It is exactly the equality $\Xi A \Xi = -\Xi \hat{B} - \hat{B} \Xi$. □

Remark 4

1. For the diffusion approximation close to any stable fixed point $\mathbf{q}$ *in Section 7.4.3, the diffusion matrix is* $V^{-1}A(\mathbf{q})$, *and the variance of the stationary linear diffusion is just* $V^{-1}\Xi^{-1}$. *Now,* Ξ *has a more rigorous and clear definition from the original stochastic chemical reaction.*

2. The diffusion approximation violates the conservation laws that are possessed by chemical reaction kinetic systems.

Chapter 8
Phase Transition and Mesoscopic Nonlinear Bistability

The phenomenon of *phase transition*, in a nutshell, is associated with a bistable, nonlinear stochastic dynamical system The system undergoes a change via a saddle-node bifurcation with *catastrophe* as $t \to \infty$, followed by the stochasticity (noise, fluctuations) tending to zero. Phase transition, thus, is a limiting behavior of non-linear stochastic dynamics. The mathematical fact that the order of taking the two limits, the stochasticity to zero and the time to infinity, cannot be switched is one of the deepest insights of mathematical science and theoretical physics [277, 7]. When considering mesoscopic chemical or biochemical reaction systems, the limit of the stochasticity to zero is represented by the Kurtz limit $V \to \infty$.

The above understanding shows why the notion of phase transition is so funda-mental a concept in classical physics of *matter*: the equilibrium state of macroscopic matter is a dynamic attractor; a state of motion that is unique under a given set of ex-ternal conditions. Bistability is truly a novel phenomenon and is the origin of almost all complex behaviors [123]. Discontinuity in phase transition is the consequence of continuous behavior in a finite system under the limit of $V \to \infty$ with a nonuniform convergence [277, 36].

8.1 A Canonical Description in Terms of a Double-Well $\varphi^{ss}(x)$

Consider a deterministic macroscopic kinetic equation

$$\frac{dx(t)}{dt} = b(x; \alpha, \beta), \tag{8.1}$$

© The Author(s), under exclusive license to Springer Nature Switzerland AG 2021
H. Qian, H. Ge, *Stochastic Chemical Reaction Systems in Biology*,
Lecture Notes on Mathematical Modelling in the Life Sciences,
https://doi.org/10.1007/978-3-030-86252-7_8

where α and β are two parameters, and its corresponding mesoscopic stochastic kinetics are represented by a stochastic differential equation

$$dX(t) = b(X; \alpha, \beta)dt + \sqrt{\varepsilon}A(X)dW(t), \qquad (8.2)$$

in which $W(t)$ is standard Brownian motion. If a system has volume V, then $\varepsilon \propto V^{-1}$. We now assume Eqn. (8.1) has two stable steady states $x_1^*(\alpha, \beta)$ and $x_2^*(\alpha, \beta)$, and $x_3^*(\alpha, \beta)$ is the saddle point located on the separatrix that divides the two basins associated with x_1^* and x_2^*. Let $\mathscr{D}_1$ and $\mathscr{D}_2$ denote the two basins of attraction, $x_1^* \in \mathscr{D}_1$ and $x_2^* \in \mathscr{D}_2$, and $x_3^* \in \overline{\mathscr{D}_1} \cap \overline{\mathscr{D}_2}$, where $\overline{\mathscr{D}_i}$ is the closure of the open set $\mathscr{D}_i$.

The stochastic dynamical system (8.2), in the limit of $\varepsilon \to 0$, or $V \to \infty$, yields (8.1). We shall use this as a caricature for Kurtz's law of large numbers. Then, as $t \to \infty$, the dynamics of (8.1) approach one of the stable fixed points x^*, which is a root of the algebraic equation $b(x; \alpha, \beta) = 0$ that defines a multivalued surface

$$x^*(\alpha, \beta) = \{x | b(x; \alpha, \beta) = 0\} = \{x_i^*(\alpha, \beta), \ i = 1, 2, 3\}. \qquad (8.3)$$

On the other hand, in the limit of $t \to \infty$ followed by $\varepsilon \to 0^+$, the $X_\varepsilon(t)$ following the stochastic differential equation (8.2) has an asymptotic expression for the stationary probability density function, i.e., the large-deviation principle, which states that when ε is sufficiently small,

$$f_\varepsilon^{ss}(x) \simeq C_\varepsilon e^{-\varphi^{ss}(x)/\varepsilon}, \ \ C_\varepsilon^{-1} = \int_{\mathbb{R}^n} e^{-\varphi^{ss}(x)/\varepsilon} dx, \qquad (8.4)$$

in which φ^{ss} is the large-deviation rate function. If the stochastic dynamics (8.2) is 1-dimensional, $x, b(x; \alpha, \beta) \in \mathbb{R}$, one explicitly obtains

$$\varphi^{ss}(x; \alpha, \beta) = -\int A^{-2}(x)b(x; \alpha, \beta)dx, \qquad (8.5)$$

which has two local minima, x_1^* and x_2^*, and a maximum, x_3^*. Generally, the function φ^{ss} always satisfies the Lyapunov property, i.e., $d\varphi^{ss}(x(t))/dt \leq 0$, in which $x(t)$ follows Eqn. (8.1) and the equality holds if and only if the right-hand side of Eqn. (8.1) vanishes. This means that φ^{ss} always has two local minima x_1^* and x_2^* and maximum x_3^* in between.

Without loss of generality, let $x_1^* < x_3^* < x_2^*$. Then $\mathscr{D}_1 = (-\infty, x_3^*)$ and $\mathscr{D}_2 = (x_3^*, +\infty)$. $\varphi^{ss}(x; \alpha, \beta)$ is a function of parameters α and β; It has two basins, or "wells", which are exactly $\mathscr{D}_1$ and $\mathscr{D}_2$, respectively. Then,

$$\lim_{\varepsilon \to 0} f_\varepsilon^{ss}(x) = \delta(x - \hat{x}), \quad \hat{x}(\alpha, \beta) = \arg\min_{x \in \mathbb{R}} \varphi^{ss}(x; \alpha, \beta), \tag{8.6}$$

i.e., $\hat{x}$ is the global minimum of $\varphi^{ss}(x)$. In contrast, $x^*(\alpha, \beta) = \{x | \nabla \varphi^{ss}(x; \alpha, \beta) = 0\}$ contains all the critical points of $\varphi^{ss}(x)$.

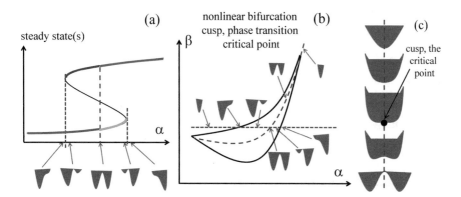

Fig. 8.1 For a mesoscopic model, if one takes its size N or $V \to \infty$ or fluctuations $\varepsilon \to 0$ first before $t \to \infty$, one obtains nonlinear dynamics with broken ergodicity. The infinite-time limit of such a deterministic system can be described by the S-shaped curve in (a), with varying α and fixed β. For different values of α and β, see the regions in (b). There is always a cusp catastrophe represented by a wedged region, inside (outside) of which the dynamical system has three (one) fixed points. For a fixed value of β while varying α, the horizontal blue dashed line crossing the boundary of the wedged region corresponds to the two vertical blue dashed lines in (a), where two saddle-node bifurcations occur. For a true thermodynamic limit, however, one first takes $t \to \infty$ for small but finite ε and obtains the system's stationary probability distribution $f_\varepsilon^{ss}(x) \simeq C_\varepsilon e^{-\varphi^{ss}(x)/\varepsilon}$, where the function $\varphi^{ss}(x)$ is the dynamic landscape, shown as green shapes in (a), (b) and (c). Then, one lets $\varepsilon \to 0$ and obtains $f_\varepsilon^{ss}(x) \to \delta(x - \hat{x})$, where $\hat{x}$ is the global minimum of the function $\varphi^{ss}(x)$. Marking the coexistence of two equal minima by the vertical red dashed line in (a), a discontinuity appears in $\hat{x}$ as a function of α, at α^*. In (b), the red dashed line represents the line of α^* values for different values of β. The cusp of the wedged region matches the critical point in phase transition theory. In fact, as magnified in (c), along the red phase transition line in (b), there is always a pitch-fork bifurcation at the critical cusp. In the thermodynamic limit, this is known as a second-order phase transition. Figure redrawn based on [221] with permission.

The solution to Eqn. (8.2) with initial value x_0, $\mathbf{X}_\varepsilon(t|x_0)$, has the following limits

$$\lim_{t \to \infty} \mathbf{X}_\varepsilon(t|x_0) = \mathbf{X}_\varepsilon^{ss}, \text{ in distribution}; \tag{8.7a}$$

$$\lim_{\varepsilon \to 0} \mathbf{X}_\varepsilon^{ss} = \hat{x}, \text{ in distribution}; \tag{8.7b}$$

$$\lim_{\varepsilon \to 0} \mathbf{X}_\varepsilon(t|x_0) = x(t|x_0), \text{ in distribution}; \tag{8.7c}$$

$$\lim_{t \to \infty} x(t|x_0) = x^*(x_0), \tag{8.7d}$$

in which $x(t|x_0)$ is the solution of Eqn. (8.1) with initial value x_0, and $x^*(x_0)$ is the stable steady state of (8.1) starting from x_0.

In a bistable system, $x^*(x_0)$ is not always the same as $\hat{x}$. The $\mathbf{X}^{ss}_{\varepsilon}$ in (8.7a) is independent of x_0 and ergodic; and $\hat{x}$ is, in general, unique except at certain critical values of (α,β). The $x^*(x_0)$ in (8.7d), however, is dependent upon x_0. With fixed β, Fig. 8.1a shows $\hat{x}(\alpha)$ in {orange, pink} colors. This function is unique for each value of α except for one critical value. In contrast, the values of $x^*(x_0;\alpha)$ are shown in {purple, pink} colors for one initial condition and {orange, turquois} colors for another initial condition. The relationship between $x^*(x_0;\alpha,\beta)$, $\hat{x}(\alpha,\beta)$, and the local and global minima of $\varphi^{ss}(x;\alpha,\beta)$ are shown by the shapes of the landscape function in green.

8.2 Nonlinear Stochastic Dynamics and Landau's Theory

Theoretical physicist Lev D. Landau (1908–1968) first proposed a theory for equilibrium phase transition in 1937. In terms of a 1-dimensional nonlinear ordinary differential equation (8.1) and its corresponding stochastic differential equation (8.2), let us consider the cubic function

$$b(x;\alpha,\beta) = -\left(x^3 + x^2 - \alpha x + \beta\right),\tag{8.8}$$

and a constant $A(x) = A$.

8.2.1 Ordinary Differential Equation With Bistability

The ordinary differential equation (8.1) $dx/dt = b(x)$ has fixed point(s), or steady state(s), as the roots of the cubic polynomial

$$b(x) = -\left(x^3 + x^2 - \alpha x + \beta\right) = 0.\tag{8.9}$$

Its solutions $x^*(\alpha,\beta)$, as a multilayer surface with a fold, is a function of α and β. In the (α,β) plane, the region in which $x^*(\alpha,\beta)$ takes three values has a wedged shape with a cusp (as illustrated in Figs. 8.1b and 8.2b). The boundary of the wedged region is given by a parametric equation with $-\infty < \xi < +\infty$,

$$\begin{cases} \alpha(\xi) = 3\xi^2 + 2\xi \\ \beta(\xi) = 2\xi^3 + \xi^2 \end{cases} \tag{8.10}$$

or, in two-branch form with $\pm$,

$$\beta_{\pm}(\alpha) = \frac{2}{9}\left(\alpha + \frac{1}{3}\right)\left(-1 \pm \sqrt{1+3\alpha}\right) - \frac{\alpha}{9}. \tag{8.11}$$

A cups is located at $\alpha = -\frac{1}{3}$ and $\beta = \frac{1}{27}$ when $\xi = -\frac{1}{3}$. At the cusp, $\alpha'(-\frac{1}{3}) = \beta'(-\frac{1}{3}) = 0$; thus, $d\beta/d\alpha = \frac{0}{0}$ is undetermined according to the parametric Eqn. (8.10). However, according to Eqn. (8.11), $d\beta_{\pm}/d\alpha = -\frac{1}{3}$ for both $\beta_+(\alpha)$ and $\beta_-(\alpha)$ as $\alpha \to (-\frac{1}{3})^+$, but only from above. $\beta_{\pm}(\alpha)$ are nonexistent for $\alpha < -\frac{1}{3}$; the second-order derivative diverges at the cusp.

8.2.2 Stochastic Differential Equation With Phase Transition

With $A(x) = A$ and letting $\varepsilon A^2/2 = V^{-1}$ in Eqn. (8.2), the corresponding Fokker–Planck equation is

$$\frac{\partial f(x,t)}{\partial t} = \frac{\partial}{\partial x}\left(\frac{1}{V}\frac{\partial f(x,t)}{\partial x} - b(x;\alpha,\beta)f(x,t)\right). \tag{8.12}$$

Its stationary solution is

$$\begin{aligned} f_V^{ss}(x) &= C_V \exp\left(V\int b(x)dx\right) \\ &= C_V \exp\left[-V\left(\frac{x^4}{4} + \frac{x^3}{3} - \frac{\alpha x^2}{2} + \beta x\right)\right], \end{aligned} \tag{8.13}$$

where C_V is a normalization constant

$$C_V^{-1} = \int_{-\infty}^{\infty} \exp\left[-V\left(\frac{x^4}{4} + \frac{x^3}{3} - \frac{\alpha x^2}{2} + \beta x\right)\right]dx. \tag{8.14}$$

We denote

$$\varphi^{ss}(x) = -\int b(x)dx = \frac{x^4}{4} + \frac{x^3}{3} - \frac{\alpha x^2}{2} + \beta x. \tag{8.15}$$

Here the global minimum of $\varphi^{ss}(x)$ is not always zero. However, the difference between this $\varphi^{ss}(x)$ and the exact large-deviation rate function is only a constant. This constant is immaterial, however, and thus, we shall call $\varphi^{ss}(x)$ the landscape function, or even sometimes the large-deviation rate function, throughout the book.

$\varphi^{ss}(x)$ has two minima separated by a maximum. In addition, the condition for the two minima being equal is a line in the $\alpha\beta$ plane:

$$\frac{\alpha}{3} + \beta + \frac{2}{27} = 0, \tag{8.16}$$

along which we have

$$\varphi^{ss}(x;\alpha) = \frac{1}{4}\left(x+\frac{1}{3}\right)^4 - \frac{1}{2}\left(\alpha+\frac{1}{3}\right)\left(x+\frac{1}{3}\right)^2 + C', \tag{8.17}$$

where C' is a constant. Note that the $\varphi^{ss}(x;\alpha)$ term in Eqn. (8.17) is an even function of $\tilde{x} = x + \frac{1}{3}$:

$$\varphi^{ss}(\tilde{x};\alpha) = \frac{\tilde{x}^4}{4} - \left(\alpha+\frac{1}{3}\right)\frac{\tilde{x}^2}{2} + C'. \tag{8.18}$$

This indicates that any cubic system can be transformed into Landau's canonical form, given in Eqn. (8.18). It has two minima of equal value when $\alpha > -\frac{1}{3}$; it turns into a single minimum at $\tilde{x} = 0$ when $\alpha < -\frac{1}{3}$. The $\varphi^{ss}(\tilde{x};\alpha)$ term has the canonical form of a pitch-fork bifurcation [255]. The line in Eqn. (8.16) is the "phase separation line"; its slope $d\beta/d\alpha = -\frac{1}{3}$ is consistent with Eqn. (8.11).

In the terminology of Landau's phase-transition theory, $\tilde{x}$ is called an order parameter, in terms of which the $b(x)$ in (8.9) takes on the simpler form

$$b(\tilde{x}) = -\left[\tilde{x}^3 - \frac{\tau}{3}\left(\tilde{x}-\frac{1}{3}\right) + \frac{J}{27}\right] = 0, \tag{8.19a}$$

in which $\tau = 3\alpha + 1$ and $J = 27\beta - 1$, and correspondingly,

$$\varphi^{ss}(\tilde{x},\tau,J) = \frac{\tilde{x}^4}{4} - \frac{\tau}{6}\left(\tilde{x}-\frac{1}{3}\right)^2 + \frac{J\tilde{x}}{27} + C'. \tag{8.19b}$$

From Eqn. (8.19), we easily see that when $\tau/9 + J/27 = \alpha/3 + \beta + 2/27 = 0$, φ^{ss} becomes an even function of $\tilde{x}$, confirming Eqns. (8.16) and (8.17). Along this line of phase transition points, e.g., the red-dashed line shown in Fig. 8.1b and 8.1c, the three fixed points are the roots of $(\tilde{x}^*)^3 - \tau\tilde{x}^*/3 = 0$. At the critical point, $\tau = 0$, and one obtains

$$\left(\frac{d\ln\tilde{x}^*}{d\ln\tau}\right)_{\tau=0} = \frac{1}{2}, \tag{8.20}$$

that is, $\tilde{x}^* \propto \tau^{\hat{\beta}}$ with exponent $\hat{\beta} = \frac{1}{2}$.

The φ^{ss} in (8.19b) is a "free energy" function in physicists' theory. Define α as "temperature" and $\tau = 3(\alpha + 1/3)$ as the reduced temperature; then $\tau = 0$ is the critical point near which, when substituting $\tilde{x}^* \propto \tau^{\hat{\beta}}$, we obtain $\varphi^{ss} \propto \tau^2$. The "heat capacity" $C = \alpha(\partial^2 \varphi^{ss}/\partial \alpha^2) \propto \tau^{\hat{\alpha}}$ with $\hat{\alpha} = 0$. Furthermore, from Eqn. (8.19a), one obtains $\tilde{x}(\tau, J)$ and let $\chi = \partial \tilde{x}/\partial J = (9\tau - 81\tilde{x}^2)^{-1}$. Then, $\chi \propto \tau^{-\gamma}$ with $\gamma = 1$. Finally, for $\tau = 0$, Eqn. (8.19a) becomes $\tilde{x}^3 + J/27 = 0$. Thus, $J \propto \tilde{x}^\delta$ with $\delta = 3$. We note that $\hat{\alpha}, \gamma, \delta$ satisfies $2 - \hat{\alpha} = \gamma \frac{\delta+1}{\delta-1}$ and that $\hat{\beta}, \gamma, \delta$ satisfies $\hat{\beta} = \frac{\gamma}{\delta-1}$. These calculations give a glimpse of the theory of mathematical singularity at a critical point in terms of scaling and exponents [73].

8.3 Symmetry Breaking, Nonlinear Catastrophe and Phase Transition

In the mathematical theory of deterministic nonlinear dynamics, symmetry breaking is intimately related to an unstable fixed point, which is involved in both problems of *saddle-node bifurcation* and *pitch-fork bifurcation*. It is important to recognize that the phenomenon of saddle-node bifurcation, and the associated catastrophe, is robust with respect to minor changes in the differential equation, while the phenomenon of pitch-fork bifurcation is not.

More specifically, the canonical ordinary differential equations (ODEs) representing the two types of bifurcations are [255]

$$\frac{dx}{dt} = \lambda - x^2 \qquad \text{saddle-node bifurcation} \tag{8.21}$$

$$\frac{dx}{dt} = x(\lambda - x^2) \qquad \text{pitchfork bifurcation} \tag{8.22}$$

where λ is the bifurcation parameter. It is easy to verify that for the system in (8.22), by employing the triple-root condition for a cubic equation, one sees that if $\varepsilon \neq 0$, the ODE $dx/dt = x(\lambda - x^2) - \varepsilon(x^2 - 1)$ will no longer have a pitchfork bifurcation but only a saddle-node type bifurcation with respect to changes in λ. On the other hand, loss of the pair of roots in (8.21) is preserved under perturbation by a generic $\varepsilon(x)$ term on the right-hand side of the equation as long as $\varepsilon(x)$ is sufficiently small. The theory of saddle-node bifurcation has a topological origin in terms of the *catastrophe theory*; it defines a symmetry breaking.

The mathematical results in Sections 8.1 and 8.2 teach us the following two important lessons:

(*i*) The phase transition does not occur near a nonlinear bifurcation point! Rather, it is "far from it" and is in between two nonlinear saddle-node bifurcation events. Actually, a local nonlinear bifurcation does not affect the global structure of a vector field; thus, it cannot have a real consequence in a physics theory on the phase transition phenomenon *per se*.

It is widely said that "quantitative differences can lead to a qualitative distinction". This is the essential feature of a bifurcation in nonlinear deterministic dynamics. We observe, however, that the qualitative topological distinction actually always involves a "definition", especially in the framework of stochastic dynamics. Qualitative distinctions, such as at the critical points of a saddle-node bifurcation, do not truly appear under stochastic conditions, even when noise tends to vanish. Similarly, at the critical point of the first-order phase transition where the two local minima of a landscape function are identical, the qualitative distinction can only be defined via a mathematical limit. While there are only very minor quantitative differences at the critical points of the first-order phase transition in a stochastic dynamical system with finite V and t, the differences are "amplified" in the limiting processes of $t \to \infty$ and $V \to \infty$, giving rise to a qualitative distinction.

(*ii*) Although a pitch-fork bifurcation in a nonlinear dynamical system is not robust in general, there is necessarily a hidden pitch-fork bifurcation associated with any **mesoscopic** catastrophe phenomenon! Eqn. (8.18) can be derived from the $\varphi^{ss}(x; \alpha, \beta)$ in (8.15), together with an emergent phase separation line (8.16). Note that the information on the phase separation line is not fully contained in function $b(x; \alpha, \beta)$; it is defined only when the landscape $\varphi^{ss}(x; \alpha, \beta)$ is known.

In the classical van der Waals theory for a nonideal gas, which predicts a phase transition from gas to liquid, the variable x is the equilibrium molar volume v of a box of gas, α and β are temperature (T) and pressure (p), respectively, and $x^*(\alpha, \beta)$ represents the roots of the van der Waals equation

$$\left(P + \frac{a}{v^2}\right)\left(v - b\right) = RT. \tag{8.23}$$

The equation $b(x; \alpha, \beta) = 0$ is known as an "equation of state" in van der Waals theory. The Maxwell construction is used in the absence of a stochastic model. In a biochemical phosphorylation feedback system, x is the fraction of phosphorylated protein, and α and β are the kinase activity and ATP phosphorylation potential, respectively [90, 91]. $b(x; \alpha, \beta) = 0$ provides an "equation of phosphorylation-dephosphorylation switching" for biochemical signaling [17].

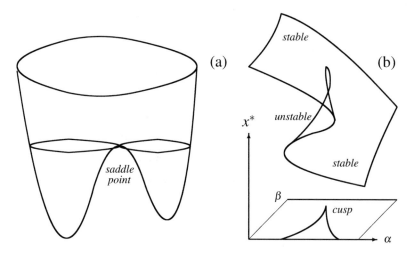

Fig. 8.2 Double wells with a saddle point and a three-layer fold with a cusp. (a) The surface represents the "potential function" $\varphi^{ss}(x)$. Moving from one well to another, the most likely path with the lowest "barrier" to overcome is through the saddle point, i.e., a mountain pass. (b) The fixed points of an ODE $dx/dt = b(x; \alpha, \beta)$ are the roots of equation $b(x; \alpha, \beta) = 0$. x^* as a function of α and β is a surface in 3-d. Folding a smooth surface into three layers, there is necessarily a cusp in the $\alpha\beta$ plane.

In Fig. 8.2b, projecting the three layer surface, which represents the root of $b(x; \alpha, \beta) = 0$, to the $\alpha\beta$ plane for the two parameters, topologist René Thom (1927–1975) had the deep insight that the region has to have a wedge shape with a cusp, as illustrated in Fig. 8.2b as well as in Fig. 8.1b [185]. Now, if one keeps β constant and varies α across the wedged region starting from the far left, as illustrated by the dashed blue line in Fig. 8.1b, the number of steady states changes from 1 to 2 to 3 then back to 2 and 1. This is seen in Fig. 8.1a. The multicolored S-shaped curve is a "bifurcation diagram", which shows the position of the steady state(s) as a smooth function of α with fixed β.

At the blue vertical lines in Fig. 8.1a, the changing number of steady states is called *saddle-node* bifurcation. For sufficiently small and sufficiently large values of α, the system has only a single steady state (fixed point). The blue dashed lines mark the critical α values at which there is the sudden appearance or disappearance of a pair of stable and unstable steady states.

The pair "bursts out of blue", thus acquiring the name "blue sky bifurcation" [255]. One of the most extensively studied examples of this type of behavior in biochemistry is forced molecular dissociation leading to noncovalent bond rupture [248].

We have described how bifurcation arises in bistable, nonlinear, deterministic systems. This theory is only valid for macroscopic systems. For a mesoscopic system with stochastic kinetics, the numbers of individuals of various species in system $\vec{N} = (n_1, n_2 \cdots)$ with volume V, are usually stochastic. The ODE perspective is only valid for infinitely large systems, i.e., introducing "concentration" $\vec{x}(t) = \vec{N}(t)/V$, and then mathematically taking $\vec{N}, V \to \infty$ to obtain a "macroscopic limiting behavior" in terms of the nonlinear dynamics for $\vec{x}(t)$. Then, in the limit as $t \to \infty$, multiple attractors are revealed. Different initial conditions will lead to different steady states. Because of this procedure, the transition between two basins of attraction, the most important consequence of fluctuations, is not possible in the deterministic analysis. There is a breakdown of ergodicity.

The *thermodynamic limit* requires a true equilibration among all the different attractors. In fact, the most important information missing in the ODE analysis is the relative probabilities for different attractors. This comes from analyses based on stochastic dynamics. A finite-size correction naturally introduces stochasticity. Depending upon the chosen representation for a system, a finite-size mesoscopic model can be either a discrete or continuous stochastic process. For example, the dynamic equation for a stochastic concentration $\mathbf{X}(t)$ can be characterized by $d\mathbf{X}(t) = b(\mathbf{X})dt + V^{-\frac{1}{2}}d\mathbf{W}(t)$, where V represents the size of the dynamical system, and $\mathbf{W}(t)$ is Brownian motion. When $V \to \infty$, the dynamics are reduced to the macroscopic limit.

In the limit as $t \to \infty$ followed by $V \to \infty$, i.e., in the true thermodynamic limit, the S-shaped curve in Fig. 8.1a is no more; only a discontinuous jump at α^*, marked by the red dashed vertical line, remains. This is reminiscent of the Maxwell construction for the van der Waals theory of a nonideal gas. Similarly, in Fig. 8.1b, the wedged region is no longer relevant; only the dashed red curve is important. This is known as a *first-order phase transition line*.

Let us now focus on the cusp in Fig. 8.1b. Moving along the red curve inside the wedged region, the landscape changes from symmetric bistable with two equal minima to a single, monostable minimum when passing the cusp. Magnified in Fig.

8.1c, this is Landau's *second-order phase transition*. In nonlinear dynamical systems theory, crossing the cusp constitutes a pitchfork bifurcation.

C. N. Yang and T. D. Lee were the first to establish a general mathematical origin for phase transition [277]. They showed that nonanalyticity is a necessary feature of any rigorous phase transition theory. Furthermore, it is related to a zero of a partition function moving from the complex plane onto the real axis in the limit as $V \to \infty$ (recall that the free energy is the logarithm of the partition function.) Recently, it was demonstrated that this same mathematical description applies to any bistable system with stochastic elements [89, 91], including a bistable, mesoscopic biochemical system. Therefore, the notion of phase transition, together with concepts such as symmetry breaking and the perspective of the "true thermodynamic limit" have broad applicability to systems exhibiting phenomena such as catastrophe, rupture, and hysteresis. It is a complementary description of bistability in the presence of stochasticity.

The notion of symmetry breaking, extensively discussed by physicists like P. W. Anderson and J. J. Hopfield [7, 123], is intimately related to nonlinear bifurcation. The existence of multistability with multiple attractors is often an emergent phenomenon itself, arising as a consequence of nonlinear interactions. Our foregoing discussions illustrate that with stochasticity, bifurcations in the true thermodynamic limit exhibit phase transitions; even more interestingly, symmetry is in a probability distribution (here, the meaning of symmetry is not strict, only indicating the bimodal nature of the distribution according to the presence of two attractors); symmetry breaking is included in a particular system as a realization of the probability.

8.4 Hierarchical Organization: Different Levels and Time Scales

One of the insights from the present chapter is that stochasticity does not completely disappear in a reasonably large, mesoscopic nonlinear population-based kinetic system. Rather, it manifests itself as a stochastic jump process on a much longer time scale among a set of discrete states. These discrete states are attractors of an interacting nonlinear dynamical system, their existence, locations, and transition times all emergent properties. Well-known examples include nonlinearity cooperativity in equilibrium physics and feedback in biological networks. In the former, cooperativity leads to crystallization; in the latter, which are nonequilibrium systems, the feedback lead to self-organization.

The emergent stochastic dynamics, of course, become the stochastic elements for a higher level system in a larger space with longer time. This is illustrated in Fig. 8.3. L. Boltzmann started the tradition of treating molecular collisions as stochastic elements in kinetic theory. Kramers showed that discrete chemical reactions can be described as Brownian motion in a force field. The stochastic population kinetic approach to chemical reactions, in terms of Delbrück–Gillespie processes, considers each chemical reaction as a stochastic process and derives the "states" and "dynamics" of cell-sized biochemical reaction *systems*.

In a famous article [7], Anderson stated that "at each level of complexity entirely new properties appear, and the understanding of the new behaviors requires research which I think is as fundamental in its nature as any other." He further listed a series of increasing levels of complexity:

$$
\begin{aligned}
&\text{few-body (elementary particle) physics} \\
&\text{many-body (condensed-matter) physics} \\
&\text{chemistry} \\
&\text{molecular biology} \\
&\text{cell biology} \\
&\text{physiology} \\
&\qquad\vdots \\
&\text{psychology} \\
&\text{social sciences} \\
&\qquad\vdots
\end{aligned}
\tag{8.24}
$$

In this hierarchy, the elementary entities at each level, e.g., atoms in chemistry and macromolecules in cell biology, obey the dynamic laws of the lower level. Then, at each level, proper and entirely new laws, concepts, and generalizations emerge, and thus different treatments and theories are necessary. Stochastic population kinetics, together with the notions of nonlinear attractors, basins of attractions, discrete transitions between basins, etc., nicely provide a mathematical description that crosses these hierarchical levels. Interestingly, this entire hierarchical organization shares many features with the organizational hierarchy among protein conformational substates, a theory advanced by H. Frauenfelder, P. G. Wolynes, and their coworkers [79]: Going downward, a protein consists of secondary structural motifs, which consist of amino acids, which in turn consist of atoms, etc. Going upward, a cell consists of a large number of macromolecules, and tissue consists of a large population of cells.

Together with the theory of proteins, the stochastic chemical reaction system theory developed in the present book serves as a paradigm for stochastic, mesoscopic, complex behavior [160]. The chemical reaction theory involves *three* of the levels described in (8.24): (*a*) few-body physics, detailing the collisions of a few water molecules with a few atoms within the reacting molecules; (*b*) many-body physics, concerning molecules A, B and C in a sea of solvent molecules; and (*c*) chemistry, whose elementary events are discrete chemical transitions such as $A \rightarrow B$. In (*a*), one is concerned with collisions that lead to high-frequency vibrations; the primary concern of (*b*) is the *mechanism* and the process by which molecules collide, interact, and form in terms of the atoms in these molecules with vibrations and diffusion, while in (*c*), the actual chemical reaction mainly involves only on a rate constant for the discrete transition. Figs. 8.3a presents schematics that illustrate this hierarchy.

The theory for the rate of transition $A \rightarrow B$, modeled as a stochastic diffusion process crossing a barrier defined by a saddle point and developed by Kramers and discussed in Section 4.6, provides a paradigm. The stochastic theory reveals different time scales in the problem: at the time scale of molecular reactions, (*a*) is so rapid that it can be essentially treated as infinitely rapid fluctuating dynamics with certain appropriate statistics. The time scale for (*b*) is of course determined by the energy and force in the molecular system, which Kramers called "a field of force", and the outcome of the theory is a discrete event of a chemical reaction whose time scale, $\sim 10^{-7}$ sec., is almost infinite on the time scale of (*a*), $\sim 10^{-12}$ sec.

At the level of cell biology, the individual biochemical reactions, such as the conformational changes of transcription factors, including covalent modifications and molecular bindings such as $A + B \rightarrow C$, are just part of rapid "noise". In terms of reaction kinetics, the steady states of a biochemical reaction system can be predicted. More importantly, single-cell phenotype switching is precisely the stochastic transitions from one biochemical steady state to another. Fig. 8.3b shows a schematic illustrating this concept. There is an obvious parallel between Fig. 8.3a and b.

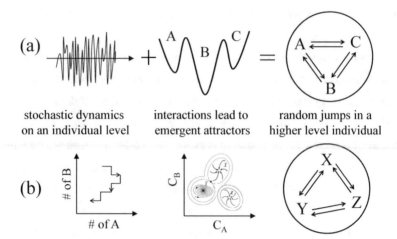

Fig. 8.3 (a) Schematics showing the rapid solvent-macromolecule collisions that, as a source of stochasticity and together with a multienergy-well landscape, gives rise to a kinetic jump process for an individual macromolecule with multiple states (shown in the circle). (b) One level higher, many interacting macromolecular individuals, each with multiple discrete states, form mesoscopic nonlinear reaction systems. In the space of copy numbers and concentrations of chemical species, such a system undergoes Monte Carlo walk due to the presence of each and every stochastic reaction with emergent multiple nonlinear attractors (C_A, C_B represent concentrations of A and B). The stochastic transitions within each macromolecule serve as "noise" leading to phenotypic-state switching at the whole-reaction system level, shown by X, Y, and Z. (a & b) The same schematic illustrates Anderson's hierarchy of complexity: there are stochastic uncertainties in the dynamics of an "individual", be it a macromolecule in an aqueous solution, a cell in tissue, an animal in an ecological environment, or a trading company in an economic system. Interactions between individuals in a system form a nonlinear dynamic system with emergent attractors. The fundamental insight of Kramers' theory is that, while the law of large numbers is at work, there will be emergent, discrete stochastic jumps beyond infinite time of deterministic nonlinear dynamics at the system level at an evolutionary time scale. At a level higher and at a much longer time scale, such a system is just a complex individual; the emergent discrete jumps, shown as transitions among the states within a stochastic individual on the right, constitute the stochastic elements for the nonlinear kinetics of a much larger system that consists of many such interacting individuals. From the first row in (a) to the second row in (b), this hierarchy repeats for complex systems, consisting of complex individuals that themselves are complex systems. Figure redrawn based on [221] with permission.

8.4.1 Symmetry Breaking As the Underpinning of Stochastic Population Kinetics

Hopfield defined "dynamical broken symmetry" as a nonlinear stochastic system that stays in one of the basins with a finite lifetime [123]: the two sides of a saddle point are considered "symmetric". In chemistry, one of the deepest concepts developed in connection to chemical reactions is the notion of the "transition state", which is precisely the saddle point of the nonlinear dynamics. We see that it is at the very transition state that the dynamics have a broken symmetry. If a molecular system is infinitely large, then this symmetry breaking is permanent: the odds of crossing Kramers' barrier are so low that it might not occur until the age of the universe. However, a macromolecule such as a protein can in fact jump among its different *conformational states* at a time scale observable in a laboratory and exhibit successive *dynamic symmetry breaking* [123]. In this latter case, a discrete-state stochastic description of the kinetics in terms of a Markov jump process, with exponential, memoryless waiting time, emerges. As the ODE description of classical dynamics, the Markov description of stochastic transitions within and between complex individuals is general. The dynamic symmetry breaking is the theoretical underpinning of our stochastic population kinetics.

Dynamic symmetry breaking is at the heart of mesoscopic behavior, and a mathematical issue is crucial to such systems. Let us quote directly from Anderson again:

> It is only as the nucleus is considered to be a many-body system — in what is often called the $V \to \infty$ limit — that such [emergent] behavior is rigorous definable.

In the thermodynamic limit in equilibrium physics of matters, the time-limit $t \to \infty$ is taken first, followed by the systems size-limit $V \to \infty$ afterward. Actually, the $t \to \infty$ limit never appears in the equilibrium statistical thermodynamics. On the other hand, a frequently observed nonlinear dynamic behavior in a macroscopic system is the taking of the $V \to \infty$ limit first for finite t. As we have shown abundantly clearly, the nonlinear, emergent dynamic behavior of a complex system can *only* be rigorously defined with the limit $V \to \infty$. In reality, both limits are simple idealizations; each limiting procedure is valid on an appropriate time scale.

Part III
Stochastic Kinetics of Biochemical Systems and Processes

In 2013, M. Karplus, M. Levitt, and A. Warshel shared the Nobel Prize in Chemistry "for the development of multiscale models for complex chemical systems." This was the culmination of a nearly half-century endeavor [163] to push the frontier of Newtonian mechanics toward understanding the macromolecular basis of biochemistry, particularly the structural and dynamic foundation of the biological functions of polypeptides and nucleic acids. One learns from these studies that the atomic motions in a macromolecule at room temperature are highly stochastic; a single enzyme molecule is already a nearly infinitely complex individual.

Similarly, the first two decades of the new millennium have witnessed a growing awareness of the "individualities" of single cells [202]. Stochastic gene expression has revealed a rich repertoire of nongenic heterogeneity that could play important roles in developmental biology, bacterial antibiotic resistance, and cancer evolution [15, 76, 168].

A single protein molecule is a "complex individual" [79], as are single cells. However, the former are only "dead" materials, but the latter are alive [13].

Equipped with the stochastic nonlinear kinetic description of biochemical reaction systems and the nonequilibrium chemical thermodynamcs that thus emerge, we now turn our attention to several key biochemical applications in Part III, which includes enzyme kinetics, macromolecular mechanics, linear and nonlinear stochastic reaction systems, and stochastic gene expression kinetics in single cells based on the central dogma of molecular biology [166].

Chapter 9
Classic Enzyme Kinetics — The Michaelis–Menten and Briggs–Haldane Theories

9.1 Enzyme: Protein as a Catalyst

Many chemical reactions inside a cell are very slow if the reactants are simply mixed alone as a multicomponent aqueous solution in a test tube. In the later part of the 19th century, people discovered that if one put ground yeast into a test tube, the reaction was significantly accelerated, its speed increasing by a factor of 10^{10} and greater! For example, the chemical conversion of dihydroxyacetone phosphate to D-glyceraldehyde 3-phosphate is approximately 10^{11} times faster in the presence of something found in yeast, an enzyme called triose-phosphate isomerase. The extraction of substances such as this to nearly 100% purity dominated the field of *biochemistry* for more than half of a century. Interested readers should consult standard textbooks on the subject [243, 208, 71, 274, 41]; they provided a great education for the first author of this book years ago.

In an *enzyme purification* process, biochemists follow "biochemical activity", by which we mean the ability to speed up a reaction or change the color of a solution. In the above yeast-based experiments, they employed all kinds of methods to separate the molecules from ground yeast into fractions by size, weight, charge, and affinity toward specific substances while keeping track of the biochemical activity. Eventually, they were able to purify a substance that formed a crystal when lyophilized. That was the beginning of biochemistry and molecular biology. Analyzing the crystals in terms of the atoms within and their relative spatial positions gave rise to the field of macromolecular structural biology.

9.2 Product Formation Rate and Double-Reciprocal Relation

The magical substance within yeast mentioned above was named "enzyme", the Greek word for "in yeast". Quantitative studies of enzymes soon followed. When mixing a specific amount of enzyme with varying concentrations of a reactant S (substrate is the jargon in biochemistry), it was discovered that the rate of product formation, J^+, was a nonlinear function of $[S]$, the concentration of reactant S. This does not conform to the law of mass action. Indeed, $\frac{1}{J^+}$ and $\frac{1}{[S]}$ exhibited a simple linear relation, as shown in Fig. 9.1. This is known as the double-reciprocal relation in enzyme kinetics.

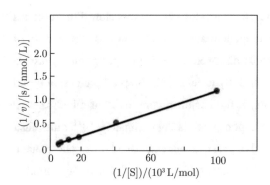

Fig. 9.1 Double-reciprocal relation of enzyme kinetics.

9.3 Michaelis–Menten (MM) Theory

One only expects the law of mass action to work for an *elementary reaction*. The enzymatic reaction, thus, must be nonelementary and involve multiple steps. In 1913, Leonor Michaelis and Maud Menten proposed a kinetic scheme, a mechanism, for how an enzymatic reaction might work [178]:

$$E + A \underset{k_{-1}}{\overset{k_1}{\rightleftharpoons}} EA \overset{k_2}{\longrightarrow} E + B, \tag{9.1}$$

in which E is the enzyme, and A is its substrate. When A is in a "pocket" of the very large enzyme molecule, where the dielectric constant is very low, the transformation of A to B, the product of the reaction, is significantly faster than it would be on its own when surrounded by water, which has a high dielectric constant. The reaction $EA \to EB$ is so fast that we can ignore the EB state and directly consider the release of B.

In chemistry, one usually uses $[A]$, $[E]$ and $[EA]$ to denote the concentrations of chemical species A, E, and EA, respectively. Without further statements, we usually assume that a biochemical reaction is in a well-mixed vessel. This is certainly not a very realistic assumption for a cell, but it provides biochemists with a good starting point. The field of enzymology, which studies enzymes in test tubes, became an important source of information on enzymes, their structures, kinetics, and functions. For a given amount of E, increasing the amount of A, eventually, will cause its concentration to greatly exceed that of E: $[A] \gg [E]$. However, since the amount of complex EA is limited by the total amount of E, E_{tot}, the second reaction in Eqn. (9.1) eventually reach its maximum value: $k_2 E_{tot}$. This is the origin of the nonlinear dependence of the rate of product formation on the substrate concentration $[A]$.

9.3.1 Michaelis–Menten Kinetic Equation

We now introduce the following notations:

$$[A] = a, \quad [E] = e, \quad [EA] = c, \quad [B] = b, \quad e + c = e_0, \quad a + b + c = a_0. \tag{9.2}$$

The Michaelis–Menten theory assumes all the reactions in (9.1) are elementary. Then, we have the differential equations according to the LMA kinetics described in Chapter 2:

$$\frac{da}{dt} = -k_1 e_0 a + (k_1 a + k_{-1})c, \tag{9.3}$$

$$\frac{dc}{dt} = k_1 e_0 a - (k_1 a + k_{-1} + k_2)c, \tag{9.4}$$

with initial values

$$a(0) = a_0, \quad b(0) = c(0) = 0, \quad e(0) = e_0. \tag{9.5}$$

Introducing the following nondimensional variables and parameters:

$$u(\tau) = a(t)/a_0, \quad v(\tau) = c(t)/e_0, \quad \tau = k_1 e_0 t, \tag{9.6}$$

the system of ordinary differential equations becomes

$$\frac{du}{d\tau} = -u + (u + K - \lambda)v, \tag{9.7a}$$

$$\varepsilon \frac{dv}{d\tau} = u - (u + K)v, \tag{9.7b}$$

$$u(0) = 1, \tag{9.7c}$$

$$v(0) = 0, \tag{9.7d}$$

where

$$\lambda = k_2/(k_1 a_0), \quad K = (k_{-1} + k_2)/(k_1 a_0), \quad \varepsilon = e_0/a_0.$$

We note that there is an ε in front of $\frac{dv}{d\tau}$. In a cell, an enzymatic reaction involved in the central metabolism usually has a much greater total substrate concentration than that of the total enzyme: $a_0 \gg e_0$. Therefore, $\varepsilon \ll 1$.

9.3.2 Singular Perturbation Problem: Examples

If $\varepsilon \ll 1$ in Eqn. (9.7b), we can simply set it to zero, obtaining

$$v = \frac{u}{u + K}, \tag{9.8}$$

which we then substitute into Eqn. (9.7a), yielding

$$\frac{du}{d\tau} = -u + \frac{u + K - \lambda}{u + K} u = -\frac{\lambda u}{u + K}. \tag{9.9}$$

Eqn. (9.9) can then be solved analytically by the method of separation of variables. However, is this valid mathematically? We do notice that with Eqn. (9.8), $v(0) = \frac{u(0)}{u(0)+K} = \frac{1}{1+K}$, which is not 0. Hence, the equation is problematic, at least for values of τ very near 0.

There are two significant applied mathematical ideas that can be implemented to further the analysis, known as *asymptotic expansion* and *singular perturbation* [184, 196]. To illustrate this problem, let us digress and first consider a simple algebraic equation, $x^2 - bx + \varepsilon = 0$, and find its roots. If we let $\varepsilon = 0$, we immediately have $x_{1,2} \approx b, 0$. In fact, one can do better by carrying out a procedure called the *perturbation method*, which assumes

$$x = \sum_{k=0}^{\infty} q_k \varepsilon^k \tag{9.10}$$

and substituting this into the quadratic equation to obtain:

$$\{q_0^2 + \varepsilon 2q_0q_1 + \varepsilon^2(q_1^2 + 2q_0q_2) + O(\varepsilon^3)\} - b\{q_0 + \varepsilon q_1 + \varepsilon^2 q_2 + O(\varepsilon^3)\} + \varepsilon = 0.$$

Rearranging the terms on the right-hand side, we have

$$\{q_0^2 - bq_0\} + \varepsilon\{2q_0x_1 - bq_1 + 1\} + \varepsilon^2\{q_1^2 + 2q_0q_2 - bq_2\} + O(\varepsilon^3) = 0. \tag{9.11}$$

Eqn. (9.11) gives a system of an infinite number of linear algebraic equations, except for the first one for q_0, with a lower triangle structure; for all those q's:

$$q_0^2 - bq_0 = 0, \quad 2q_0q_1 - bq_1 + 1 = 0, \quad q_1^2 + 2q_0q_2 - bq_2 = 0, \quad \cdots, \tag{9.12}$$

and we obtain

$$q_0^{(1)} = b, \quad q_0^{(2)} = 0;$$
$$q_1^{(1)} = -\frac{1}{b}, \quad q_1^{(2)} = \frac{1}{b};$$
$$q_2^{(2)} = -\frac{1}{b^3}, \quad q_2^{(2)} = \frac{1}{b^3};$$
$$\cdots$$

These equations can be compared with Taylor's expansions of the exact result for the two roots based on the quadratic equation:

$$\frac{b \pm \sqrt{b^2 - 4\varepsilon}}{2} \approx \frac{b \pm b(1 - 2\varepsilon/b^2 - 2\varepsilon^2/b^4)}{2}$$
$$= \left\{ \left(b - \frac{\varepsilon}{b} - \frac{\varepsilon^2}{b^3} \right), \ \left(\frac{\varepsilon}{b} + \frac{\varepsilon^2}{b^3} \right) \right\}. \tag{9.13}$$

The perturbation method was the main methodology for engineering computations before the arrival of the computer.

However, how does one determine the approximate roots of $\varepsilon x^2 + bx + c = 0$? If we let $\varepsilon = 0$, we have $x = -c/b$; thus, we lost one of the two roots of the quadratic equation. Let us see where it has gone:

$$x_{1,2} = \frac{-b \pm \sqrt{b^2 - 4\varepsilon c}}{2\varepsilon}$$

$$\approx \left\{ -\left(\frac{c}{b} + \varepsilon \frac{c^2}{b^3}\right), \; -\left(\frac{b}{\varepsilon} - \frac{c}{b} - \varepsilon \frac{c^2}{b^3}\right) \right\}. \tag{9.14}$$

The missing root, which is singular with respect to ε, goes to ∞. Such an abrupt change in the behavior of the solution, i.e., the loss of a root, is known as a singular perturbation problem.

9.3.3 Singular Perturbation Theory: Outer and Inner Solutions and Their Matching

The above example suggests that one can proceed with letting $\varepsilon = 0$ but with caution, since there will be something missing. By the method of separation of variables, the Eqn. (9.9) we have obtained can be solved together with the initial value $u(0) = 1$ to further obtain

$$u + K \ln u = 1 - \lambda \tau. \tag{9.15}$$

u decreases with τ and approaches 0 when $\tau \to \infty$. At $\tau = 0$, $u(0) = 1$, $u'(0) = -\frac{\lambda}{1+K} < 0$, $u''(0) = -\frac{\lambda K}{(1+K)^2} < 0$. For large τ, $u''(\tau) > 0$.

Now let us see what happens to $v(\tau)$. According to Eqn. (9.8)

$$v(\tau) = \frac{u(\tau)}{u(\tau) + K}. \tag{9.16}$$

Thus, $v(0) = \frac{1}{1+K}$ and $v'(0) = \frac{K}{(1+K)^2}$. $v(\tau)$ does not start at 0, as required by the initial condition. Actually we never had the opportunity to use the initial value.

This is a characteristic symptom of singular perturbed ordinary differential equations when one naively lets $\varepsilon = 0$. A more careful inspection of Eqn. (9.7) reveals that at $\tau = 0$, $\frac{dv}{d\tau} = \frac{1}{\varepsilon}$ is very large. If we let $\varepsilon = 0$, then the time derivative is infinite. One needs a good way to investigate such fast-increasing kinetics.

We can go to a shorter time scale as follows. Instead of working with time τ, we can work with time $\sigma = \frac{\tau}{\varepsilon}$ and rewrite the differential equations in Eqn. (9.7) in terms of

$$\sigma = \frac{\tau}{\varepsilon}, \quad U(\sigma) = u(\varepsilon \sigma), \quad V(\sigma) = v(\varepsilon \sigma). \tag{9.17}$$

Then, the equations in (9.7) become

$$\frac{dU}{d\sigma} = \varepsilon(-U + (U + K - \lambda)V),$$ (9.18a)

$$\frac{dV}{d\sigma} = U - (U + K)V$$ (9.18b)

$$U(0) = 1$$ (9.18c)

$$V(0) = 0$$ (9.18d)

There is still an ε term, but the problem is no longer singular. It can be solved by a regular perturbation method, in which the lowest order solution is obtained by setting $\varepsilon = 0$.

This gives us

$$U(\sigma) = U(0) = 1, \quad V(\sigma) = \frac{1 - e^{-(1+K)\sigma}}{1 + K}.$$ (9.19)

Therefore, the time scale for the fast-increasing $v(t)$ is $\frac{1+K}{\varepsilon}$.

When $\sigma \to \infty$, $V(\sigma) \to \frac{1}{1+K}$ and $U(\sigma) \to 1$. This matches the $v(0)$ and $u(0)$ obtained at the slow time scale.

The solution at time scale τ is called the outer solution to the singular perturbation problem, and the solution at time scale σ is called the inner solution. The solutions match at $\sigma = \infty$ and $\tau = 0$, which gives us a complete quantitative description for the kinetics of the MM model. In mathematical terms, one has a "uniformly valid" solution for $\tau \in [0, \infty)$:

$$u(\tau) = \begin{cases} 1 & \tau \sim \varepsilon \\ \tilde{u}(\tau) & \tau \gg \varepsilon \end{cases}, \quad v(\tau) = \begin{cases} \dfrac{1 - e^{-(1+K)\tau/\varepsilon}}{1 + K} & \tau \sim \varepsilon \\ \dfrac{\tilde{u}(\tau)}{\tilde{u}(\tau) + K} & \tau \gg \varepsilon \end{cases},$$ (9.20)

in which $\tilde{u}(\tau)$ is implicitly defined as the solution in Eqn. (9.15).

The approximated result can be improved by seeking higher order results in ε. This is the entire subject within applied mathematics known as singular perturbation and matched asymptotics [196].

9.3.4 MM kinetics, Saturation and Bimolecular Reactions

The above discussions show that for large time scale problems, the outer solution to the MM equation is quite accurate approximation. This is the justification for using the standard MM equation, i.e., Eqn. (9.9), in realistic bioengineering computations:

$$\frac{d[B]}{dt} = -\frac{d[A]}{dt} = \frac{k_2 e_0 [A]}{K_m + [A]}, \tag{9.21}$$

for enzyme kinetics. The parameter $K_m = \frac{k_{-1} + k_2}{k_1}$ is called the Michaelis constant for the enzyme with respect to the substrate molecule. The MM equation (9.21) is not valid if the total concentration of the enzyme e_0 is not much smaller than the total concentration of substrate a_0. In this case, $\frac{d[B]}{dt} \neq -\frac{d[A]}{dt}$.

The hyperbolic function in the right-hand side of Eqn. (9.21) has two distinct regimes: one that increases with substrate concentration $[A]$ when $[A] \ll K_m$ and one that is independent of the substrate concentration when $[A] \gg K_m$. In the former regime, we have effectively a bimolecular reaction

$$\frac{d[B]}{dt} = -\frac{d[A]}{dt} \approx \frac{k_2}{K_m} e_0 [A] \quad \text{when } [A] \ll K_m \tag{9.22}$$

with an effective bimolecular rate constant k_2/K_m. In the latter regime, the reaction is first order with respect to the enzyme concentration:

$$\frac{d[B]}{dt} = -\frac{d[A]}{dt} \approx k_2 e_0 \quad \text{when } [A] \gg K_m. \tag{9.23}$$

For a given amount of enzyme, the latter is zeroth order with respect to substrate A.

9.4 Briggs–Haldane Reversible Enzymatic Reaction

Briggs and Haldane generalized the kinetic scheme in Eqn. (9.1) to reversible enzymatic reactions [31]:

$$E + A \underset{k_{-1}}{\overset{k_1}{\rightleftharpoons}} EA \underset{k_{-2}}{\overset{k_2}{\rightleftharpoons}} E + B. \tag{9.24}$$

Again, assuming that both the substrate A and product B have concentrations much greater than that of the enzyme, one can perform the mathematical analysis parallel to that in Sec. 9.3 and obtain

$$-\frac{d[A]}{dt} = \frac{d[B]}{dt} = \frac{\dfrac{V_1 [A]}{K_1} - \dfrac{V_2 [B]}{K_2}}{1 + \dfrac{[A]}{K_1} + \dfrac{[B]}{K_2}}, \tag{9.25}$$

in which

$$K_1 = \frac{k_{-1}+k_2}{k_1}, \; K_2 = \frac{k_2+k_{-1}}{k_{-2}}, \; V_1 = k_2 e_0, \; V_2 = k_{-1} e_0. \qquad (9.26)$$

For an enzymatic reaction at the chemical equilibrium between A and B, we have

$$-\frac{d[A]}{dt} = \frac{d[B]}{dt} = 0 \Leftrightarrow \frac{k_1 k_2 [A]^{eq}}{k_{-1} k_{-2} [B]^{eq}} = 1 \Leftrightarrow \left(\frac{K_2 V_1}{K_1 V_2}\right) = \frac{[B]^{eq}}{[A]^{eq}}, \qquad (9.27)$$

as expected from the detailed balance condition. The last relation in (9.27) is called the Haldane equation.

9.5 Allosteric Cooperativity

The hyperbolic curve from the quasisteady-state treatment of MM kinetics, i.e., the outer solution in the singular perturbation analysis, figured prominently in the previous section. There are other reaction mechanisms that also produce such a dependence on substrate concentration. Let us consider the simplest protein-substrate *binding* reaction:

$$P + S \underset{k_{-1}}{\overset{k_1}{\rightleftharpoons}} PS. \qquad (9.28)$$

What is the fraction of protein in PS-complex form in equilibrium? The answer to this question follows a simple calculation: given $k_1 [S]^{eq} [P]^{eq} = k_{-1} [PS]^{eq}$, due to the equilibrium condition, and $[P]^{eq} + [PS]^{eq} = P_{tot}$, we have

$$\frac{[PS]^{eq}}{P_{tot}} = \frac{K_{eq}[S]^{eq}}{1 + K_{eq}[S]^{eq}}, \quad K_{eq} = \frac{k_1}{k_{-1}}. \qquad (9.29)$$

K_{eq} is called the *binding constant* in biochemistry. If we denote $[PS]^{eq}/P_{tot} = f$, the fraction of saturation, and $x = [S]^{eq}/[S]_{\frac{1}{2}}$, where $[S]_{\frac{1}{2}}$ is the amount of $[S]^{eq}$ when $f = \frac{1}{2}$, then we have $[S]_{\frac{1}{2}} = K_{eq}^{-1}$, and

$$f = \frac{x}{1+x}. \qquad (9.30)$$

However, in the 1950s, measurements on the binding of oxygen to purified hemoglobin showed that

$$f = \frac{x^\nu}{1+x^\nu}, \qquad (9.31)$$

with $\nu \approx 2.6$. Curves like this are called *sigmoidal* shaped, and the exponent ν is often called Hill's coefficient, after English biophysicist Archibald V. Hill (1886–1977). Experimentalists often obtain Hill's coefficient, n_h, from the slope in the middle point of the saturation curve, $f(x)$, in a log-log plot:

$$n_h = 2 \left(\frac{\mathrm{d} \ln f}{\mathrm{d} \ln x} \right)_{f=\frac{1}{2}}. \tag{9.32}$$

$n_h > 1$ is called positive cooperativity, and $n_h < 1$ is called negative cooperativity.

What is the possible biomolecular mechanism that produces a sigmoidal-shaped curve? A possible simple model is

$$P + S \underset{k_{-1}}{\overset{k_1}{\rightleftharpoons}} PS, \quad PS + S \underset{k_{-2}}{\overset{k_2}{\rightleftharpoons}} PS_2. \tag{9.33}$$

This gives

$$\frac{[PS]^{eq}}{[P]^{eq}[S]^{eq}} = \frac{k_1}{k_{-1}}, \quad \frac{[PS_2]^{eq}}{[P]^{eq}([S]^{eq})^2} = \frac{k_2}{k_{-2}} \frac{[PS]^{eq}}{[P]^{eq}[S]^{eq}} = \frac{k_1 k_2}{k_{-1} k_{-2}}. \tag{9.34}$$

Therefore,

$$f = \frac{[PS]^{eq} + 2[PS_2]^{eq}}{2([P]^{eq} + [PS]^{eq} + [PS_2]^{eq})} = \frac{K_1[S]^{eq} + 2K_1 K_2([S]^{eq})^2}{2 + 2K_1[S]^{eq} + 2K_1 K_2([S]^{eq})^2}, \tag{9.35}$$

where $K_1 = \frac{k_1}{k_{-1}}$, $K_2 = \frac{k_2}{k_{-2}}$.

Hill's coefficient $n_h = \frac{2}{1 + \sqrt{K_1/(4K_2)}}$. Hence, if $K_2 > K_1/4$, then $n_h > 1$ is positive cooperativity, and if $K_2 < K_1/4$, then $n_h < 1$ is negative cooperativity. In fact, when $K_2 = K_1/4$, Eqn. (9.35) reduces to Eqn. (9.29) with $K_{eq} = K_1/2 = 2K_2$. Protein P in (9.33) has two binding sites for substrate S, which are not independent if $K_2 \neq K_1/4$. If there are N binding sites, then n_h can be as large as N.

9.5.1 Combinatorics of Identical and Independent Subunits

We see that when $K_2 = K_1/4$, the above mechanism in Eqn. (9.33) is reduced to the simple binding in Eqn. (9.28). This turns out to be a very general and important statistical result that we shall discuss.

Let us consider a protein P with N subunits each contains a binding site for substrate S. Let us use α_i and β_j for the rate constants for the binding:

$$PS_{k-1} + S \underset{\beta_k}{\overset{\alpha_k}{\rightleftharpoons}} PS_k, \quad k = 1, \cdots, N. \tag{9.36}$$

Therefore, the fraction of saturation curve

$$f = \frac{\sum_{k=0}^{N} k[PS_k]^{eq}}{N\left(\sum_{k=0}^{N}[PS_k]^{eq}\right)}$$

$$= \frac{\frac{\alpha_1}{\beta_1}z + 2\frac{\alpha_1\alpha_2}{\beta_1\beta_2}z^2 + \cdots + N\frac{\alpha_1\cdots\alpha_N}{\beta_1\cdots\beta_N}z^N}{N\left(1 + \frac{\alpha_1}{\beta_1}z + \cdots + \frac{\alpha_1\cdots\alpha_N}{\beta_1\cdots\beta_N}z^N\right)}$$

$$= \frac{\partial \ln Q(z)}{N\partial \ln z}$$

in which $z = [S]^{eq}$ and $Q(z)$ is called a binding polynomial:[1]

$$Q(z) = 1 + \frac{\alpha_1}{\beta_1}z + \cdots + \frac{\alpha_1\cdots\alpha_N}{\beta_1\cdots\beta_N}z^N. \qquad (9.37)$$

If all the subunits in the protein are identical and independent, and each on its own has binding constant $K_{eq} = \frac{k_1}{k_{-1}}$, where k_1 and k_{-1} are the rate constants, as shown in Eqn. (9.28), then a simply combinatorial calculation leads to $\alpha_k = k_1(N - k + 1)$ and $\beta_k = k_{-1}k$. Then, we have the binding polynomial

$$Q(z) = \sum_{k=0}^{N} \binom{N}{k} (K_{eq}z)^k = (1 + K_{eq}z)^N. \qquad (9.38)$$

The corresponding fraction of saturation curve $f = \frac{K_{eq}z}{1+K_{eq}z}$, which is the same as Eqn. (9.30).

For example, when $N = 2$, we have for identical and independent subunits:

$$\alpha_1 = 2k_1, \quad \beta_1 = k_{-1}, \quad \alpha_2 = k_1, \quad \beta_2 = 2k_{-1}.$$

This is the condition for $K_2 = \frac{K_1}{4}$ with $K_2 = \frac{k_1}{2k_{-1}}$ and $K_1 = \frac{2k_1}{k_{-1}}$. In contrast, allosteric cooperativity means that the binding of the second substrate, also called the ligand, is influenced by the binding of the first substrate that leads to a change in the protein structure. If the influence is positive, we say there is positive cooperativity; and if the influence is negative, we say the cooperativity is negative.

Positive cooperativity is similar to positive feedback. It produces a sigmoidal-shaped binding curve [217]. In particular, if $K_1 \approx 0$ and $K_2 \approx \infty$ such that K_1K_2 is finite, then we obtain the binding curve in Eqn. (9.31) with $v = 2$. The binding of the

---

[1] Biophysical chemists discovered this mathematical approach independently on their own [275]; but it is really only a part of J. W. Gibbs' theory of *grand partition function*. This was well recognized by Terrell L. Hill (1917–2014) and John A. Schellman (1924–2014), see the opening page of Chapter 4 in [118] and [239].

first ligand is very difficult; but when it is bound, the second ligand binds extremely easy. Thus, there is a positive cooperativity.

9.5.2 *Mathematical Definition of Molecular Cooperativity*

The above analysis gives us a mathematical definition of a cooperative binding system: for N identical and independent subunits, each of which binds to a ligand, we have the rate constants that satisfy the condition

$$\frac{i\alpha_i}{(N-i+1)\beta_i} = \text{constant.} \tag{9.39}$$

Any deviation from this relation necessarily means there are nonindependent subunits. Thus, some kind of cooperativity must be present.

We see that a noncooperative system has a binomial distribution for a population with k bound ligands. This is intimately related to the concept of *independent and identically distributed* (i.i.d.) random variables. The above discussion is based on the same mathematics.

Chapter 10
Single-Molecule Enzymology and Driven Biochemical Kinetics with Chemostat

The new millennium has witnessed a resurgence of interest in the theory of enzyme kinetics due to several developments in biochemical research: The foremost was the systems approach to cell biology, which demands quantitative representations of cellular enzymatic reactions in terms of Michaelis–Menten (MM)-like kinetics. Second, advances in single-molecule enzymology since the late 1990s have generated exquisite information on protein dynamics in connection to enzyme catalysis [173, 276]. Third, the most relevant to the present chapter, the theoretical advances in our understanding of open, driven biochemical reaction systems in terms of the stochastic theory of the nonequilibrium steady state (NESS) [213, 86, 100, 220, 258].

Single-molecule enzymology and enzymatic reactions inside cells have showed the necessity of characterizing enzyme reactions in terms of stochastic models. On the theory side, an alternative to the traditional approach to enzyme kinetics that is particularly applicable to stochastic single-molecule enzyme measurements was developed in parallel to the measurements themselves. In this chapter, motivated by single-molecule experiments, we will introduce the concepts of and derive mathematical expressions for *cycle fluxes*, *cycle completion times*, and *stepping probabilities*. Furthermore, we shall discuss the classic Haldane equality in light of the NESS as well as an unexpected generalization of this equality. The generalized Haldane equation is part of a collection of new results in nonequilibrium statistical physics called fluctuation theorems [230, 135].

In Sections 10.3, 10.4, and 10.5 we shall give three concrete examples of the application of NESS theory to single-molecule enzymology. All operate, at the indi-

© The Author(s), under exclusive license to Springer Nature Switzerland AG 2021
H. Qian, H. Ge, *Stochastic Chemical Reaction Systems in Biology*,
Lecture Notes on Mathematical Modelling in the Life Sciences,
https://doi.org/10.1007/978-3-030-86252-7_10

vidual enzyme level, as "living biochemical processes" that exhibit nonequilibrium phenomena that are also essential to their biological functions. It is widely believed in current molecular biology that the biochemical function of a protein molecule is derived from its structure. While this is certainly true in a general sense, our examples explicitly show how function(s) of a molecule can change depending upon the NESS in which the molecular system is situated. The three examples we shall study are:

(1) A same-enzyme modifier that switches between being an activator and an inhibitor of an enzyme [134];

(2) A dynamic cooperativity that gives rise to an enzymatic response sharper than that of MM, without using the traditional mechanism of allosteric cooperativity; and

(3) A kinetic proofreading mechanism in which the specificities of different ligands that associate with the same enzyme is not determined by their equilibrium affinities [122, 191].

All these examples together illustrate the importance of nonstructural-based molecular regulations in NESS. Specifically, we shall show that all these phenomena will disappear in chemical equilibrium; thus, they depend critically upon nonequilibrium kinetic processes with energy dissipation, rather than merely the structural properties of macromolecules themselves.

10.1 Single-Molecule Irreversible Michaelis–Menten Enzyme Kinetics

10.1.1 Product Arrival As a Renewal Process

The traditional Michaelis–Menten kinetics

$$E + S \underset{k_{-1}}{\overset{k_1}{\rightleftharpoons}} ES \xrightarrow{k_2} E + P, \qquad (10.1)$$

when there is only a single enzyme, and when one considers only a single turnover event from the left to the right of (10.1), can be rewritten as $E \rightleftharpoons ES \longrightarrow E^+$. If we use $p_E(t)$, $p_{ES}(t)$, and $p_{E^+}(t)$ as the probabilities of the three states, respectively, then we have

$$\begin{cases} \dfrac{\mathrm{d}p_E(t)}{\mathrm{d}t} = -k_1[S]p_E + k_{-1}p_{ES}, \\[2mm] \dfrac{\mathrm{d}p_{ES}(t)}{\mathrm{d}t} = k_1[S]p_E - (k_{-1}+k_2)p_{ES}, \\[2mm] \dfrac{\mathrm{d}p_{E^+}(t)}{\mathrm{d}t} = k_2 p_{ES}, \end{cases} \tag{10.2}$$

with initial conditions $p_E(0) = 1$, $p_{ES}(0) = 0$. Furthermore, $p_{E^+}(t) = P\{\mathbf{T} \leq t\}$, where $\mathbf{T}$ is the waiting time for the arrival of the next product, or the *cycle time* of the enzyme.

The linear ODE system in Eqn. (10.2) has three eigenvalues: 0, $-\lambda_1$ and $-\lambda_2$:

$$\lambda_{1,2} = \frac{(k_1[S]+k_{-1}+k_2) \pm \sqrt{(k_1[S]+k_{-1}+k_2)^2 - 4k_1k_2[S]}}{2}. \tag{10.3}$$

Both $\lambda_1, \lambda_2 > 0$. The solution to Eqn. (10.2), thus, is

$$p_{E^+}(t) = a_0 + a_1 e^{-\lambda_1 t} + a_2 e^{-\lambda_2 t}, \tag{10.4}$$

in which the different as are determined from the conditions

$$p_{E^+}(0) = 0, \quad \frac{\mathrm{d}p_{E^+}}{\mathrm{d}t}(0) = k_2 p_{ES}(0) = 0, \quad \text{and} \quad p_{E^+}(\infty) = p_E(0) = 1. \tag{10.5}$$

This yields

$$p_{E^+}(t) = 1 + \frac{\lambda_2 e^{-\lambda_1 t} - \lambda_1 e^{-\lambda_2 t}}{\lambda_1 - \lambda_2}, \tag{10.6}$$

and the probability density function (pdf) for $\mathbf{T}$

$$f_T(t) = \frac{\mathrm{d}}{\mathrm{d}t}p_{E^+}(t) = \frac{\lambda_1\lambda_2\left(e^{-\lambda_1 t} - e^{-\lambda_2 t}\right)}{\lambda_2 - \lambda_1}. \tag{10.7}$$

If the magnitude of one λ is much larger than that of the other, say $\lambda_2 \gg \lambda_1$, then we have

$$f_T(t) \approx \lambda_1 e^{-\lambda_1 t}, \tag{10.8}$$

which is a single exponential. This is what biochemists called *only one rate-limiting step*. In this case, the arrival of successive production steps is a simple Poisson process with mean waiting time λ_1^{-1}. It has also been called a *Poisson stepper*.

In general, however, the waiting time distribution is not a single exponential; its precise shape contains more information on the detailed enzyme kinetic mechanism, as shown for example in Eqn. (10.1). The successive arrivals of products form a *renewal process*, by which we mean a sequence of repeated events with a random but statistically identical waiting time, which is sampled independently from the same $\mathbf{T}$ with distribution $f_T(t)$.

10.1.2 Mean Waiting Time and Steady-state Flux

From the probability density function for $\mathbf{T}$ in Eqn. (10.7), one can easily compute the mean waiting time for the MM mechanism:

$$\mathbb{E}[\mathbf{T}] = \int_0^\infty t f_T(t) \mathrm{d}t = \frac{\lambda_1 + \lambda_2}{\lambda_1 \lambda_2}. \tag{10.9}$$

We note that $\lambda_1 + \lambda_2 = k_1[S] + k_{-1} + k_2$ is the trace of the matrix

$$\begin{pmatrix} -k_1[S] & k_{-1} \\ k_1[S] & -(k_{-1} + k_2) \end{pmatrix},$$

which represents the linear system in Eqn. (10.2), and $\lambda_1 \lambda_2 = k_1 k_2[S]$ is its determinant. Hence,

$$\mathbb{E}[\mathbf{T}] = \frac{k_1[S] + k_{-1} + k_2}{k_1 k_2[S]} = \frac{1}{k_2} + \frac{k_{-1} + k_2}{k_1 k_2[S]}. \tag{10.10}$$

This is exactly the "double-reciprocal relation" of the Michaelis–Menten kinetics, where $K_m = (k_{-1} + k_2)/k_1$ and $V_{max} = k_2$ for a single enzyme.

While the arrival time interval follows $f_T(t)$, the single-enzyme molecule randomly jumps between states E and ES. This two-state Markov process has a steady-state, or stationary, probability distribution

$$p_E^{ss} = \frac{k_{-1} + k_2}{k_1[S] + k_{-1} + k_2}, \quad p_{ES}^{ss} = \frac{k_1[S]}{k_1[S] + k_{-1} + k_2}. \tag{10.11}$$

Therefore, the steady-state flux

$$J^{ss} = k_1[S]p_E^{ss} - k_{-1}p_{ES}^{ss} = k_2 p_{ES}^{ss} = \frac{k_1 k_2[S]}{k_1[S] + k_{-1} + k_2}. \tag{10.12}$$

Comparing Eqns. (10.10) and (10.12), we note that $1/J^{ss} = \mathbb{E}[\mathbf{T}]$: J^{ss} is the number of product arrivals per unit time, and $\mathbb{E}[\mathbf{T}]$ is the mean duration between consecutive product arrivals. The equation in fact is general for any renewal process; it is known as the *elementary renewal theorem*.

There is a different method for directly computing the mean, or expected value $\mathbb{E}[\mathbf{T}]$, from the reaction scheme in Eqn. (10.1). We note that by adding the mean time step by step:

$$\mathbb{E}[\mathbf{T}] = \frac{1}{k_1[S]} + \frac{1}{k_{-1}+k_2} + \frac{k_{-1}}{k_{-1}+k_2}\mathbb{E}[\mathbf{T}] + \frac{k_2}{k_{-1}+k_2}0. \tag{10.13}$$

From this, we obtain the same $\mathbb{E}[\mathbf{T}]$ given in Eqn. (10.10). This derivation is very insightful: it shows that if there is only one unbound enzyme form E, then no matter how complex the kinetics for the bound enzymes ES_1, ES_2, etc., $\mathbb{E}[\mathbf{T}] = 1/(k_1[S]) + c_1\mathbb{E}[\mathbf{T}] + c_2$, where c_1 and c_2 are independent of $[S]$. Hence, $\mathbb{E}[\mathbf{T}] = a/[S] + b$ has the same reciprocal relation as the simplest Michaelis–Menten mechanism, and $J^{ss} = \frac{[S]}{a+b[S]}$ is a hyperbolic function of $[S]$. This observation partly explains why the hyperbolic enzyme rate has been so widely observed empirically. To obtain something nonhyperbolic, the unbound form of an enzyme has to have at least two different conformations, E_1 and E_2, with slow conversions between them.

The same insight also suggests to us a simple example for a nonhyperbolic enzyme rate: an enzyme with two different states, nonexchangable at the time scale of interest. Hence, one simply treats them as a mixture with a fraction c in state 1 and a fraction $(1-c)$ in state 2. Then

$$J^{ss} = \frac{cV_{max}^{(1)}[S]}{K_M^{(2)}+[S]} + \frac{(1-c)V_{max}^{(1)}[S]}{K_M^{(2)}+[S]}, \tag{10.14}$$

where $0 < c < 1$. It is easy to verify that as a function of $[S]$, J^{ss} is not hyperbolic; in other words, $1/J^{ss}$ is not a linear function of $1/[S]$. However, the curvature of J^{ss} is always negative. So there is no positive "cooperativity".

10.2 Single-molecule Reversible Michaelis–Menten Enzyme Kinetics

We now consider a three-step mechanism for an enzymatic reaction in which the conversion of a substrate S into a product P occurs at the catalytic site of the enzyme:

$$E + S \underset{k_{-1}}{\overset{k_1^0}{\rightleftharpoons}} ES \underset{k_{-2}}{\overset{k_2}{\rightleftharpoons}} EP \underset{k_{-3}^0}{\overset{k_3}{\rightleftharpoons}} E + P. \tag{10.15}$$

If there is only one enzyme molecule, then from the enzyme perspective, the kinetics are stochastic and cyclic, as shown in Fig. 10.1(a), with pseudo first-order rate constants $k_1 = k_1^0 c_S$ and $k_{-3} = k_{-3}^0 c_P$, where c_S and c_P are the sustained concentrations of substrate S and product P, respectively, in a steady state.

At chemical equilibrium, the concentrations of S and P satisfy $c_P/c_S = k_1^0 k_2 k_3 / \left(k_{-1} k_{-2} k_{-3}^0\right)$, i.e.,

$$\frac{k_1 k_2 k_3}{k_{-1} k_{-2} k_{-3}} = 1.$$

This is the detailed balance condition, also known as the thermodynamic "box" in elementary chemistry. However, if c_S and c_P are maintained at some constant levels that are not at chemical equilibrium, as metabolite concentrations are in living cells, then the enzyme reaction is in an open system, which will eventually reach a NESS. This is the scenario for most enzyme kinetics in a living, homeostatic cell.

In this case,

$$\frac{k_1 k_2 k_3}{k_{-1} k_{-2} k_{-3}} = \gamma \neq 1, \tag{10.16}$$

and in fact $\triangle\mu = k_B T \ln\gamma$ is the chemical potential difference between P and S, independent of the enzyme (see Section 2.2).

From the perspective of a single enzyme molecule, the probabilities of the states follow a master equation

$$\frac{dP_E(t)}{dt} = -(k_1 + k_{-3})P_E(t) + k_{-1}P_{ES}(t) + k_3 P_{EP}(t),$$

$$\frac{dP_{ES}(t)}{dt} = k_1 P_E(t) - (k_{-1} + k_2)P_{ES}(t) + k_{-2}P_{EP}(t),$$

$$\frac{dP_{EP}(t)}{dt} = k_{-3}P_E(t) + k_2 P_{ES}(t) - (k_{-2} + k_3)P_{EP}(t). \tag{10.17}$$

The steady-state probabilities for states E, ES and EP are easily computed by setting the time derivative to zero and noting that $P_E + P_{ES} + P_{EP} = 1$ for the total probability:

$$P_E^{ss} = \frac{k_2 k_3 + k_{-1}k_3 + k_{-1}k_{-2}}{\mathscr{D}},$$

$$P_{ES}^{ss} = \frac{k_1 k_3 + k_{-2}k_{-3} + k_1 k_{-2}}{\mathscr{D}},$$

$$P_{EP}^{ss} = \frac{k_1 k_2 + k_2 k_{-3} + k_{-1}k_{-3}}{\mathscr{D}},$$

in which the denominator

$$\mathscr{D} = k_1 k_2 + k_2 k_3 + k_3 k_1 + k_{-1}k_{-3} + k_{-2}k_{-3} + k_{-1}k_{-2} + k_1 k_{-2} + k_2 k_{-3} + k_3 k_{-1}.$$

Then, the clockwise steady-state cycle flux in Fig. 10.1(a), which is precisely the enzyme turnover rate, e.g., the number of turnovers per time, of $S \to P$ in the reaction scheme (10.15), $J^{ss} = P_E^{ss}k_1 - P_{ES}^{ss}k_{-1} = P_{ES}^{ss}k_2 - P_{EP}^{ss}k_{-2} = P_{EP}^{ss}k_3 - P_E^{ss}k_{-3}$, which follows

$$J^{ss} = \frac{k_1 k_2 k_3 - k_{-1}k_{-2}k_{-3}}{\mathscr{D}} = J^{ss}_{+} - J^{ss}_{-}, \tag{10.18}$$

where

$$J^{ss}_{+} = \frac{k_1 k_2 k_3}{\mathscr{D}}, \quad J^{ss}_{-} = \frac{k_{-1}k_{-2}k_{-3}}{\mathscr{D}}, \tag{10.19}$$

are the forward and backward cycle fluxes, respectively.

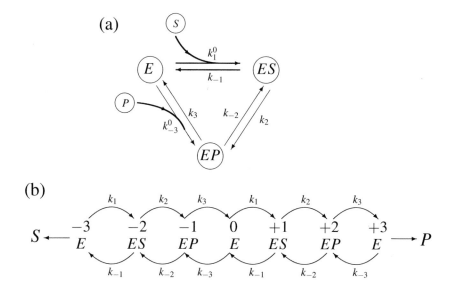

Fig. 10.1 (a) Kinetic scheme of a simple reversible enzyme reaction, in which k_1^0 and k_{-3}^0 are second-order rate constants. From the perspective of a single enzyme molecule, the reaction is unimolecular and cyclic, with first-order and pseudo first-order rate constants $k_1 = k_1^0 c_S$, $k_{-3} = k_{-3}^0 c_P$. (b) Kinetic scheme for computing the cycle completion times $\mathbf{T}$, $\mathbf{T}_{+}$ and $\mathbf{T}_{-}$. To distinguish the forward and backward cycles, cyclic reaction (a) is transformed into a one-dimensional random walk in (b). See text for details. Figure redrawn based on [100] with permission.

The net cycle flux is just the Michaelis–Menten steady-state flux in Fig. 10.1(a), i.e.,

$$v = \frac{V_S \frac{c_S}{K_{mS}} - V_P \frac{c_P}{K_{mP}}}{1 + \frac{c_S}{K_{mS}} + \frac{c_P}{K_{mP}}},$$

with maximal velocities

$$V_S = \frac{k_2 k_3}{k_{-2} + k_2 + k_3}, \quad V_P = \frac{k_{-1}k_{-2}}{k_{-2} + k_2 + k_{-1}}, \tag{10.20}$$

and Michaelis constants

$$K_{mS} = \frac{k_{-1}k_{-2} + k_{-1}k_3 + k_2 k_3}{k_1^0(k_{-2} + k_2 + k_3)}, \quad K_{mP} = \frac{k_{-1}k_{-2} + k_{-1}k_3 + k_2 k_3}{(k_{-2} + k_2 + k_{-1})k_{-3}^0}. \tag{10.21}$$

In addition, J_+^{ss} and J_-^{ss} are the average numbers of forward and backward cycles per time, respectively, due to ergodicity [137], i.e.,

$$J^{ss} = \lim_{t \to \infty} \frac{1}{t} v(t), \quad J_+^{ss} = \lim_{t \to \infty} \frac{1}{t} v_+(t), \quad J_-^{ss} = \lim_{t \to \infty} \frac{1}{t} v_-(t), \tag{10.22}$$

where $v_+(t)$ and $v_-(t)$ are the number of occurrences of forward and backward cycles up to time t, and $v(t) = v_+(t) - v_-(t)$.

Before closing this section, it is important to highlight that the quantity γ can be approximated by $v_+(t)/v_-(t)$ in a single-molecule experiment when time t is sufficiently large, because $\gamma = J_+^{ss}/J_-^{ss}$. One also has $J_+^{ss} = J_-^{ss}$, i.e., $\gamma = 1$, if and only if the enzymatic reaction is at chemical equilibrium.

10.2.1 Mean Waiting Cycle Times

In single-molecule enzyme kinetic studies, the most salient feature of either the substrate turnover time courses or the enzyme cyclic trajectories is that they are stochastic. The stochasticity, however, is exhibited in the amount of time "waiting" for a chemical reaction, a discrete event that occurs via a "fluctuating" diffusion encounter and thermal activation.

The amount of time actually taken by a transition itself, e.g., the atomic and molecular movements that accomplish a chemical structural change, is often on the order of subpicoseconds, which is considered instantaneous in enzyme kinetics in an aqueous solution. Therefore, in a single-molecule experiment, the observed state as a function of time corresponds to the "stochastic waiting time" for a reaction. Once statistical time course data is in hand, the most straightforward analysis of the trajectories clearly reveals the two distributions of the on-time and off-time. Therefore, in our theoretical model, we shall first define the waiting cycle times and then calculate their means, variances, and ultimately distributions.

Starting from the free enzyme state E, three kinds of waiting cycle times can be defined. Let $\mathbf{T}$ represent the waiting time for the completion of a forward or a backward cycle, $\mathbf{T}_+$ represent the waiting time for only the occurrence of a forward cycle, and $\mathbf{T}_-$ represent the waiting time for only the occurrence of a backward cycle. Obviously, of these three random variables, $\mathbf{T}$ is just the smaller of $\mathbf{T}_+$ and $\mathbf{T}_-$.

The problem of computing the mean waiting time $\langle \mathbf{T} \rangle$ is in fact the same as in the mean first-passage-time (MFPT) problem (Fig. 10.1b) for a random walk. This is a perfect application of the classical problem of probability to single-enzyme kinetics.

Let τ_i be the mean first hitting time for state 3 or -3 in Fig. 10.1(b) starting from state i. Obviously, $\langle \mathbf{T} \rangle = \tau_0$ and $\tau_3 = \tau_{-3} = 0$.

Then, we need to derive the equations for $\{\tau_i\}$. Starting from state i, the system will first wait for an average exponential time $\frac{1}{q_{i,i-1}+q_{i,i+1}}$, where q_{ij} is just the reaction constant from state i to state j, then jump to either state $i-1$ or state $i+1$ according to the ratio of their reaction constants. Hence, τ_i would be the sum of $\frac{1}{q_{i,i-1}+q_{i,i+1}}$ and the probability weighted mean first hitting time starting from $i-1$ or $i+1$.

Hence, τ_i satisfies the following equations with boundary $\mathbb{E}[\mathbf{T}] = \tau_0$ and $\tau_3 = \tau_{-3} = 0$:

$$\tau_{-2} = \frac{1}{k_{-1}+k_2} + \frac{k_{-1}}{k_{-1}+k_2} \times 0 + \frac{k_2}{k_{-1}+k_2}\tau_{-1},$$

$$\tau_{-1} = \frac{1}{k_{-2}+k_3} + \frac{k_{-2}}{k_{-2}+k_3}\tau_{-2} + \frac{k_3}{k_{-2}+k_3}\tau_0,$$

$$\tau_0 = \frac{1}{k_{-3}+k_1} + \frac{k_{-3}}{k_{-3}+k_1}\tau_{-1} + \frac{k_1}{k_{-3}+k_1}\tau_1,$$

$$\tau_1 = \frac{1}{k_{-1}+k_2} + \frac{k_{-1}}{k_{-1}+k_2}\tau_0 + \frac{k_2}{k_{-1}+k_2}\tau_2,$$

$$\tau_2 = \frac{1}{k_{-2}+k_3} + \frac{k_{-2}}{k_{-2}+k_3}\tau_1 + \frac{k_3}{k_{-2}+k_3} \times 0. \tag{10.23}$$

Through a simple calculation, one can obtain

$$\mathbb{E}[\mathbf{T}] = \frac{1}{J_+^{ss}+J_-^{ss}}, \tag{10.24}$$

where J_+^{ss} and J_-^{ss} are given in Eqn. (10.19). Similarly, another mean waiting cycle time $\mathbb{E}[\mathbf{T}_+]$, which is the mean time to complete the forward cycle in Fig. 10.1(a) either before or after completing cycling in the opposite direction, can be obtained as the solution of equations identical to Eqn. (10.23) but with different boundary conditions. Let τ_{i+} be the mean time first hitting state 3, whether before or after the time hitting state -3 in Fig. 10.1(b), starting from state i. Obviously, $\mathbb{E}[\mathbf{T}_+] = \tau_{0+}$, $\tau_{3+} = 0$ and $\tau_{-3+} = \tau_{0+}$.

Then,

$$\mathbb{E}[\mathbf{T}_+] = \frac{1}{J_+^{ss}},$$

where J_+^{ss} is given in Eqn. (10.19). With almost the same derivation, one can compute $\mathbb{E}[\mathbf{T}_-]$, which is the mean time to complete the backward cycle in Fig. 10.1(a), whether before or after complete cycling in the opposite direction. It immediately follows that

$$\mathbb{E}[\mathbf{T}_-] = \frac{1}{J_-^{ss}},$$

where J_-^{ss} is given in Eqn. (10.19). These relations between mean waiting time and steady flux can also be derived through elementary renewal theorem.

In fact, one can obtain the expression for $\mathbb{E}[\mathbf{T}_-]$, e.g., Eqn. (10.19), directly based on the expression for $\mathbb{E}[\mathbf{T}_+]$, e.g., Eqn. (10.19), according to the symmetry of the random walk in Fig. 10.1(b): $(k_1, k_{-1}, k_2, k_{-2}, k_3, k_{-3}) \rightarrow (k_{-3}, k_3, k_{-2}, k_2, k_{-1}, k_1)$.

We see that $\mathbb{E}[\mathbf{T}_+] = \mathbb{E}[\mathbf{T}_-]$ if and only if this system is at chemical equilibrium because of $\gamma = \mathbb{E}[\mathbf{T}_-]/\mathbb{E}[\mathbf{T}_+]$. Consequently, γ can also be measured by the ratio of averaged forward and backward waiting cycle times up to time t in the single-molecule experiment, which is different from its previous measurement $\gamma = \frac{v_+(t)}{v_-(t)}$. Nevertheless, as indicated in Eqn. (10.22), by applying the elementary renewal theorem, the two methods are asymptotically identical because $\mathbb{E}[\mathbf{T}_+]\rangle \approx t/v_+(t)$ and $\mathbb{E}[\mathbf{T}_-] \approx t/v_-(t)$ when t is large.

10.2.2 Stepping Probabilities

Let $v_+(t)$ and $v_-(t)$ be the number of forward and backward cycles, respectively being completed in the time interval $[0,t]$. Then, stepping frequencies

$$f^+(t) = \frac{v_+(t)}{v_+(t)+v_-(t)}, \; f^-(t) = \frac{v_-(t)}{v_+(t)+v_-(t)}, \tag{10.25}$$

can be obtained statistically in an experiment. According to Eqn. (10.22), one can obtain the eventual stepping probabilities

$$p^+ \triangleq \lim_{t\to\infty} f^+(t) = \frac{J_+^{ss}}{J_+^{ss}+J_-^{ss}} = \frac{k_1 k_2 k_3}{k_1 k_2 k_3 + k_{-1} k_{-2} k_{-3}}, \tag{10.26a}$$

$$p^- \triangleq \lim_{t\to\infty} f^-(t) = \frac{J_-^{ss}}{J_+^{ss}+J_-^{ss}} = \frac{k_{-1} k_{-2} k_{-3}}{k_1 k_2 k_3 + k_{-1} k_{-2} k_{-3}}. \tag{10.26b}$$

It is necessary to highlight here that the stepping frequencies in Eqn. (10.25) are random variables that depend on the trajectories, while their variances, e.g., fluctuations, vanish as t tends to infinity. Hence, the eventual stepping probabilities p^+ and p^- are independent of the trajectories due to ergodicity.

Interestingly, the forward stepping probability can also be defined in terms of cycle completion time as $p^+ \triangleq P_{\{E\}}\{T_+ < T_-\}$, which describes the probability that the enzyme first completes a forward cycle, starting from the initial free enzyme state E, before a backward cycle. Similarly, the backward stepping probability can be defined as $p^- \triangleq P_{\{E\}}\{T_- < T_+\}$.

This equivalence can be seen explicitly through translating this problem to the corresponding random walk shown in Fig. 10.1(b). Let p_{i+} be the probability of hitting state 3 before state -3 in Fig. 10.1(b), starting from state i. Obviously, $p_{3+} = 1$ and $p_{-3+} = 0$.

Again applying the strong Markov property of Markov chains as we did in the precious section, $\{p_{i+}\}$ satisfies the following equations:

$$p_{-2+} = \frac{k_{-1}}{k_{-1}+k_2} \times 0 + \frac{k_2}{k_{-1}+k_2} p_{-1+},$$

$$p_{-1+} = \frac{k_{-2}}{k_{-2}+k_3} p_{-2+} + \frac{k_3}{k_{-2}+k_3} p_{0+},$$

$$p_{0+} = \frac{k_{-3}}{k_{-3}+k_1} p_{-1+} + \frac{k_1}{k_{-3}+k_1} p_{1+},$$

$$p_{1+} = \frac{k_{-1}}{k_{-1}+k_2} p_{0+} + \frac{k_2}{k_{-1}+k_2} p_{2+},$$

$$p_{2+} = \frac{k_{-2}}{k_{-2}+k_3} p_{1+} + \frac{k_3}{k_{-2}+k_3} \times 1.$$

Through a similar calculation, one can obtain

$$p^+ = P_{\{E\}}\{T_+ < T_-\} = p_{0+} = \frac{k_1 k_2 k_3}{k_1 k_2 k_3 + k_{-1}k_{-2}k_{-3}},$$

and

$$p^- = P_{\{E\}}\{T_+ > T_-\} = 1 - P_{\{E\}}\{T_+ < T_-\} = \frac{k_{-1}k_{-2}k_{-3}}{k_1 k_2 k_3 + k_{-1}k_{-2}k_{-3}}.$$

Consequently,

$$p^+ = \frac{J_+^{ss}}{J_+^{ss}+J_-^{ss}} = \frac{\mathbb{E}[T]}{\mathbb{E}[T_+]}, \quad p^- = \frac{J_-^{ss}}{J_+^{ss}+J_-^{ss}} = \frac{\mathbb{E}[T]}{\mathbb{E}[T_-]},$$

and the chemical potential difference

$$\triangle \mu = k_B T \ln \gamma = k_B T \ln \frac{p^+}{p^-} = k_B T \log \frac{J_+^{ss}}{J_-^{ss}} = k_B T \ln \left(\frac{\mathbb{E}[\mathbf{T}_-]}{\mathbb{E}[\mathbf{T}_+]} \right), \tag{10.27}$$

which follows $p^+ = p^-$ if and only if the enzyme reaction is at chemical equilibrium.

10.2.3 Haldane Equality and Its Generalization

In enzyme kinetics, the Haldane equality is between the forward Michaelis constant K_M, the forward maximum velocity V_{max}, and their backward counterparts. Using the results in Eqs. (10.20) and (10.21), we see that

$$\frac{V_S/K_{mS}}{V_P/K_{mP}} = \frac{k_1^0 k_2 k_3}{k_{-1} k_{-2} k_{-3}^0}. \tag{10.28}$$

The right-hand side is independent of the enzyme. In other words, $(V_S/K_{mS})c_S = \gamma (V_P/K_{mP})c_P$. The Haldane equality in Eqn. (10.28), thus, is simply an alternative expression for Eqn. (10.27) since

$$\left(1 + \frac{c_S}{K_{mS}} + \frac{c_P}{K_{mP}} \right) \frac{K_{mS}}{V_S c_S} = \mathbb{E}[\mathbf{T}_+],$$

and

$$\left(1 + \frac{c_S}{K_{mS}} + \frac{c_P}{K_{mP}} \right) \frac{K_{mP}}{V_P c_P} = \mathbb{E}[\mathbf{T}_-]!$$

An interesting observation from single-molecule enzyme kinetics-generalized Haldane equality can be made: although the mean values of cycle times $\mathbf{T}_-$ and $\mathbf{T}_+$ can be very different, the entire probability distributions of $\mathbf{T}_-$ and $\mathbf{T}_+$, under the condition that the corresponding cycle is completed before the completion of the opposite cycle, are identical, and both are the same as that of $\mathbf{T}$. In fact, waiting cycle time $\mathbf{T}$ is *independent* of whether enzyme E completes a forward cycle or a backward cycle, although the probability of these two cycles might be rather different (i.e., $\gamma \neq 1$).

We introduce a one-to-one "quasi-time-reversal" mapping r for the trajectory of the simple kinetics in Fig. 10.1(a), which belongs to the event $\{\mathbf{T}_+ < \mathbf{T}_-\}$, mapped to its "quasi-time-reversal" mapping [100].

For each trajectory $\omega = \{\omega_t : t \geq 0,\ \omega_0 = \{E\}\}$ belonging to the set $\{\mathbf{T}_+ < \mathbf{T}_-\}$, let $\mathbf{T}^*$ be the *last time* when it leaves the $\{E\}$ before finishing a forward cycle in Fig. 10.1(a). Then, its "quasi-time-reversal" trajectory $r\omega = \{(r\omega)_t : t \geq 0\}$ is defined as follows:

i) when time t is before or equal to $\mathbf{T}^*$, then one just copies ω to $r\omega$, i.e., $(r\omega)_t = \omega_t$;

ii) when time t is between $\mathbf{T}^*$ and $\mathbf{T}_+$, then one maps the real time-reversal trajectory of ω with respect to the time interval $[\mathbf{T}^*, \mathbf{T}_+]$ to $r\omega$, i.e., $(r\omega)_t = \omega_{\mathbf{T}^* + \mathbf{T}_+ - t}$;

iii) when time t is greater than $\mathbf{T}_+$, then one can also simply copy ω to $r\omega$ as in i).

See Fig. 10.2 for an illustrative example. As noted in the figure caption, $\mathbf{T}^*$ is denoted as the *last time* when the trajectory leaves the state $\{E\}$ before finishing a forward cycle $E \to ES \to EP \to E$. Then, *the ratio* of the probability density of the above trajectory to its "quasi-time-reversal" probability density below is

$$\gamma = \frac{k_1 k_2 k_{-2} k_2 k_3}{k_{-3} k_{-2} k_2 k_{-2} k_{-1}} = \frac{k_1 k_2 k_3}{k_{-1} k_{-2} k_{-3}} = \gamma.$$

Now, it is indispensable to explain why we construct the above mapping like this.

1) The number of $E \to ES$, $ES \to EP$ and $EP \to E$ steps in the original trajectory ω that belong to $\{\mathbf{T}_+ < \mathbf{T}_-\}$ is one more than that in the corresponding "quasi-time-reversal" trajectory $r\omega$ that belong to $\{\mathbf{T}_+ > \mathbf{T}_-\}$, while the number of $ES \to E$, $EP \to ES$ and $E \to EP$ steps in ω is one less than that in $r\omega$;

2) The dwell time upon reaching each state of the trajectory ω and its corresponding "quasi-time-reversal" trajectory $r\omega$ is mapped quite well such that the differences between ω and $r\omega$ are only exhibited upon their sequences of states.

Consequently, the most important observation is that *the ratio* of the probability density of each trajectory ω in $\{\mathbf{T}_+ < \mathbf{T}_-\}$ with respect to its "quasi-time-reversal" trajectory $r\omega$ in $\{\mathbf{T}_+ > \mathbf{T}_-\}$ is invariable, which is, surprisingly, always equal to the constant $\gamma = k_1 k_2 k_3 / (k_{-1} k_{-2} k_{-3})$.

Furthermore, map r is a one-to-one correspondence between the trajectory sets $\{\mathbf{T}_+ < \mathbf{T}_-\}$ and $\{\mathbf{T}_+ > \mathbf{T}_-\}$. More specifically, for each $t \geq 0$, map r is also actually a one-to-one correspondence between the trajectory sets $\{\mathbf{T}_+ = t < \mathbf{T}_-\}$ and $\{\mathbf{T}_+ > \mathbf{T}_- = t\}$.

Therefore, for each $t \geq 0$,

$$P_{\{E\}}\{\mathbf{T}_+ = t, \mathbf{T}_+ < \mathbf{T}_-\} = \gamma P_{\{E\}}\{\mathbf{T}_- = t, \mathbf{T}_- < \mathbf{T}_+\},$$

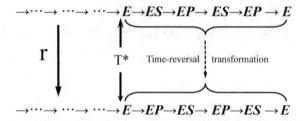

Fig. 10.2 A schematic illustration of the "quasi-time-reversal" mapping. $\mathbf{T}^*$ is the *last time*, the moment when the system leaves state $\{E\}$ before finishing a forward cycle $E \to ES \to EP \to E$; then, one can map the real time-reversal trajectory ω with respect to the time interval $[\mathbf{T}^*, \mathbf{T}_+]$ to $r\omega$. See text for details. Figure redrawn based on [100] with permission.

and

$$p^+ = \mathrm{P}_{\{E\}}\{\mathbf{T}_+ < \mathbf{T}_-\} = \gamma \mathrm{P}_{\{E\}}\{\mathbf{T}_- < \mathbf{T}_+\} = \gamma p^-,$$

which has already been proven in the previous section, Eqn. (10.27).

Denote the conditional probability density of T_+ given $\{\mathbf{T}_+ < \mathbf{T}_-\}$ as $\Theta_+(t)\mathrm{d}t = \mathrm{P}_{\{E\}}\{t \le \mathbf{T}_+ < t + \mathrm{d}t | \mathbf{T}_+ < \mathbf{T}_-\}$, and the conditional probability density of $\mathbf{T}_-$ given $\{\mathbf{T}_- < \mathbf{T}_+\}$ as $\Theta_-(t)\mathrm{d}t = \mathrm{P}_{\{E\}}\{t \le \mathbf{T}_- < t + \mathrm{d}t | \mathbf{T}_- < \mathbf{T}_+\}$. Hence,

$$
\begin{aligned}
\Theta_+(t)\mathrm{d}t &= \mathrm{P}_{\{E\}}\{t \le \mathbf{T}_+ < t + \mathrm{d}t | \mathbf{T}_+ < \mathbf{T}_-\} \\
&= \frac{\mathrm{P}_{\{E\}}\{t \le \mathbf{T}_+ < t + \mathrm{d}t, \mathbf{T}_+ < \mathbf{T}_-\}}{\mathrm{P}_{\{E\}}\{\mathbf{T}_+ < \mathbf{T}_-\}} \\
&= \frac{\gamma \mathrm{P}_{\{E\}}\{t \le \mathbf{T}_- < t + \mathrm{d}t, \mathbf{T}_- < \mathbf{T}_+\}}{\gamma \mathrm{P}_{\{E\}}\{\mathbf{T}_- < \mathbf{T}_+\}} \\
&= \mathrm{P}_{\{E\}}\{t \le \mathbf{T}_- < t + \mathrm{d}t | \mathbf{T}_- < \mathbf{T}_+\} = \Theta_-(t)\mathrm{d}t, \ \forall t.
\end{aligned}
$$

And also denote the probability density of $\mathbf{T}$ as $\Theta(t)\mathrm{d}t = \mathrm{P}_{\{E\}}\{t \le \mathbf{T} < t + \mathrm{d}t\}$, so

$$\Theta(t) = \Theta_+(t)p^+ + \Theta_-(t)p^- = \Theta_+(t) = \Theta_-(t).$$

A very important corollary consequently follows: the distribution of waiting cycle time T is *independent* of whether enzyme E completes a forward cycle or a backward cycle, although the probability of these two cycles might be rather different, i.e.,

$$P_{\{E\}}\{t \leq \mathbf{T} < t + dt, \mathbf{T}_+ < \mathbf{T}_-\} = P\{t \leq \mathbf{T}_+ < t + dt, \mathbf{T}_+ < \mathbf{T}_-\}$$
$$= \Theta_+(t)dt\, p^+$$
$$= \Theta(t)dt\, p^+, \tag{10.29}$$

and

$$P_{\{E\}}\{t \leq \mathbf{T} < t + dt, \mathbf{T}_+ > \mathbf{T}_-\} = P\{t \leq \mathbf{T}_- < t + dt, \mathbf{T}_+ < \mathbf{T}_-\}$$
$$= \Theta_-(t)dt\, p^-$$
$$= \Theta(t)dt\, p^-. \tag{10.30}$$

Furthermore, we have

$$\mathbb{E}[\mathbf{T}_+|\mathbf{T}_+ < \mathbf{T}_-] = p^+ \mathbb{E}[\mathbf{T}],$$

$$\mathbb{E}[\mathbf{T}_-|\mathbf{T}_- < \mathbf{T}_+] = p^- \mathbb{E}[\mathbf{T}],$$

and

$$\mathbb{E}[\mathbf{T}_+|\mathbf{T}_+ < \mathbf{T}_-] = \mathbb{E}[\mathbf{T}_-|\mathbf{T}_- < \mathbf{T}_+] = \mathbb{E}[\mathbf{T}],$$

which means that even in cases far from equilibrium ($\gamma \gg 1$), the dwell times for each forward cycle and those for each backward cycle are identical, although their frequencies may be very different ($p^+ \gg p^-$). This novel equality was first discovered in the context of motor proteins, both experimental and theoretical.

Now to close this section, we shall present an interesting corollary regarding the entropy production rate e_p (see Section 6). We have already shown that

$$e_p = (J_+^{ss} - J_-^{ss})\ln\gamma, \tag{10.31}$$

where $\ln\gamma = \log(J_+^{ss}/J_-^{ss})$ is the entropy production of the single cycle $E \rightarrow ES \rightarrow EP \rightarrow E$, and $J_+^{ss} - J_-^{ss}$ is the net cycle flux: the number of cycles per time.

Applying Eqn. (10.31) to the waiting cycle times, e_p can be expressed as

$$e_p = \left(\frac{1}{\mathbb{E}[\mathbf{T}_+]} - \frac{1}{\mathbb{E}[\mathbf{T}_-]}\right)\ln\gamma$$
$$= \left(\frac{p^+}{\mathbb{E}[\mathbf{T}]} - \frac{p^-}{\mathbb{E}[\mathbf{T}]}\right)\ln\gamma$$
$$= (p^+ - p^-) \times \text{mean epr}, \tag{10.32}$$

where

$$\text{mean epr} = \frac{1}{\mathbb{E}[\mathbf{T}]}\ln\gamma = \frac{1}{\mathbb{E}[\mathbf{T}]}\ln\frac{J_+^{ss}}{J_-^{ss}} = \frac{1}{\mathbb{E}[\mathbf{T}]}\ln\frac{\mathbb{E}[\mathbf{T}_-]}{\mathbb{E}[\mathbf{T}_+]} = \frac{1}{\mathbb{E}[\mathbf{T}]}\ln\frac{p^+}{p^-}$$

is regarded as the time-averaged entropy production rate of the cycle $E \to ES \to EP \to E$.

10.2.4 Fluctuation Theorems

There is a deep connection between the generalized Haldane equation in the previous section and the fluctuation theorems widely studied in statistical physics in recent years [43, 132, 88, 230]. In this section, we will give a brief account of two fluctuation theorems: one for the stochastic number of substrate cycles $v(t)$ and one for the fluctuating chemical work done to sustain the NESS $W(t) = v(t)\triangle\mu/k_BT = v(t)\log\gamma$.

Regarding the probability distribution of the nonstationary process $v(t) = v_+(t) - v_-(t)$, one has

$$P\{v(t) = k\} = \sum_{n-r=k} P\{v_+(t) = n, v_-(t) = r\}$$

$$= \sum_{n-r=k}^{\infty} P\{v_+(t) + v_-(t) = n+r\}C_{n+r}^n(p^+)^n(p^-)^r,$$

and

$$P\{v(t) = -k\} = \sum_{n-r=k}^{\infty} P\{v_+(t) + v_-(t) = n+r\}C_{n+r}^r(p^+)^r(p^-)^n.$$

Note that p^+ and p^- are the stepping probabilities, and $C_{n+r}^r = C_{n+r}^n = (n+r)!/(n!\cdot r!)$ is the standard binomial combinatorial factor.

Since $p^+/p^- = \gamma$, we have

$$\frac{P\{v(t) = k\}}{P\{v(t) = -k\}} = \gamma^k,$$

which is called the transient fluctuation theorem for $v(t)$.

Therefore, the expected value

$$\mathbb{E}\left[e^{-\lambda v(t)}\right] = \sum_{k=-\infty}^{\infty} P\{v(t) = k\}e^{-k\lambda} = \sum_{k=-\infty}^{\infty} P\{v(t) = -k\}e^{-k(\lambda - \log\gamma)}$$

$$= \mathbb{E}\left[e^{-(\log\gamma - \lambda)v(t)}\right], \tag{10.33}$$

which is the fluctuation theorem known as Kurchan–Lebowitz–Spohn-type symmetry. If one lets $\lambda = \log \gamma$, then the special case of Kawasaki equality arises,

$$\mathbb{E}\left[e^{-W(t)}\right] = 1. \tag{10.34}$$

In the previous section, most of the results were obtained through solving a system of linear equations that resemble Eqn. (10.23), the Kolmogorov backward master equations. The results have been extended to the n-step cycle in [86], according the elementary renewal theorem in probability theory and the theory of continuous-time Markov processes. The key step is still the same "quasi-time-reversal" mapping r introduced in the previous section. Recently, the generalized Haldane equality, also named cycle symmetry, has been generalized to cases with multiple cycles and diffusion processes on a circle [135, 99].

10.3 Modifier Activation-inhibition Switching in Enzyme Kinetics

Reversible modifiers of an enzyme are ligands that can form a complex with the enzyme through a binding reaction, resulting in changed catalytic properties for the enzyme. Reversible enzyme modifiers play a crucial role both in the regulation of metabolic pathways inside a cell and in the studies of enzymatic catalyses and functions. Moreover, they have found widespread application in pharmacology and toxicology, as well as in industry and agriculture. A modifier is called an activator or an inhibitor according to its ability to increase or decrease the catalytic rate of an enzyme, respectively.

In standard molecular biology textbooks, the activity of an activator/inhibitor of an enzyme is widely assumed to be an attribute of the modifying substance determined by its molecular structure. In this section, we shall show how the same ligand molecule can be either one depending upon the *context* of the biochemical reaction system: the NESS [223, 136, 216]. Even though kinetics are a rather classic problem in biochemical literature, it was not fully understood until the 2010s through the theory of NESS. We carry out our discussion with a simple example.

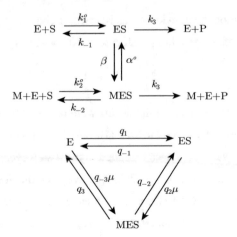

Fig. 10.3 The upper enzyme reaction system has modifier M, second-order rate constant k_1^0 and α^0 and third-order rate constant k_2^0. In the absence of the modifier, i.e., its concentration $\mu \equiv [M] = 0$, the steady-state velocity of the enzymatic reaction is the Michaelis–Menten hyperbolic $\frac{k_1 k_3 [S]}{k_1 [S] + k_{-1} + k_3}$. In the lower three-state single enzyme conformation representation for the system: $q_1 = k_1^0 [S]$, $q_{-1} = k_{-1} + k_3$, $q_2 = \alpha^0$. $q_{-2} = \beta$, $q_3 = k_{-2} + k_3$, $q_{-3} = k_2^0 [S]$.

10.3.1 Case Study of a Simple Example

Consider a simple enzyme-catalyzed reaction that is regulated by a modifier M: there are parallel catalysis pathways with and without the modifier, as shown in the upper panel of Fig. 10.3. Note that the system in Fig. 10.3 contains irreversible steps; hence, ES and E would never reach an equilibrium due to the presence of a flux. However, a NESS among states E, ES and MES can be reached in a single enzyme just as in the standard Michaelis–Menten kinetics, which also contain an irreversible step. The lack of backward steps $E + P \rightarrow EP$ and $M + E + P \rightarrow MEP$ is achieved if one assumes a very rapid depletion of product P; the sustained concentration of product P thus vanishes, i.e., $[P] = 0$. The same justification applies for all irreversible enzyme kinetics.

A molecule M is called an activator (inhibitor) of an enzyme if the enzyme-catalyzed reaction velocity is greater (smaller) in the presence of M. We shall show that in certain kinetic systems, however, the same molecule can act as either an activator or an inhibitor depending upon its concentration $[M]$. There can be a switching from one to another with changing $[M]$. This phenomenon is called kinetics-based activation-inhibition switching.

From the standpoint of a single enzyme, there are two types of kinetic cycles in Fig. 10.3: One substrate binding cycle $M + E + S \rightleftharpoons M + ES \rightleftharpoons MES \rightleftharpoons M + E + S$, and two catalytic cycles $E + S \rightleftharpoons ES \rightleftharpoons E + P$ and $M + E + S \rightleftharpoons MES \rightleftharpoons M + E + P$. The former should obey the detailed balance condition, $k_2^0 \beta k_{-1} = k_1^0 \alpha^0 k_{-2}$, while the latter does not since there is a continuous $S \rightarrow P$ turnover.

Because the rate constants for both $ES \rightarrow E + S$ and $MES \rightarrow M + E + S$ are the same, k_3, the rate of catalytic reaction $v = k_3 (p_{ES} + p_{MES})$. To show the activation-inhibition switching as an NESS phenomenon, we shall first consider the "control" case in which E, ES and MES are at equilibrium with each other, when

$$[ES]^{eq} = \frac{k_1^0 [E]^{eq} [S]^{eq}}{k_{-1}} \quad \text{and} \quad [MES]^{eq} = \frac{\alpha^0 [M]^{eq} [ES]^{eq}}{\beta}.$$

Then, the corresponding equilibrium turnover rate

$$v^{eq} = k_3 \left(\frac{\frac{k_1^0 [S]^{eq}}{k_{-1}} + \frac{k_1^0 [S]^{eq} \alpha^0 [M]^{eq}}{k_{-1} \beta}}{1 + \frac{k_1^0 [S]^{eq}}{k_{-1}} + \frac{k_1^0 [S]^{eq} \alpha^0 [M]^{eq}}{k_{-1} \beta}} \right). \tag{10.35}$$

We see that this rate always increases with $[M]^{eq}$. Hence, the modifier in this model is clearly an activator, never becoming an inhibitor in the equilibrium state.

The NESS of the kinetic system in the upper panel of Fig. 10.3, with irreversible steps, can be readily solved. This involves a mathematical "trick" that combines rate constant k_3 with k_{-1} and k_{-2}, as shown in the lower panel.[1] Hence, the total enzyme-substrate complex (TES)

$$p_{TES}^{ss}(u) = (p_{ES} + p_{MES})^{ss} = \frac{A + Bu + Cu^2}{D + Eu + Fu^2}, \tag{10.36}$$

in which u is the concentration of the modifier M,

---

[1] When combining steps like this, it is very important to make a distinction between the mathematical trick and the physical reality: for some particular values of k, the resulting tri-state Markov chain in the lower panel of Fig. 10.3 could have $q_1 q_2 q_3 = q_{-1} q_{-2} q_{-3}$. That *does not* mean the physical problem is an equilibrium; rather, this arises because the mathematical simplification hides certain details. To articulate this distinction, Vellela and Qian [265] introduced the notion of *chemical detailed balance* and *mathematical detailed balance*. A Markov chain with the latter could still be a useful mathematical model for a nonequilibrium chemical system. However, it can be shown that a mathematical model with chemical detailed balance is necessarily in the mathematical detailed balance condition.

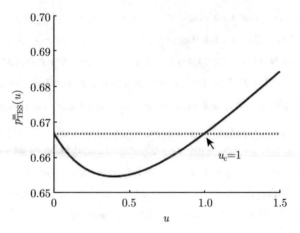

Fig. 10.4 Modifier activation-inhibition switching. $q_1 = q_2 = q_3 = 2, q_{-1} = q_{-2} = q_{-3} = 1, A = 6, B = 6, C = 2, D = 9, E = 10, F = 2$. Figure redrawn based on [100] with permission.

$$A = q_1(q_3 + q_{-2}), \quad B = q_{-2}q_{-3} + q_1q_2 + q_{-1}q_{-3}, \tag{10.37a}$$

$$C = q_2q_{-3}, \quad D = (q_1 + q_{-1})(q_3 + q_{-2}), \tag{10.37b}$$

$$E = q_{-2}q_{-3} + q_1q_2 + q_{-1}q_{-3} + q_2q_3, \quad F = q_2q_{-3}. \tag{10.37c}$$

We see that

$$p_{TES}^{ss}(0) = \frac{A}{D} = \frac{k_1}{k_1 + k_{-1}} < p_{TES}^{ss}(\infty) = \frac{C}{F} = 1$$

and

$$\left(p_{TES}^{ss}\right)'(0) = \frac{B}{A} - \frac{E}{D} = \frac{q_{-1}(q_{-2}q_{-3} + q_1q_2 + q_{-1}q_{-3}) - q_1q_2q_3}{q_1(q_3 + q_{-2})(q_1 + q_{-1})}. \tag{10.38}$$

If $\left(p_{TES}^{ss}\right)'(0) < 0$, then there is an inhibition-to-activation switching with increasing $[M]$. One can in fact obtain the critical u_c value (the switching point):

$$\frac{A + Bu_c + Cu_c^2}{D + Eu_c + Fu_c^2} = \frac{A}{D}.$$

This yields

$$u_c = \frac{DC - AF}{AE - DB} = \frac{F}{A}\left(\frac{\frac{C}{F} - \frac{A}{D}}{\frac{E}{D} - \frac{B}{A}}\right) > 0.$$

An example is shown in Fig. 10.4.

10.3.2 Modifier With a More General Mechanism — An In-depth Study

As early as 1953, Botts and Morales [29] considered a mechanism for a general modifier with reversible binding kinetics. The four-state catalytic reaction system shown in Fig. 10.5 is central to their mechanism. Fig. 10.5 is also kinetically iso-morphic to the fluctuating (hysteretic) enzyme model of C. Frieden [81]. Many pre-vious works have studied the steady-state as well as transient kinetics of this general modifier mechanism.

The rate of product formation associated with the general modifier mechanism is shown in Fig. 10.5, in which E, S, R and P represent the enzyme, the substrate, the regulator (modifier) and the product, respectively, and ES, ER and ERS are three complexes, whose meanings are self-evident. k_1^0, q_1^0 are second-order rate constants. This mechanism can produce hyperbolic inhibition or activation or a combination of the two. All of the simple mechanisms for inhibition and activation (apart from product inhibition) are special cases of this general mechanism.

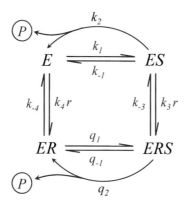

Fig. 10.5 Kinetic scheme of an enzyme reaction with a general modifier. The pseudo first-order rate constants $k_1 = k_1^0 s$ and $q_1 = q_1^0 s$. s and r are the substrate (not shown) concentration $[S]$ and the regulator (not shown) concentration $[R]$, respectively.

Similar to the previous section, the rate of product formation from this enzyme regulation system at equilibrium is

$$v^{eq}(r) = k_2 p_{ES}^{eq} + q_2 p_{ERS}^{eq} = \frac{k_2 \frac{k_1}{k_{-1}} + q_2 \frac{k_1 k_3 r}{k_{-1} k_{-3}}}{1 + \frac{k_1}{k_{-1}} + \frac{k_1 k_3 r}{k_{-1} k_{-3}} + \frac{k_4 r}{k_{-4}}}, \qquad (10.39)$$

in which $r = [R]$, $k_1 = k_1^0[S]$ and $q_1 = q_1^0[S]$. In the derivation of Eqn. (10.39), we have used the detailed balance condition in the substrate-binding cycle: $k_1 k_3 q_{-1} k_{-4} = k_{-1} k_{-3} q_1 k_4$. Again, for fixed $[S]^{eq}$, $v^{eq}(r)$ is always monotonic with respect to r.

Moreover, we have

$$v^{eq}(0) = \frac{k_2 \frac{k_1}{k_{-1}}}{1 + \frac{k_1}{k_{-1}}},$$

and

$$v^{eq}(\infty) = \frac{q_2 \frac{k_1 k_3}{k_{-1} k_{-3}}}{\frac{k_1 k_3}{k_{-1} k_{-3}} + \frac{k_4}{k_{-4}}},$$

Hence, when $[v^{eq}(\infty) - v^{eq}(0)] > 0$, the modifier acts as an activator; otherwise, it acts as an inhibitor. This is true for all values of r. Both $v^{eq}(\infty)$ and $v^{eq}(0)$ could be measured directly and are functions of rate constants ks and qs, which are intrinsic properties of the molecules E and R and their interaction. Hence, whether R is an equilibrium activator or inhibitor is determined by not only the molecular structure but also the concentration of S.

In a NESS, from a single-enzyme perspective, the kinetics described in Fig. 10.5 can be represented as a Markov chain $\xi(t)$ with a finite discrete state space $\mathscr{S} = \{E, ES, ER, ERS\}$ in continuous-time t. The rate matrix Q of the Markov transition probability of the process can be readily obtained from Fig. 10.5:

$$Q = \begin{pmatrix} -(k_1 + k_4 r) & k_1 & k_4 r & 0 \\ k_{-1} + k_2 & -(k_{-1} + k_2 + k_3 r) & 0 & k_3 r \\ k_{-4} & 0 & -(k_{-4} + q_1) & q_1 \\ 0 & k_{-3} & q_{-1} + q_2 & -(q_{-1} + q_2 + k_{-3}) \end{pmatrix},$$

where the rows and columns are indexed by $\mathscr{S}$, and r again is the modifier concentration. We will only study the steady state and assume that the concentration of substrate s and the concentration of product p are kept constant in some way. In fact $p = 0$ again as in the previous section. This assumption implies that the single, irreversible enzyme reaches a NESS.

Let $p^{ss} = (p_E^{ss}, p_{ES}^{ss}, p_{ER}^{ss}, p_{ERS}^{ss})$ be the invariant probability distribution of Markov process $\xi(t)$. Then, in the steady state, the rate of product formation is

$$v(r) = k_2 p_{ES}^{ss} + q_2 p_{ERS}^{ss}.$$

The rate of product formation $v(r)$ is dependent upon the concentration of the modifier r, via p^{ss}, which is obtained by solving the system of linear equations $p^{ss}Q = 0$. By obtaining the expression of $v(r)$ in terms of r and examining the influence of r on $v(r)$, the kinetics can be analyzed.

We shall not present the mathematical details here but wish only to give a brief summary of the phenomenon. For a comprehensive account, please see [136].

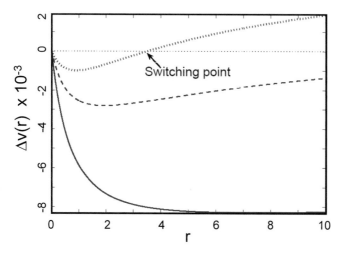

Fig. 10.6 Three types of kinetic behaviors in the general enzyme regulation mechanism shown in Fig. 10.5. $\Delta v(r) = v(r) - v(0)$. Parameters: $k_1 = 2$; $k_2 = 0.1$; $k_{-1} = 2 - k_2$; $q_1 = 1$; $q_2 = 0.085$ (solid), 0.102 (dashed) and 0.11 (dotted); $q_{-1} = 1 - q_2$; $k_3 = 1$; $k_{-3} = 2$; $k_4 = 2$; $k_{-4} = 1$. Here γ is set to 1/4. Figure redrawn based on [100] with permission.

Hyperbolic behavior. Under certain conditions, the steady-state velocity of product formation $v(r)$ in a NESS exhibits an approximately hyperbolic dependence on r, the concentration of modifier R. In this case, the modifier behaves in the same way as it does at equilibrium steady state, as shown by the solid line in Fig. 10.6.

Bell-shaped behavior. The velocity $v(r)$ in the NESS could also exhibit a bell-shaped dependence on $[R]$. In this case, the quantity $\Delta v(r) = v(r) - v(0)$ is always positive or negative depending on whether $v(\infty)$ is greater or smaller than $v(0)$, respectively. Although both the hyperbolic-behaving and bell-shaped-behaving modifiers are overall activators or inhibitors for all possible values of modifier concentration $[R]$, the bell-shaped-behaving modifier would make the enzyme activity exceed its limit value for the saturated rate, as seen in the dashed line in Fig. 10.6.

Switching behavior. More interestingly, in this case, the quantity $\Delta v(r)$ may change sign somewhere in the range of the modifier concentration. The role played by the modifier will convert from an activator to an inhibitor or vice versa. Therefore, at NESS, the effect of activation or inhibition should not be viewed as an intrinsic property of the modifier. It instead depends on the concentrations of modifier $[R]$. This is shown by the dotted line in Fig. 10.6.

10.4 Fluctuating Enzymes and Dynamic Cooperativity

Many experimental measurements have shown that most enzymes have a great deal of conformational fluctuation, also called dynamic disorder. The term *dynamic disorder* refers to the behavior by which a reaction rate constant fluctuates as a function of time. One of the mechanisms underlying this phenomenon is that a reactant has fluctuations among multiple conformations, each with a different reaction rate constant. One of the surprising consequences of this, when an enzyme operates in a living cellular environment (see below), is that the enzyme catalysis can exhibit positive cooperativity, giving rise to a "sigmoidal dependence" of the product formation rate on the substrate concentration in the steady state [81, 214]. We now study this phenomenon.

10.4.1 A Simple Model for Fluctuating Enzymes

As we have shown, if there is only one unbound enzyme state E, then no matter how complex the reaction scheme within the ES state, the Michaelis–Menten kinetics are valid. Therefore, we consider the case of having two unbound enzyme states E_1 and E_2, both of which can result in binding substrate S and forming ES.

For simplicity, we assume that E_1S and E_2S are essentially the same. Thus, we have the simplest kinetic model for a fluctuating enzyme

$$E_1 \underset{k_{-1}}{\overset{k_1}{\rightleftharpoons}} E_2, \ E_1 + S \underset{\beta_1}{\overset{\alpha_1}{\rightleftharpoons}} ES, \ E_2 + S \underset{\beta_2}{\overset{\alpha_2}{\rightleftharpoons}} ES,$$

$$ES \overset{\alpha_3}{\longrightarrow} E_1 + P, \ ES \overset{\alpha_4}{\longrightarrow} E_2 + P. \tag{10.40}$$

Wong and Hanes proposed a similar model in 1962. In 1970, Frieden introduced the concept of hysteretic enzymes [81], while Ricard and his colleagues championed the concept of mnemonic enzymes (see [214] for a more recent account in light of single-molecule enzymology). As we shall show, while both hysteretic and monomeric enzymes are consequences of slow conformational disorder, they in fact say something different: one concept emphasizes transient kinetics and the other relates to a driven NESS.

Similar to the previous section, we assume a sustained $[P]^{ss} = 0$ here, an hence an enzyme would reach a nonequilibrium steady state (NESS) with nonzero flux. When an enzyme is in equilibrium with detailed balance, i.e., with the thermodynamic constrain $k_1 \alpha_2 \beta_1 = k_{-1} \alpha_1 \beta_2$, however,

$$\theta(s) = \frac{[ES]}{E_{tot}} = p_{ES} = \frac{\frac{k_1 \alpha_2 [S]^{eq}}{k_{-1} \beta_2}}{1 + \frac{k_1}{k_{-1}} + \frac{k_1 \alpha_2 [S]^{eq}}{k_{-1} \beta_2}}.$$

This is always hyperbolic.

In the nonequilibrium steady state, from a single-enzyme perspective, we have a tristate Markov system, as shown in Fig. 10.7(A).We shall assume that the fluctuating rates between E_1 and E_2, k_1 and k_{-1}, are small. Otherwise, if they are fast, then this model is reduced to simple Michaelis–Menten kinetics with association rate constant $(k_{-1}\alpha_1 + k_1\alpha_2)/(k_{-1}+k_1)$ and dissociation rate constant $\beta_1 + \beta_2$.

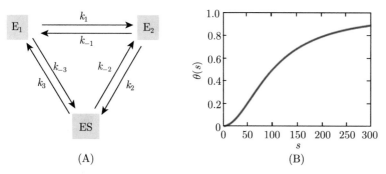

(A)   (B)

Fig. 10.7 (A) A three-state Markov system representing the simplest fluctuating enzyme with two distinctly different unbound states E_1 and E_2, where $k_2 = \alpha_2[S]$, $k_3 = \beta_1 + \alpha_3$, $k_{-2} = \beta_2 + \alpha_4$ and $k_{-3} = \alpha_1[S]$. (B) Positive dynamic cooperativity in simple model (A) with parameters $k_1 = 0, k_{-1} = 1, \alpha_1 = 0.01, k_{-2} = 100, k_3 = 0$, and $\alpha_2 = 100$. Figure redrawn based on [100] with permission.

A simple calculation gives

$$p_E = \frac{[E]}{E_{tot}} = \frac{k_2 k_3 + k_{-1} k_3 + k_{-1} k_{-2}}{\mathscr{D}},$$

$$p_{E^*} = \frac{[E^*]}{E_{tot}} = \frac{k_1 k_3 + k_{-2} k_{-3} + k_1 k_{-2}}{\mathscr{D}},$$

$$p_{ES} = \frac{[ES]}{E_{tot}} = \frac{k_1 k_2 + k_2 k_{-3} + k_{-1} k_{-3}}{\mathscr{D}},$$

and

$$\mathscr{D} = k_2 k_3 + k_{-1} k_3 + k_{-1} k_{-2} + k_1 k_3 + k_{-2} k_{-3} + k_1 k_{-2} + k_1 k_2 + k_2 k_{-3} + k_{-1} k_{-3}.$$

Hence

$$\theta(s) = \frac{[ES]}{E_{tot}} = \frac{ds + cs^2}{a + bs + cs^2}, \tag{10.41}$$

where $a = k_{-1} k_3 + k_{-1} k_{-2} + k_1 k_3 + k_1 k_{-2}$, $b = \alpha_2 k_3 + k_{-2} \alpha_1 + k_1 \alpha_2 + k_{-1} \alpha_1$, $c = \alpha_1 \alpha_2$ and $d = k_1 \alpha_2 + k_{-1} \alpha_1$.

The steady-state velocity of the enzyme-catalyzed reaction is $v = (\alpha_3 + \alpha_4)\theta(s)$. It contains terms involving $[S]^2$. Then, it is possible that the catalyzed reaction velocity of the fluctuating enzyme will exhibit a sigmoidal dependence on the substrate concentration s. This is known as dynamic cooperativity and is in contrast to allosteric cooperativity, which requires multiple binding sites within an enzyme. This phenomenon is also known as mnemonic behavior. Both dynamic disorder and the breakdown of detailed balance are necessary for a monomeric enzyme to show this interesting behavior in NESS.

Applying the mathematical method in Sec. 10.4.2 below, we obtain $(k_{-1} k_{-2} \alpha_1 - k_1 \alpha_2 k_3)(\alpha_2 - \alpha_1)$ for $ac - d(b - d)$ in Eqn. (10.41). Hence, positive cooperativity corresponds to $k_1 \alpha_2 k_3 < k_{-1} k_{-2} \alpha_1$ (Fig. 10.7B) and negative cooperativity corresponds to $k_1 \alpha_2 k_3 > k_{-1} k_{-2} \alpha_1$ when $\alpha_2 > \alpha_1$. When $\alpha_2 = \alpha_1$, the catalytic capability is the same for both of the different enzyme conformations, so there will always be a hyperbolic mechanism.

Similar result holds for more general cases (Fig. 10.8), such as

$$\theta(s) = \frac{[ES] + [E^* S]}{E_{tot}} = \frac{ds + cs^2}{a + bs + cs^2},$$

where

$$a = k_{-1}k_{-2}k_{-3} + k_{-1}k_{-2}k_4 + k_{-1}k_3k_4 + k_1k_{-2}k_{-3} + k_1k_{-2}k_4 + k_1k_3k_4,$$

$$b = k_2k_3k_4 + k_{-2}k_{-3}k_{-4} + k_1k_2k_3 + k_{-1}k_{-2}k_{-4} + k_{-1}k_3k_{-4}$$

$$\quad + k_1k_2k_{-3} + k_1k_2k_4 + k_{-1}k_{-3}k_{-4},$$

$$c = k_2k_3k_{-4} + k_2k_{-3}k_{-4},$$

$$d = k_1k_2k_3 + k_{-1}k_{-2}k_{-4} + k_{-1}k_3k_{-4} + k_1k_2k_{-3} + k_1k_2k_4 + k_{-1}k_{-3}k_{-4}.$$

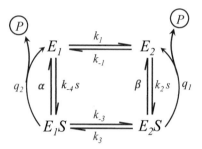

Fig. 10.8 The canonical four-state model for fluctuating enzymes, where $k_4 = \alpha + q_2$ and $k_{-2} = \beta + q_1$.

Again applying the mathematical method in Sec. 10.4.2, one obtains $ac - d(b - d) = (k_{-1}k_{-2}k_{-3}k_{-4} - k_1k_2k_3k_4)(k_2k_{-3} + k_2k_4 + k_2k_3 - k_{-3}k_{-4} - k_{-2}k_{-4} - k_3k_{-4})$. Similar conclusions as in the previous simple model can be reached.

10.4.2 Mathematical Method For Analyzing Dynamic Cooperativity

The dependence of fractional saturation $\theta(s)$ on the substrate concentration is always described by the ratio between a polynomial numerator and a polynomial denominator, whose order is either the same as or less than the number of different conformations:

$$\theta(s) = \frac{\sum_{i=1}^{n} a_i s^i}{\sum_{i=0}^{n} b_i s^i}.$$

In particular, for the two-conformational models discussed in the previous sections, the polynomial order is two:

$$\theta(s) = \frac{ds + cs^2}{a + bs + cs^2}. \tag{10.42}$$

Rewriting Eqn. (10.42) in terms of the reciprocal fractional saturation and reciprocal substrate concentration gives Eqn. (10.43):

$$\frac{1}{\theta(s)} = \frac{a(\frac{1}{s})^2 + b\frac{1}{s} + c}{d\frac{1}{s} + c}. \tag{10.43}$$

Eqn. (10.43) is a second-order polynomial divided by a first-order polynomial and therefore is called a "2/1" function of $\frac{1}{s}$.

Dividing by the denominator gives

$$\frac{1}{\theta(s)} = \frac{bd - ac}{d^2} + \frac{a}{d}\frac{1}{s} + \frac{c(ac - (b-d)d)}{d^2(\frac{d}{s} + c)}. \tag{10.44}$$

The first two terms on the right-hand side of Eqn. (10.44) represent the Michaelis–Menten relation between θ and s which approached asymptotically at low substrate concentrations. The third term represents the deviation from Michaelis–Menten kinetics at high substrate concentration when $\frac{1}{s}$ is no longer large enough to make the third term negligible. The value of $ac - d(b - d)$, the numerator of the third term, is a measure of the deviation from the Michaelis–Menten relationship. When this expression is greater than, equal to, or less than zero, apparent positive cooperativity, Michaelis–Menten behavior, or apparent negative cooperativity is observed, respectively.

10.5 Kinetic Proofreading and Specificity Amplification

Noise, stochasticity, and specificity are among the most important emerging concepts in current molecular cell biology, particularly in connection with cellular processes such as gene regulation and signal transduction. Through evolution, biological organisms have acquired a repertoire of mechanisms to counteract stochasticity and thus improve the accuracy of their informational processing. Kinetic proofreading theory, first developed in the 1970s [122, 191], provides a concrete example of how a cellular biochemical network can function as an error reduction device, suppressing noise and improving biochemical specificity between macromolecules without relying on molecular structural modification. The molecular mechanisms of the proofreading have been extensively elucidated for DNA polymerase in terms of its exonuclease activity and for protein synthesis in terms of ribosomes structure and kinetics. However, having the right molecular structures and biochemical reaction networks are not sufficient for the error reduction mechanism to function. The central idea of the Hopfield–Ninio theory is the necessity of a free energy source in the form of the chemical potential gradient, either from GTP/GDP or from other enzymatic cofactors. Biochemical error reduction requires a continuous free energy expenditure: high-grade chemical energy is transformed into low-grade heat accompanied by increasing entropy [214]. The role of free energy has been conspicuously absent in the general discussions on biochemical specificity and error reduction in cell biology.

With respect to kinetic proofreading, here we address a fundamental question: for a given amount of available free energy to a cell, in the form of the ATP/ADP (or GTP/GDP) ratio, what is the thermodynamic limitation on the error reduction and specificity amplification?

10.5.1 Minimal Error Rate Predicted By the Hopfield–Ninio Model

While this model was developed for high fidelity in biosynthetic processes, the kinetic model and the idea within have a much broader cellular applications. Here we shall frame our discussion in terms of the receptor-ligand association. The upper triangle of Fig. 10.9 shows a schematic of the 3-state, cyclic receptor-ligand binding model. We shall denote the equilibrium association constant for the ligand-receptor complex as K_a. Then, the equilibrium between the empty receptor R and the activated complex RL^* is $[RL^*]^{eq}/[R]^{eq} = K_a[L] = k^0_{-3}[L]/k_3 = k_{-3}/k_3$, where k_3 and k^0_{-3} are the dissociation and association rate constants, respectively. We shall use k_{-3} to denote the pseudo first-order rate constant $k^0_{-3}[L]$.

We now quote a few results from standard thermodynamics. The equilibrium, standard-state free energy of hydrolysis

$$\Delta G^0_{DT} = -RT\ln\left(\frac{[T]^{eq}}{[D]^{eq}}\right).$$

Furthermore, the equilibrium ATP and ADP concentrations are related to the rate constants:

$$\frac{[T]^{eq}}{[D]^{eq}} = \frac{k^0_{-2}[RL^*]^{eq}}{k^0_2[RL]^{eq}} = \frac{k_{-1}k^0_{-2}k_{-3}}{k_1 k^0_2 k_3}, \tag{10.45}$$

in which the second equality is due to the equilibrium relation $[RL]^{eq}/[RL^*]^{eq} = k_1 k_3/(k_{-1}k_{-3})$. Hence, the free energy of ATP hydrolysis in a cell,

$$\Delta G_{DT} = \Delta G_{DT}^0 + RT\ln\left(\frac{[T]}{[D]}\right) = RT\ln\left(\frac{k_1 k_2 k_3}{k_{-1}k_{-2}k_{-3}}\right),$$

where $[T]$ and $[D]$ are the cellular ATP and ADP concentrations, which are not at equilibrium if the cell is alive: $\Delta G_{DT} > 0$.

As the point of departure from the original work of Hopfield, we introduce a parameter, $\gamma = e^{\Delta G_{DT}/RT}$, representing the available free energy from each ATP hydrolysis:

$$\gamma = \frac{k_1 k_2 k_3}{k_{-1}k_{-2}k_{-3}}. \tag{10.46}$$

The free energy in a living cell is from the sustained physiological level of ATP ($\sim$8 mM) and ADP ($\sim$10 M). It is not from the phosphate bond of the ATP molecule: "The Pacific Ocean could be filled with an equilibrium mixture of ATP, ADP and Pi, but the ATP would have no capacity to do work"[188]. When a cardiac myocyte experiences ischemia, its ATP/ADP ratio and the cellular energy decreases.

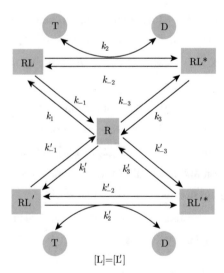

Fig. 10.9 A simple kinetic model of receptor-ligand binding coupled with hydrolysis reaction $T \rightleftharpoons D$. $k_{-3} = k_{-3}^0[L]$, $k_1 = k_1^0[L]$ and $k_{-3}' = k_{-3}'^0[L']$, $k_1' = k_1'^0[L']$. Figure redrawn based on [100] with permission.

In Fig. 10.9, let us consider two possible ligands L and L' at equal concentration. We assume they are structurally related so that they have the same k_1, k_2, k_{-2} and k_{-3}.

In a test tube at equilibrium (i.e., $[T] = [T]^{eq}$ and $[D] = [D]^{eq}$),

$$\frac{k_1^0 k_2^0 k_3}{k_{-1} k_{-2}^0 k_{-3}^0} = \frac{k_1'^0 k_2'^0 k_3'}{k_{-1}' k_{-2}'^0 k'0_{-3}} = \frac{[D]^{eq}}{[E]^{eq}},$$

hence

$$\gamma = \frac{k_1 k_2 k_3}{k_{-1} k_{-2} k_{-3}} = \frac{k_1' k_2' k_3'}{k_{-1}' k_{-2}' k_{-3}'} = 1.$$

Therefore, the ratio of the two affinities at equilibrium is

$$f = \frac{\frac{[RL^*]}{[R][L]}}{\frac{[RL'^*]}{[R][L']}} = \frac{\frac{k_{-3}}{k_3}}{\frac{k_{-3}}{k_3'}} = \theta.$$

Their affinities with the receptor, however, are different due to $\frac{k_{-1}'}{k_{-1}} = \frac{k_3'}{k_3} = \theta$, where $\theta < 1$. That is, L' has a higher affinity to the receptor than L; the receptor is more specific for L' than for L.

In Hopfield's model for biosynthesis, f represents the expected error rate of an incorrect amino acid being incorporated into a protein. Here, f represents the error rate of activation due to nonspecific binding.

In living cells, due to the hydrolysis reaction in Fig. 10.9, $RL + T \rightleftharpoons RL^* + D$, $[D]$ and $[T]$ are not at their equilibrium concentrations. The error rate therefore depends on how much energy is available, i.e., f is a function of γ. Here $\gamma \neq 1$, but $\gamma = \frac{k_1 k_2 k_3}{k_{-1} k_{-2} k_{-3}} = \frac{k_1' k_2' k_3'}{k_{-1}' k_{-2}' k_{-3}'}$ still hold as the thermodynamic constrain.

For the model in Fig. 10.9,

$$f = \theta \frac{(k_1 k_2 + k_2 k_{-3} + k_{-1} k_{-3})\,(k_2 k_3 + \theta k_3 k_{-1} + k_{-1} k_{-2})}{(k_2 k_3 + k_3 k_{-1} + k_{-1} k_{-2})\,(k_1 k_2 + k_2 k_{-3} + \theta k_{-1} k_{-3})}, \tag{10.47}$$

using relation Eqn. (10.46) to eliminate $k_{-1} k_{-2}$ in Eqn. (10.47), one obtains

$$f(\gamma) = \theta \frac{(k_1 k_2 + k_2 k_{-3} + k_{-1} k_{-3})\left(k_2 k_{-3} + \theta k_{-1} k_{-3} + \frac{k_1 k_2}{\gamma}\right)}{\left(k_2 k_{-3} + k_{-1} k_{-3} + \frac{k_1 k_2}{\gamma}\right)(k_1 k_2 + k_2 k_{-3} + \theta k_{-1} k_{-3})}.$$

One useful inequality is presented here. For $\gamma > 1$, $\theta < 1$, and nonnegative variables a, b, and c, we have $\frac{(a+b+c)}{\left(a+\frac{b}{\gamma}+c\right)} \frac{\left(a\theta+\frac{b}{\gamma}+c\right)}{(a\theta+b+c)} \geq \left(\frac{1+\sqrt{\gamma\theta}}{\sqrt{\gamma}+\sqrt{\theta}}\right)^2$, where the equality holds true when $c = 0$ and $\frac{b}{a} = \sqrt{\gamma\theta}$.

Then, we have the minimal error rate

$$f_{min}(\gamma) = \theta \left(\frac{1+\sqrt{\gamma\theta}}{\sqrt{\gamma}+\sqrt{\theta}}\right)^2 \tag{10.48}$$

for all possible rate constants with a given γ when $\frac{k_{-3}}{k_1} = 0$ and $\frac{k_1 k_2}{k_{-1}k_{-3}} = \sqrt{\gamma\theta}$. These two relations imply the inequalities

$$k_{-1} \gg k_2, \quad k_1 \gg k_{-3}, \quad \frac{k_1}{k_{-3}}\frac{k_2}{k_{-1}} > \theta, \quad k_3 > k_{-2}. \tag{10.49}$$

The conditions in Eqn. (10.49) can be described, à la Hopfield, as the "wrong substrate arriving at RL^* must come typically through step 2 rather than 3" and "the rate of loss of molecules RL^* must be dominantly by path 3" [122]. $k_{-1} \gg k_2$ and $k_1 \gg k_{-3}$ imply that step 1 is in rapid equilibrium for an optimal error reduction. When $\gamma \gg \theta^{-1}$, i.e., there is sufficient amount of energy available, f_{min} approaches θ^2. This is the celebrated result of Eqn. (10.48), which offers a more complete, quantitative description of how much the error can be reduced with a finite amount of energy available.

10.5.2 Absolute Thermodynamic Limit on the Error Rate With Finite Available Free Energy.

The f_{min} obtained above is confined within the kinetic scheme given in Fig. 10.9. We now seek to provide an estimation of the absolute lower bound of the error rate with a given amount of energy γ for any possible kinetic scheme, i.e., a true thermodynamic limit irrespective of the detailed "wiring diagram".

The competition between a cognate ligand L and a noncognate competitor L' for a receptor R can be written into a single biochemical reaction as

$$L + RL'^* \rightleftharpoons L' + RL^*. \tag{10.50}$$

Let the equilibrium constant for the reaction be θ. Then, with an equal concentration of L and L', the free energy difference between RL^* and RL'^*, ΔG^{eq}, is zero in a chemical equilibrium:

$$\Delta G^{eq} = -RT\ln\theta + RT\ln\frac{[RL^*]^{eq}}{[RL'^*]^{eq}} = 0. \tag{10.51}$$

In living cells, this reaction is coupled to an "energy source" with NESS free energy $RT\ln\gamma$. This sets a upper bound for the amount of free energy that can be utilized when there is a tight coupling between the chemical energy source and the "ligand exchange" reaction in Eqn. (10.50). Assuming there is no waste of energy in the coupling, the maximum contribution yields

$$\Delta G^{ness} = -RT\ln\theta + RT\ln\frac{[RL^*]}{[RL'^*]} = -RT\ln\gamma < 0.$$

This yields a minimal error rate independent of the kinetic scheme:

$$\frac{[RL^*]}{[RL'^*]} = \frac{\theta}{\gamma}. \tag{10.52}$$

Checking the result in Eqn. (10.48) against (10.52), we have:

$$\theta\left(\frac{1+\sqrt{\gamma\theta}}{\sqrt{\gamma}+\sqrt{\theta}}\right)^2 = \frac{\theta}{\gamma}\left(\frac{1+\sqrt{\gamma\theta}}{1+\sqrt{\theta/\gamma}}\right)^2 > \frac{\theta}{\gamma},$$

when $\gamma > 1$.

Chapter 11
Stochastic Linear Reaction Kinetic Systems

In Chapter 10, we discussed the single-molecular enzyme kinetics, which consists of a set of unimolecular, or pseudo first-order reactions representing conformational transitions of an individual enzyme molecule. Not all the linear chemical kinetics are unimolecular reactions, however. One of the simplest counter examples is a constant-rate synthesis reaction. The stochastic Markov models based on single-molecule conformational states discussed in the previous chapter are not applicable to kinetic systems in which the number of molecules can fluctuate with time. In this chapter, we study the general theory for kinetic systems with linear reactions. Because of linearity, the expected value of a linear function is a function of the expected value. Therefore, there is a close relationship between macroscopic deterministic kinetics and mesoscopic stochastic kinetics.

Historically, there have been many motivations for studying the general kinetics of unimolecular reaction networks: heterogeneous catalysis is one example [273], and the biophysical study of single-membrane channel proteins is another. Both predate the studies of single-molecule enzymology in an aqueous environment. We have seen from Chapters 2 and 10 that the master equation that describes the probabilities of the stochastic kinetics of single macromolecules is essentially identical to first-order, unimolecular chemical kinetics in terms of concentration: one needs only to reinterpret the concentration of the molecule in state k, normalized by the total concentration, as the probability of a single molecule being in state k. Such a simple relation between stochastic kinetics and macroscopic kinetics no longer holds true for either nonlinear chemical reactions or linear open reaction systems in which the total number of molecules are not conserved over time. For open chemical systems and nonlinear chemical reaction systems, the stochastic chemical kinetic theory has to be represented very differently, in terms of the Delbrück–Gillespie processes (DGP) introduced in Chapter 5.

Recall that in a DGP, the *states* in the stochastic model no longer correspond to the states of the molecules: they become discrete, integer numbers of the participating molecules in different states. It is instructive to know that many years ago, in the study of fluid mechanics, a similar *change of representation* occurred, from (Lagrangian) tracking of the states of point masses according to their positions and velocities to (Eulerian) counting of the number of point masses with a particular position and velocity. J. W. Gibbs went through a similar process when he was formulating chemical thermodynamics. In fact, a paradox named after him arose due to this change in representation, and scholars continue to debate the issue [19, 21]. The stochastic approach of counting the molecules has been nicely discussed more recently by D. T. Gillespie [104]. Mino et al. [179] suggested the terminology of *state-tracking* versus *number-counting*. Amazingly, even in quantum physics, there was also such a change in representation, known as the second quantization.

In this chapter, we will first study two- and three-state stochastic kinetic systems in detail. Then, in Sections 11.3 and 11.4 we will address the systems of linear reactions with and without conservation of the number of molecules. Section 11.3 seems to overlap with that in Chapter 10 on enzyme kinetics, but the emphasis here will be on how the number-counting approach is related to the state-tracking approach via the concept of i.i.d. (independent and identically distributed) and multinomial distribution. The nonlinear reaction systems will be discussed in Chapter 12.

In the language of Gibbsian equilibrium statistical mechanics, an isothermal system with a fluctuating total number of particles is called a *grand canonical ensemble*. In biophysical chemistry, the formulation of a single macromolecule with multiple binding sites and a fluctuating number of bound ligands with constant ligand concentrations in the solution is the essence of J. Wyman's theory of *binding polynomials* [275] and R. A. Alberty's *semigrand* system [3]. The rediscovery of what is essentially the grand partition function for this type of problem by various physical biochemists testifies to the naturalness of the mathematical method [118]. Part of the material covered in the present chapter can be found in [83, 115]. In connection to the applications of Gibbsian equilibrium statistical thermodynamics to molecular biophysics, readers are also referred to excellent textbooks on biophysical chemistry [20, 14].

11.1 Stochastic Two-state Unimolecular Reaction

The simplest chemical reaction is the unimolecular isomerization reaction, also called conformational change in the literature on proteins and enzymes:

$$A \underset{k_{-1}}{\overset{k_1}{\rightleftharpoons}} B \tag{11.1}$$

where the forward and backward rate constants are k_1 and k_{-1}, respectively. A and B could be the two possible conformations of a six-carbon ring (cyclohexane) molecule that can be in either a "boat" or a "chair" conformation. Then, the kinetic equations for their macroscopic concentrations are

$$\frac{dc_A(t)}{dt} = -k_1 c_A + k_{-1} c_B,$$
$$\frac{dc_B(t)}{dt} = k_1 c_A - k_{-1} c_B, \tag{11.2}$$

in which $c_A(t)$ and $c_B(t)$ are the concentrations of A and B, respectively, at time t. One can solve this equation with initial conditions $c_A(0)$ and $c_B(0)$ and obtain

$$c_A(t) = \left(\frac{k_1 c_A(0) - k_{-1} c_B(0)}{k_1 + k_{-1}} \right) e^{-\lambda t} + \frac{k_{-1}[c_A(0) + c_B(0)]}{k_1 + k_{-1}},$$
$$c_B(t) = \left(\frac{k_{-1} c_B(0) - k_1 c_A(0)}{k_1 + k_{-1}} \right) e^{-\lambda t} + \frac{k_1[c_A(0) + c_B(0)]}{k_1 + k_{-1}}, \tag{11.3}$$

in which $\lambda = k_1 + k_{-1}$. The two equations in (11.2) can be expressed in a compact form in terms of a matrix and vectors: $\dot{\mathbf{x}} = \mathbf{Q}\mathbf{x}(t)$, where

$$\mathbf{Q} = \begin{bmatrix} -k_1 & k_{-1} \\ k_1 & -k_{-1} \end{bmatrix}, \quad \mathbf{x} = \begin{bmatrix} c_A \\ c_B \end{bmatrix}, \tag{11.4}$$

The 2×2 matrix $\mathbf{Q}$ has two eigenvalues, $\lambda_1 = 0$ and $\lambda_2 = -k_1 - k_{-1}$, with corresponding eigenvectors

$$\mathbf{v}_1 = \begin{bmatrix} k_{-1} \\ k_1 \end{bmatrix}, \quad \mathbf{v}_2 = \begin{bmatrix} 1 \\ -1 \end{bmatrix}. \tag{11.5}$$

Then, the solution in (11.3) has a vector form

$$\mathbf{x}(t) = \left(\frac{c_A(0) + c_B(0)}{k_1 + k_{-1}} \right) e^{\lambda_1 t} \mathbf{v}_1 + \left(\frac{k_1 c_A(0) + k_{-1} c_B(0)}{k_1 + k_{-1}} \right) e^{\lambda_2 t} \mathbf{v}_2. \qquad (11.6)$$

Single-molecule experiments in biophysics now allow people to monitor macro-molecules one at a time, follow the state of each molecule as a function of time and observe discrete state transitions. To mathematically represent such an experiment, we note that in the Eqn. (11.2), the $c_A(t) + c_B(t) = c_A(0) + c_B(0) = c_{tot}$ for all the time t. Since all the molecules are assumed to be noninteracting, e.g., they behave statistically independently, $c_A(t)/c_{tot}$ and $c_B(t)/c_{tot}$ can be interpreted as the "probability of a molecule in the A and B states", respectively. Indeed, when one only observes a single molecule [181], the concentrations no longer make sense. Rather, one talks about the probabilities of the molecule being in state A and state B, $p_A(t)$ and $p_B(t)$, respectively. The single molecule continuously jumps back-and-forth between the two conformational states. When it is in A, its probability of going to B is like radioactive decay with an exponential distribution, i.e., a linear rate equation with only $A \to B$,

$$\frac{dp_A}{dt} = -k_1 p_A, \quad \text{and} \quad \frac{dp_B}{dt} = k_1 p_A.$$

Similarly, if the molecule is in B, then with only $B \to A$,

$$\frac{dp_A}{dt} = k_{-1} p_B, \quad \text{and} \quad \frac{dp_B}{dt} = -k_{-1} p_B.$$

Therefore, combining these two processes, when the molecule has probabilities $p_A(t)$ and $p_B(t)$ in states A and B at time t, respectively, we have the set of equations:

$$\frac{dp_A}{dt} = -k_1 p_A + k_{-1} p_B,$$
$$\frac{dp_B}{dt} = k_1 p_A - k_{-1} p_B. \qquad (11.7)$$

This is the model for unimolecular reactions in terms of a two-state, continuous-time Markov process. Matrix $\mathbf{Q}$ is called the transition rate matrix (or infinitesimal transition probability) for the Markov process.

The $\mathbf{Q}$ matrix in (11.4) has the following two essential properties:

(1) all of its off-diagonal elements are nonnegative,

(2) all the rows sum to zero.

The matrix has a clear probabilistic meaning, as a transition probability from time 0 to time t:

$$e^{Qt} = \begin{bmatrix} p_A(t|A) & p_B(t|A) \\ p_A(t|B) & p_B(t|B) \end{bmatrix},$$ (11.8)

where $p_\alpha(t|\beta)$, $\alpha,\beta = A,B$, is the probability of the molecule being in state α at time t, given it is in state β at time 0. It is called the transition probability matrix of the continuous-time Markov process.

11.1.1 Approaching a Stationary Distribution

One could be interested in the behavior of e^{Qt}. Clearly, when $t = 0$, it is an identity matrix, as should be expected. What it will be when $t \to \infty$? To determine this, let us assume that we can diagonalize matrix Q into a diagonal Λ: $Q = B\Lambda B^{-1}$ where B is the transformation matrix, which consists of the eigenvectors of Q. Then,

$$e^{Qt} = Be^{\Lambda t}B^{-1} = B\begin{pmatrix} e^{\lambda_1 t} & 0 \\ 0 & e^{\lambda_2 t} \end{pmatrix}B^{-1},$$ (11.9)

where $\lambda_1 = 0$ and $\lambda_2 = -(k_1 + k_{-1})$,

$$B = \begin{pmatrix} 1 & \frac{k_1}{k_1+k_{-1}} \\ 1 & -\frac{k_{-1}}{k_1+k_{-1}} \end{pmatrix}, \quad B^{-1} = \begin{pmatrix} \frac{k_{-1}}{k_1+k_{-1}} & \frac{k_1}{k_1+k_{-1}} \\ 1 & -1 \end{pmatrix}.$$ (11.10)

Therefore,

$$\lim_{t\to\infty} e^{Qt} = \begin{pmatrix} \frac{k_{-1}}{k_1+k_{-1}} & \frac{k_1}{k_1+k_{-1}} \\ \frac{k_{-1}}{k_1+k_{-1}} & \frac{k_1}{k_1+k_{-1}} \end{pmatrix}.$$ (11.11)

The left eigenvector of Q associated with the zero eigenvalue,

$$\left(\frac{k_{-1}}{k_1+k_{-1}}, \frac{k_1}{k_1+k_{-1}} \right)$$

is the stationary distribution: (p_A^{ss}, p_B^{ss}). Stationary means that if one starts with this distribution, it does not change with time; and if one starts with any other distribution, the system approaches this distribution as time tends to infinity.

We now introduce the concept of stationary stochastic processes. If we use the stationary distribution as the initial value and e^{Qt} as the transition probability, then we obtain a stochastic process that statistically no longer changes with time. However, if one samples this stochastic process, one will still obtain a fluctuating, "noisy" trajectory. Experimentally, this is where one can learn a great deal about a process just by "watching" it fluctuate without any perturbation.

In the *stationary stochastic process*, the molecule spends a fraction p_A^{ss} of time in state A and a fraction p_B^{ss} of time in state B. To demonstrate this, we note that the lifetime of the molecule in state A (or B) is an exponential distribution with pdf $k_1 e^{-k_1 t}$ (or $k_{-1} e^{-k_{-1} t}$). Hence, the mean lifetime (also known as the sojourn time) is $1/k_1$ for state A and $1/k_{-1}$ for state B:

$$p_A^{ss} = \frac{\frac{1}{k_1}}{\frac{1}{k_1} + \frac{1}{k_{-1}}}, \quad p_A^{ss} = \frac{\frac{1}{k_{-1}}}{\frac{1}{k_1} + \frac{1}{k_{-1}}}.$$

11.1.2 Identical, Independent Molecules With Unimolecular Reactions

How is Eqn. (11.2) related to Eqn. (11.7)? To answer this question, let us consider a system of N identical but independent molecules, each following stochastic kinetics. The most important implication of unimolecular reactions is that the behavior of each molecule is statistically independent of that of any other molecules. Therefore, if an individual molecule has probabilities $p_A(t)$ and $p_B(t)$ of being in states A and B, then the number of molecules, among the N total, in state A and state B, $n_A(t)$ and $n_B(t)$, has a binomial distribution:

$$P\{n_A(t) = \ell\} = \frac{N!}{\ell!(N-\ell)!} \left(p_A(t)\right)^\ell \left(p_B(t)\right)^{N-\ell}, \tag{11.12}$$

in which the notion $P\{\cdots\}$ reads as "the probability of $\cdots$". The expected value of the binomial distributed random variables $n_A(t)$ and $n_B(t)$ are $N p_A(t)$ and $N p_B(t)$, respectively. Their expected values, $\mathbb{E}[n_A(t)]$ and $\mathbb{E}[n_B(t)]$, therefore satisfy Eqn. (11.2). The variance of the binomial distribution is $N p_A(t) p_B(t)$. Therefore, the relative variances are

$$\frac{\text{var}[n_A(t)]}{\mathbb{E}^2[n_A(t)]}, \frac{\text{var}[n_B(t)]}{\mathbb{E}^2[n_B(t)]} \propto N^{-1}, \tag{11.13}$$

which are negligibly small for large N. Therefore, for a macroscopic reactions in a vessel, Eqn. (11.2) is exact.

Actually, for a single molecule, we can introduce a random variable m, which equals 1 when the molecule is in state A and 0 when the molecule is in state B.

Then, random variable $m(t)$ has expected value $p_A(t)$ and variance $p_A(t)[1 - p_A(t)]$. The above $n_A(t)$ can be expressed as the sum of N independent and identically distributed (i.i.d.)

$$n_A(t) = m_1(t) + m_2(t) + \cdots + m_N(t).$$

Then, according to the theory of elementary probability, the expected value for the sum is $N p_A(t)$, and its variance is $N p_A(t)[1 - p_A(t)]$.

11.1.3 Autocorrelation and Autocovariance Functions

Different from a steady state in deterministic dynamics, when all the concentrations are at fixed values, the stochastic stationary state is a very rich object: all the concentrations are continuously fluctuating, with a probability distribution independent of time. More interestingly, a stochastic stationary state has an autocorrelation function that actually contains all the information on kinetics!

The autocorrelation function is formally related to the covariance between $m(t)$ to $m(t + \tau)$. To compute this quantity, we need to know their joint probability distribution, which means the probability of a molecule being in state A at time t and in state A at time $t + \tau$. Since in a stationary state,

$$P\{m(t) = 1, m(t + \tau) = 1\} = P\{m(\tau) = 1 | m(0) = 1\} \times p_A^{ss}, \qquad (11.14)$$

one needs to calculate the conditional probability in (11.14). This is the solution to Eqn. (11.7), or (11.2), with initial conditions $p_A(0) = 1$ and $p_B(0) = 0$:

$$P\{m(\tau) = 1 | m(0) = 1\} = \left(\frac{k_1}{k_1 + k_{-1}} \right) e^{-\lambda \tau} + \frac{k_{-1}}{k_1 + k_{-1}}. \qquad (11.15)$$

This is indeed the $p_A(\tau | A)$ in (11.8). Therefore, one has the autocovariance function in the stationary state

$$\mathrm{cov}\big[m(t), m(t + \tau) \big] = \mathbb{E}\big[m(\tau) m(0) \big] - \mathbb{E}^2[m(t)] = \left(\frac{k_1 k_{-1}}{(k_1 + k_{-1})^2} \right) e^{-\lambda \tau}. \qquad (11.16)$$

When $\tau = 0$, this autocovariance is the variance of stationary $m(t)$. For the system with N i.i.d. molecules in a stationary state, the expected value $\mathbb{E}\big[n_A^{ss} \big] = N p_A^{ss}$, and its autocovariance function is

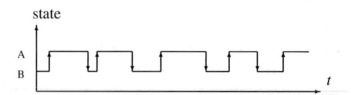

state

A

B

t

Fig. 11.1 A typical stochastic Markov jump process with two states A and B. The lifetime in state A follows an exponential distribution $k_1 e^{-k_1 t}$, and the lifetime in state B follows $k_{-1} e^{-k_{-1} t}$.

$$\mathrm{cov}\left[n_A(t+\tau), n_A(t)\right] = N \, \mathrm{cov}\left[m(t), m(t+\tau)\right]. \tag{11.17}$$

11.1.4 Statistical Analysis of Stochastic Trajectories

There are two different ways to model and understand a stochastic process: one follows the actual stochastic dynamics (i.e., a realization of the stochastic process), and the other deals with the probabilities of the dynamics. The master equation (11.7) is the latter. What is the former, then?

Fig. 11.1 is a typical "trajectory" of the two-state reaction in Eqn. (11.7), jumping between two states A and B stochastically. This kind of data are what biophysicists obtain from membrane protein single-channel recordings and soluble single-molecule enzymology.

There are several standard, statistical treatments for such data. The goal for a statistical treatment is to obtained quantities that are independent of the "realization" and thus are intrinsic to the dynamics, for example, the A and B lifetimes. One can actually compute the autocovariance function from an long trajectory. Let $X(t)$ be a measurement of the state of the molecule as a stationary process; the sample autocovariance function

$$\hat{G}(\tau) = \frac{1}{T} \int_0^T X(t) X(t+\tau) dt - \left(\frac{1}{T} \int_0^T X(t) dt \right)^2. \tag{11.18}$$

For example, $X(t)$ could be the conductance of a channel and/or the fluorescence of an enzyme.

For the two-state single molecule, what does one expect to obtain for the limit of $G(\tau)$ as T goes to infinity? We note that the two terms in Eqn. (11.18), as $T \to \infty$, are in fact the estimations of the expected value and the autocovariance

$$\frac{1}{T} \int_0^T X(t)\,dt \to \mathbb{E}\big[\mathbf{X}^{ss}(t)\big], \tag{11.19a}$$

$$\frac{1}{T} \int_0^T X(t)X(t+\tau)\,dt \to \mathbb{E}\big[\mathbf{X}^{ss}(0)\mathbf{X}^{ss}(\tau)\big]. \tag{11.19b}$$

To calculate theoretically the right-hand side of Eqn. (11.19), let us assume that X_A and X_B are the signals for states A and B, respectively. Then, we have

$$\mathbb{E}\big[\mathbf{X}^{ss}(t)\big] = p_A^{ss} X_A + p_B^{ss} X_B = \frac{k_{-1}X_A + k_1 X_B}{k_1 + k_{-1}}, \tag{11.20}$$

where p_A^{ss} and p_B^{ss} are the stationary state probabilities for states A and B, respectively. In addition, based on the matrix elements in (11.8)

$$\mathbb{E}\big[\mathbf{X}^{ss}(0)\mathbf{X}^{ss}(\tau)\big] = \sum_{\alpha=A,B}\sum_{\beta=A,B} X_\alpha X_\beta p_\alpha(\tau|\beta)p_\beta^{ss}$$

$$= \big(X_A p_A^{ss}, X_B p_B^{ss}\big)e^{\mathbf{Q}\tau}\begin{pmatrix} X_A \\ X_B \end{pmatrix}. \tag{11.21}$$

The element of $e^{\mathbf{Q}\tau}$ is already contained in Eqn. (11.6). Let us now compute it by matrix manipulation. We note that $\mathbf{Q}^2 = -(k_1 + k_{-1})\mathbf{Q}$. Therefore,

$$e^{\mathbf{Q}\tau} = \sum_{k=0}^\infty \frac{(\mathbf{Q}\tau)^k}{k!} = \mathbf{I} + \mathbf{Q}\tau - \frac{(k_1+k_{-1})}{2!}\mathbf{Q}\tau^2 + \frac{(k_1+k_{-1})^2}{3!}\mathbf{Q}\tau^3 + \cdots$$

$$= \mathbf{I} + \Big(1 - e^{-(k_1+k_{-1})\tau}\Big)\frac{\mathbf{Q}}{k_1 + k_{-1}}. \tag{11.22}$$

Note that the exponential of an $n \times n$ matrix can always be expressed in an n-term sum. Because of the Cayley–Hamilton theorem, one can always express the matrix power in terms of a polynomial of order n.

Therefore, for a two-state Markov system, $\mathbb{E}\big[\mathbf{X}^{ss}(0)\mathbf{X}^{ss}(\tau)\big]$ as a function of τ has only a single exponential term:

$$\mathbb{E}\big[\mathbf{X}^{ss}(0)\mathbf{X}^{ss}(\tau)\big] = a + b^{-(k_1+k_{-1})\tau}, \tag{11.23}$$

in which

$$a + b = \mathbb{E}\left[\mathbf{X}^{ss}(0)\mathbf{X}^{ss}(0)\right] = \frac{k_{-1}X_A^2 + k_1 X_B^2}{k_1 + k_{-1}},$$

$$a = \mathbb{E}^2\left[\mathbf{X}^{ss}(t)\right] = \left(\frac{k_{-1}X_A + k_1 X_B}{k_1 + k_{-1}}\right)^2,$$

$$b = (X_A - X_B)^2 \frac{k_1 k_{-1}}{(k_1 + k_{-1})^2} = (X_A - X_B)^2 p_A^{ss} p_B^{ss}.$$

Therefore, as $T \to \infty$,

$$\hat{G}(\tau) \to G(\tau) = (X_A - X_B)^2 p_A^{ss} p_B^{ss} e^{-(k_1 + k_{-1})\tau}. \tag{11.24}$$

Note that $G(\tau)$ decays faster than both lifetimes A and B.

We note that the autocovariance function $G(\tau)$ has essentially the same functional form as the linear relaxation of the macroscopic kinetics. There is a very simple and intuitive explanation. Note that $\mathbb{E}\left[\mathbf{X}(0)\mathbf{X}(\tau)\right]$ can be written as

$$\mathbb{E}\left[\mathbf{X}(0)\mathbf{X}(\tau)\right] = \mathbb{E}^{X_0}\left[\mathbb{E}^{X_\tau}\left[\mathbf{X}(\tau)|\mathbf{X}(0)\right]\mathbf{X}(0)\right], \tag{11.25}$$

in which $\mathbb{E}^{X_0}[\cdots]$ and $\mathbb{E}^{X_\tau}[\cdots]$ are the expectations of random variables $\mathbf{X}(0)$ and $\mathbf{X}(\tau)$, respectively. This is best shown in the Fig. 11.2.

For every given $X(0)$, the conditional expectation $\mathbb{E}^{X_\tau}[\mathbf{X}(\tau)|\mathbf{X}(0)]$ gives the average relation of the system back to equilibrium, starting at $\mathbf{X}(0)$. If all the $\mathbf{X}(0)$ are in the linear region near the equilibrium, all the kinetics will be represented as the "linear combination" of all the $e^{\lambda_k \tau}$, where λ_k are the eigenvalues of the linear system.

11.2 A Three-state Unimolecular Reaction

Let us now consider the following unimolecular chemical reaction:

$$A \underset{k_{-1}}{\overset{k_1}{\rightleftharpoons}} B \underset{k_{-2}}{\overset{k_2}{\rightleftharpoons}} C \tag{11.26}$$

There are two approaches to this problem: one is based on only a single molecule with a simple 3-state Markov process, and the other is based on N total molecules and a birth-death process. We will show that these two approaches are identical in the case of a unimolecular (linear) reaction. This will not be the case for nonlinear chemical reactions.

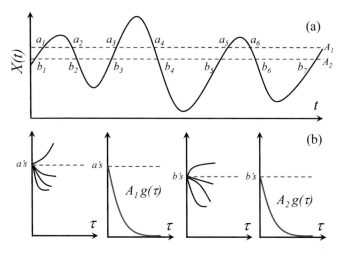

Fig. 11.2 Given a realization of a stationary Markov signal $X(t)$ shown in (a), for every given value, say A_1, one finds $X(t) = A_1$ at different time points $a_1, a_2, \cdots$, and then, due to the stationary Markov property, averages all the subsequent $X_i(\tau) = X(a_i + \tau)$'s as in (b) and obtains the relaxation function in red: $\mathbb{E}[X(a+\tau)|X(a) = A]$. For a linear system, one expects that the average has the general form $\mathbb{E}[X(a+\tau)|X(a) = A] = Ag(\tau)$. One does this for all values of A_1, A_2, etc. and then the autocovariance function in (11.25) is obtained by averaging over all the different $X(a)$'s. This illustrates why the relaxation function and autocorrelation function have the same form for a linear system.

Consider N i.i.d. molecules, each of which follows the kinetics in (11.26). Let $p(\ell, m;t) = P\{n_A(t) = \ell, n_C(t) = m\}$. The birth-death process associated with reaction scheme (11.26) for N total molecules is

$$
\frac{dp(\ell, m;t)}{dt} = (\ell + 1)k_1 p(\ell, m) + (N - \ell - m + 1)k_{-1} p(\ell - 1, m)
$$
$$
+ (N - \ell - m + 1)k_2 p(\ell, m - 1) + (m + 1)k_{-2} P(\ell, m + 1)
$$
$$
- [k_1 \ell + (k_{-1} + k_2)(N - \ell - m) + k_{-2} m] p(\ell, m). \tag{11.27}
$$

Note that $n_B(t) = N - n_A(t) - n_C(t)$.

With some algebra, one can easily verify that the $p(\ell, m;t)$ in Eqn. (11.26) can be expressed in terms of $p_A(t)$ and $p_C(t)$, which satisfy

$$
\frac{dp_A}{dt} = -k_1 p_A + k_{-1}(1 - p_A - p_C),
$$
$$
\tag{11.28}
$$
$$
\frac{dp_C}{dt} = -k_{-2} p_C + k_2(1 - p_A - p_C).
$$

In fact,

$$p(\ell,m;t) = \frac{N!}{\ell!(N-\ell-m)!m!}(p_A(t))^\ell(1-p_A(t)-p_C(t))^{N-\ell-m}(p_C(t))^m.$$

$$(11.29)$$

The distribution in (11.29) is a *multinomial* (see Sec. 3.1.5). It is a generalization of the binomial distribution in (11.12). This result is completely expected
since the N particles are identical and statistically independent. $p_A(t)$, $p_C(t)$ and
$p_B(t) = 1 - p_A(t) - p_C(t)$ are the solutions to Eqn. (11.26) with initial condition
$p_A(0) = n_A(0)/N$, $p_B(0) = n_B(0)/N$, and $p_C(0) = n_C(0)/N$.

11.2.1 Diffusion Approximation

Even though the diffusion approximation to a DGP is problematic in general, as we
shall learn in Chapter 12, it is a good approximation for linear kinetics. For a large N,
we can introduce continuous variables $x = n_A/N$, $y = n_C/N$, $(n_A + 1)/N = x + dx$,
and $(n_B + 1)/N = y + dy$. Then Eqn. (11.27) can be approximated by the linear
partial differential equation

$$\frac{\partial f(x,y;t)}{\partial t} = \frac{1}{2N}\left\{\frac{\partial^2}{\partial x^2}\left(k_1 x + k_{-1}(1-x-y)\right)f(x,y)\right.$$

$$\left. + \frac{\partial^2}{\partial y^2}\left(k_{-2}y + k_2(1-x-y)\right)f(x,y)\right\} + \frac{\partial}{\partial x}\left[\left(k_1 x - k_{-1}(1-x-y)\right)f\right]$$

$$+ \frac{\partial}{\partial y}\left[\left(k_{-2}y - k_2(1-x-y)\right)f(x,y)\right] + O\left(N^{-2}\right),$$

$$(11.30)$$

in which $f(x,y;t)$ is the probability density function for the fractions of molecules
in the A state and in the C state as real-valued random variables $\mathbf{x}(t)$ and $\mathbf{y}(t)$:

$$f(x,y,;t)dxdy = P\{x < \mathbf{x}(t) \le x+dx, y < \mathbf{y}(t) \le y+dy\}.$$

$$(11.31)$$

Since the conservation of total probability holds, the boundary conditions for the
PDE are no flux with the reflecting boundaries defined by the region in $\mathbb{R}^2$, $\Gamma : x \ge$
$0, y \ge 0, x+y \le 1$.

Note that if N is large, one can further neglect the diffusive terms on the rhs of
(11.30), and the equation then will be identical to a pair of ODEs for x and y, as in
a deterministic mass-action kinetic model:

$$\frac{dx}{dt} = -\left(k_1 x - k_{-1}(1 - x - y)\right),$$

$$\frac{dy}{dy} = -\left(k_{-2} y - k_2(1 - x - y)\right). \tag{11.32}$$

For a linear reaction system, this large N result is the same as that of the equations for the expected values of $n_A(t), n_B(t), n_C(t)$. This is not generally true for nonlinear reactions, where the expected values do not satisfy a set of self-contained, ordinary differential equations.

We shall continue to derive the differential equations for the expected values and variances of $\mathbf{x}$ and $\mathbf{y}$ following the diffusion approximation. As an exercise, the readers should determine the exact results directly from the CME and compare them with those obtained below.

The expected values for stochastic $\mathbf{x}(t)$ and $\mathbf{y}(t)$ in (11.30) are

$$\mathbb{E}[\mathbf{x}(t)] = \int_\Gamma x f(x,y;t) dxdy, \quad \mathbb{E}[\mathbf{y}(t)] = \int_\Gamma y f(x,y;t) dxdy. \tag{11.33}$$

Multiplying x or y on both sides of (11.30) and integrating, assuming the integration and $\partial/\partial t$ can exchange order, we obtain from Eqn. (11.30):

$$\frac{d\mathbb{E}[\mathbf{x}(t)]}{dt} = -k_1 \mathbb{E}[\mathbf{x}(t)] + k_{-1}\left(1 - \mathbb{E}[\mathbf{x}(t)] - \mathbb{E}[\mathbf{y}(t)]\right),$$

$$\frac{d\mathbb{E}[\mathbf{y}(t)]}{dt} = -k_{-2}\mathbb{E}[\mathbf{y}(t)] + k_2\left(1 - \mathbb{E}[\mathbf{x}(t)] - \mathbb{E}[\mathbf{y}(t)]\right), \tag{11.34}$$

which are identical to (11.28), the kinetic equation for single-molecule probabilities, and (11.32), the equation based on macroscopic mass-action kinetics. The solutions to these equations are $\langle \mathbf{x}(t)\rangle = p_A(t)$ and $\langle \mathbf{y}(t)\rangle = p_C(t)$.

In addition to the expected values $\mathbb{E}[\mathbf{x}(t)]$ and $\mathbb{E}[\mathbf{y}(t)]$, more information concerning the stochastic kinetics is contained in (11.30). One can also derive differential equations for variances and the covariance

$$\sigma_{\xi\eta}(t) = \int_\Gamma (\xi - \langle\xi\rangle)(\eta - \langle\eta\rangle) f(x,y;t) dxdy, \tag{11.35}$$

where $\xi, \eta = x$ or y.

The obtained equations are as follows:

$$\frac{d\sigma_{xx}(t)}{dt} = \frac{k_1 \mathbb{E}[\mathbf{x}] + k_{-1}(1 - \mathbb{E}[\mathbf{x}] - \mathbb{E}[\mathbf{y}])}{N} - 2k_1\sigma_{xx} - 2k_{-1}(\sigma_{xx} + \sigma_{xy}) \tag{11.36a}$$

$$\frac{d\sigma_{yy}(t)}{dt} = \frac{k_{-2}\mathbb{E}[\mathbf{y}] + k_2(1 - \mathbb{E}[\mathbf{x}] - \mathbb{E}[\mathbf{y}])}{N} - 2k_{-2}\sigma_{yy} - 2k_2(\sigma_{xx} + \sigma_{xy}) \quad (11.36b)$$

$$\frac{d\sigma_{xy}(t)}{dt} = \frac{d\sigma_{yx}(t)}{dt} = -(k_1 + k_{-1} + k_2 + k_{-2})\sigma_{xy} - k_2\sigma_{xx} - k_{-1}\sigma_{yy}. \quad (11.36c)$$

It can be verified that the solutions to these equations are

$$\sigma_{xx}(t) = \frac{p_A(t)(1 - p_A(t))}{N}, \quad \sigma_{yy}(t) = \frac{p_C(t)(1 - p_C(t))}{N}, \quad \sigma_{xy}(t) = -\frac{p_A(t)p_C(t)}{N}.$$
$$(11.37)$$

Diffusion approximation is also known as *linear noise approximation*. The results in the present section show that it is a valid approach to linear stochastic kinetics. In fact, it is quite accurate up to the second moments.

11.3 Stochastic Theory for Closed Unimolecular Reaction Networks

With the above examples for systems with two and three states, we are now in a position to study the general linear kinetic system with a fluctuating number of molecules. To set the ground rules for the next section on grand canonical unimolecular networks, here we will discuss closed unimolecular reaction systems in general. We are particularly interested in the relation between the number-counting approach and the state-tracking approach in Chapter 10. In the chemical kinetic literature, there have been numerous occasions in which investigators did not fully appreciate this connection [120, 234, 55].

The term *open* has several different meanings that deserve to be clarified: in a strictly thermodynamic sense, a closed system has to have detailed balance, and thus no energy pumped in, as well as a conserved number of molecules. In the present chapter, the term "open" is understood as open to the exchange of molecules. Therefore, the term *closed* in the present section refers to a unimolecular network, with or without detailed balance but with a fixed total number of molecules.

Following Chapter 3.3, the dynamics of a single macromolecule with K states in a closed system are represented by a Markov jump process among different states with the master equation

$$\frac{dp_i(t)}{dt} = \sum_{j=1, j\neq i}^{K} \left(p_j(t)q_{j,i} - p_i(t)q_{i,j} \right), \quad 1 \leq i \leq K, \quad (11.38)$$

where $p_i(t)$ represents the probability of the molecule being in state i at time t and $q_{i,j}$ represents the transition probability from state i to j in instantaneous time. In this chapter, we will assume that the systems being studied are irreducible so that all pairs of states can communicate with each other, directly or indirectly. Chemically, this means that the reaction system does not contain isolated subreaction systems. It is also useful to note that the $j \neq i$ in (11.38) is not necessary: when $i = j$, the two terms in the summand cancel.

Let there be N such fixed single molecules in a closed chemical reaction system. If we use $N_i(t)$ to represent the number of molecules in state i at time t, then

$$\sum_{i=1}^{K} N_i(t) = N.$$

This "canonical" system is actually a collection of noninteracting single molecules. It is an i.i.d. ensemble. The joint probability distribution for the $N_i(t)$'s,

$$p(n_1, n_2, \ldots, n_K; t) = P\{N_1(t) = n_1, \cdots, N_K(t) = n_K\},$$

is related to the combinatorial problem of computing the number of ways to arrange the N molecules into the K states with n_i being in state i:

$$
\begin{aligned}
\binom{N}{n_1, n_2, \ldots, n_K} &= \binom{N}{n_1}\binom{N-n_1}{n_2} \cdots \binom{n - n_1 - n_2 - \cdots - n_{K-1}}{n_K} \\
&= \frac{N!}{n_1!(N-n_1)!} \frac{(N-n_1)!}{n_2!(N-n_1-n_2)!} \cdots \frac{(N - n_1 - \cdots - n_{K-1})!}{n_K!} \\
&= \frac{N!}{n_1! n_2! \cdots n_K!},
\end{aligned}
\tag{11.39}
$$

which is known as the multinomial coefficient. Note that the N molecules are independent of each other, which implies that the joint probability of having n_i molecules in each state is

$$p(n_1, n_2, \ldots, n_K; t) = \frac{N!}{n_1! n_2! \cdots n_K!} (p_1(t))^{n_1} (p_2(t))^{n_2} \cdots (p_K(t))^{n_K}, \tag{11.40}$$

where the $p_i(t)$ are the solutions of (11.38). This joint probability is a multinomial distribution. Eqns. (11.12) and (11.29) are special cases with $N = 2$ and $N = 3$, respectively.

An immediate consequence of (11.40) is that $\mathbb{E}[N_i(t)] = Np_i(t)$ and $\text{Var}[N_i(t)] = Np_i(t)[1 - p_i(t)]$. We note that when dealing with N_i, one can effectively consider

N_i and $N_i' = N - N_i$ as a two-state system. Actually, the marginal distribution of N_1, from (11.40), is

$$
\sum_{n_2,\cdots,n_K} p(n_1,\cdots,n_K;t) = \sum_{n_2,\cdots,n_K} \frac{N!}{n_1!n_2!\cdots n_K!} \left(p_1(t)\right)^{n_1} \left(p_2(t)\right)^{n_2} \cdots \left(p_K(t)\right)^{n_K}
$$

$$
= \frac{N!}{n_1!(N-n_1)!} \left(p_1(t)\right)^{n_1} \sum_{n_2+\cdots+n_K=N-n_1} \frac{(N-n_1)!}{n_2!\cdots n_K!} \left(p_2(t)\right)^{n_2} \cdots \left(p_K(t)\right)^{n_K}
$$

$$
= \frac{N!}{n_1!(N-n_1)!} \left(p_1(t)\right)^{n_1} \left(1-p_1(t)\right)^{N-n_1}.
$$

For sufficiently large N and following Section 4.4, if we denote the mole fraction of the molecules in state i, $v_i \equiv N_i/N$, then their joint probability distribution

$$
f(v_1,\cdots,v_K) \simeq \exp\left\{ -N \sum_{i=1}^{K} v_i \ln\left(\frac{v_i}{p_i}\right) \right\}. \tag{11.41}
$$

A chemical kinetic system with detailed balance is expected to achieve equilibrium, i.e., the forward and backward fluxes of a given reaction must be equal [272, 164, 125, 119]. Therefore, if π_i represents the probability of being in state i at equilibrium, we have $\mathbb{E}[N_i^{eq}] = N\pi_i$, and

$$
\frac{\mathbb{E}[N_i^{eq}]}{\mathbb{E}[N_j^{eq}]} = \frac{q_{j,i}}{q_{i,j}} = K^{eq} \tag{11.42}
$$

where K^{eq} is the equilibrium constant of the reaction between states i and j, and $q_{i,j}$ should be identified as the rate constant for the unimolecular chemical reaction between species $X_i \to X_j$. This implies that for any cycle of reactions, $X_{i_0} \to X_{i_1} \to \cdots \to X_{i_m} \to X_{i_0}$, within the system, the chemical reaction rate constants must satisfy [119, 137]

$$
\frac{q_{i_0,i_1} q_{i_1,i_2} \cdots q_{i_m,i_0}}{q_{i_1,i_0} q_{i_2,i_1} \cdots q_{i_0,i_m}} = 1, \tag{11.43a}
$$

or equivalently the production of the equilibrium constants around the reaction cycle, e.g., the thermodynamic "box" in elementary chemistry

$$
K_{i_0 i_1}^{eq} K_{i_1 i_2}^{eq} \cdots K_{i_m i_0}^{eq} = 1. \tag{11.43b}
$$

We see the logic of chemical reaction kinetics follows the same logic as in the theory of the Markov process in Chapter 6.

The concept of the *semigrand canonical ensemble* in chemical kinetics refers to a multicomponent system in which one of the components has a conserved number while other components have fixed chemical potential and thus with fluctuations of numbers. It is also known as an osmotic ensemble because of its close relationship with laboratory measurements using a dialysis tube with a semipermeable membrane. The multiple components with fixed chemical potentials, *e.g.* fixing external concentrations in reality, can themselves be with or without detailed balance. The former will be an equilibrium dialysis; the latter will be a chemostat.

In the previous chapters, however, all these effects were absorbed into pseudo first-order transition rates. If these external species are not in their chemical equilibrium, then the detailed balance condition will be broken. Thus far we have not studied explicit exchange with the surrounding *material reservoir* under chemostatic control. This is the focus of the next section.

11.4 Linear Reaction Systems with Open Boundaries

We now introduce the Markov model for systems driven by an explicit chemostatic material reservoir; such a system has an open boundary. That is, while each and every molecule undergoes a set of unimolecular transitions, the total number of i.i.d. molecules vary stochastically due to the fluctuating influx and efflux at the boundary. To focus on the exchanges with the environment with a nonzero chemical potential difference, we assume the reactions inside the system are strictly detail balanced, i.e., the reaction rates satisfy the Kolmogorov cycle condition. The chemostat only comes from the boundary. Thus, we consider a system similar to the closed system in Sec. 11.3 but with the addition of a source at state 0 and a sink at state $K+1$, communicating with the K internal species through linear reactions, and keep the number of molecules in state 0 equal to n_0 and the number in state $K+1$ equal to n_{K+1}.

11.4.1 A Simple Chemical System With Synthesis and Degradation

The simple chemical reactions given in Eqns. (11.1) and (11.26) have no input or output; they are a *closed* chemical reaction and system of reactions. A cell clearly cannot be a closed chemical system with detailed balance; that would mean death.

Now let us consider the simplest open chemical system with a source and a sink

$$\text{source} \xrightarrow{J} X \xrightarrow{k} \text{sink};\tag{11.44}$$

the source has a steady influx of J, in the units of number of molecules per unit time, and X is degraded with the usual first-order, i.e., exponential, kinetics.

Let $\mathbf{N}$ be an integer random variable representing the number of X molecules in the system. Then, we have its probability $p_n(t) = P\{\mathbf{N}(t) = n\}$

$$\frac{dp_n(t)}{dt} = -nkp_n(t) + (n+1)kp_{n+1}(t) + Jp_{n-1}(t) - Jp_n(t),\tag{11.45}$$

for $n = 0, 1, 2, \cdots$.

The steady state probability distribution for the system can be obtained by setting

$$\frac{dp_n(t)}{dt} = 0$$

for all n. Note that for p_0^{ss}, we have $kp_1^{ss} - Jp_0^{ss} = 0$. Therefore, we obtain

$$\frac{p_{n+1}^{ss}}{p_n^{ss}} = \frac{J}{nk}, \quad p_n^{ss} = \frac{1}{n!}\left(\frac{J}{k}\right)^n e^{-J/k}.\tag{11.46}$$

This is a Poisson distribution with mean number of X molecules of J/k, which is exactly what one expects from the traditional law of mass action model:

$$\frac{dc_X}{dt} = -kc_X + J, \quad c_X^{ss} = \frac{J}{k}.\tag{11.47}$$

11.4.1.1 Expected Value and Variance

For the linear reaction, one can also derive the deterministic kinetic equations for the expected value $\mathbb{E}[\mathbf{N}(t)]$ and even variance $\mathrm{Var}[\mathbf{N}(t)]$ directly from Eqn. (11.45) as follows.

$$\langle n(t) \rangle \equiv \mathbb{E}[\mathbf{N}(t)] = \sum_{n=0}^{\infty} np_n(t), \quad \langle n^2(t) \rangle \equiv \mathbb{E}[\mathbf{N}^2(t)] = \sum_{n=0}^{\infty} n^2 p_n(t),$$

$$\begin{aligned}
\frac{d}{dt}\langle n(t) \rangle &= \frac{d}{dt}\mathbb{E}[\mathbf{N}(t)] \\
&= \sum_{n=0}^{\infty} n\Big(-nkp_n(t) + (n+1)kp_{n+1}(t) + Jp_{n-1}(t) - Jp_n(t) \Big) \\
&= J - k\langle n(t) \rangle,
\end{aligned}\tag{11.48}$$

and

$$\frac{d}{dt}\langle n^2(t)\rangle = \sum_{n=0}^{\infty} n^2 \Big(-nkp_n(t)+(n+1)kp_{n+1}(t)+Jp_{n-1}(t)-Jp_n(t)\Big)$$

$$= \sum_{n=0}^{\infty} \big[n^2(n+1)-(n+1)^3\big]kp_{n+1}(t)+\big[n^2-(n-1)^2\big]Jp_{n-1}(t)$$

$$= k\Big[-2\langle n^2(t)\rangle+\langle n(t)\rangle\Big]+\Big[2\langle n(t)\rangle+1\Big]J, \tag{11.49}$$

therefore,

$$\frac{d}{dt}\text{Var}[\mathbf{N}(t)] = \frac{d}{dt}\langle(\Delta n(t))^2\rangle$$

$$= \frac{d}{dt}\langle n^2(t)\rangle-2\langle n(t)\rangle\frac{d}{dt}\langle n(t)\rangle$$

$$= k\Big[-2\langle n^2(t)\rangle+\langle n(t)\rangle\Big]+\Big[2\langle n(t)\rangle+1\Big]J-2\langle n(t)\rangle\Big(J-k\langle n(t)\rangle\Big)$$

$$= J+k\langle n(t)\rangle-2k\langle(\Delta n(t))^2\rangle, \tag{11.50}$$

in which we denote $n(t)-\langle n(t)\rangle = \Delta n(t)$. One sees that from Eqn. (11.48) and (11.50), we have steady state $\langle n\rangle^{ss} = \langle(\Delta n)^2\rangle^{ss} = J/k$, in agreement with the Poisson distribution in (11.46).

11.4.1.2 Time-dependent Solution to (11.45)

One can obtain the time-dependent solution to Eqn. (11.45), the distribution $p_n(t)$ in terms of its generating function

$$G(s,t) = \sum_{n=0}^{\infty} p_n(t)s^n.$$

which satisfies

$$\frac{\partial G(s,t)}{\partial t} = \sum_{n=0}^{\infty} \Big(-nkp_n(t)+(n+1)kp_{n+1}(t)+Jp_{n-1}(t)-Jp_n(t)\Big)s^n$$

$$= J(s-1)G(s,t)-k(s-1)\frac{\partial}{\partial s}G(s,t). \tag{11.51}$$

Its solution is

$$G(s,t) = G\Big(1-(1-s)e^{-kt},0\Big)\exp\Big[-\lambda(1-s)\Big(1-e^{-kt}\Big)\Big], \tag{11.52}$$

in which $\lambda = J/k$, and $G(s,0)$ is the generating function for the initial $p_n(0)$. One immediately has

$$\langle n(t) \rangle = \frac{\partial G(1,t)}{\partial s} = \langle n(0) \rangle e^{-kt} + \lambda \left(1 - e^{-kt}\right), \tag{11.53}$$

$$\langle n(t)(n(t)-1) \rangle = \frac{\partial G^2(1,t)}{\partial s^2}$$

$$= \langle n(0)(n(0)-1) \rangle e^{-2kt} + 2\lambda \langle n(0) \rangle e^{-kt} \left(1 - e^{-kt}\right) + \lambda^2 \left(1 - e^{-kt}\right)^2,$$

$$\mathrm{Var}[n(t)] = \mathrm{Var}[n(0)]e^{-2kt} + \langle n(0) \rangle e^{-kt} \left(1 - e^{-kt}\right) + \lambda \left(1 - e^{-kt}\right). \tag{11.54}$$

11.4.2 K-state Open Linear Reaction System

The above example shows that in the steady state, there is a continuous, sustained flux from the source to the sink. It is a nonequilibrium steady state (NESS). This is a general feature of open, linear kinetic systems, which eventually approach a stochastic stationary state that is in general not a chemical equilibrium. We can see it captures the essential characteristics of a living biochemical system under a single or multiple chemostats.

The total number of molecules in systems with an open boundary is itself an integer-valued random variable; it fluctuates. Actually, the number of molecules in each state i at time t is a random variable $N_i(t)$. Following the terminology in Gibbs' ensemble theory, we call this type of Markov model *grand canonical* since it closely resembles the grand canonical ensemble. Then, the probability

$$P\{N_1(t) = n_1, N_2(t) = n_2, \cdots, N_K(t) = n_K\} \equiv p(n_1, \ldots, n_K; t)$$

satisfies a master equation:

$$\frac{\mathrm{d}}{\mathrm{d}t} p(n_1, \ldots, n_K; t) = \tag{11.55a}$$

$$K \text{ terms} \begin{cases} p(n_1 - 1, n_2, \ldots, n_K; t)(q_{0,1}n_0 + q_{K+1,1}n_{K+1}) \\ + p(n_1, n_2 - 1, \ldots, n_K; t)(q_{0,2}n_0 + q_{K+1,2}n_{K+1}) \\ \vdots \\ + P(n_1, n_2, \ldots, n_K - 1; t)(q_{0,K}n_0 + q_{K+1,K}n_{K+1}) \end{cases} \tag{11.55b}$$

$$K \text{ terms} \begin{cases} +p(n_1+1,n_2,\ldots,n_K;t)(q_{1,0}+q_{1,K+1})(n_1+1) \\ +p(n_1,n_2+1,\ldots,n_K;t)(q_{2,0}+q_{2,K+1})(n_2+1) \\ \vdots \\ +p(n_1,n_2,\ldots,n_K+1;t)(q_{K,0}+q_{K,K+1})(n_K+1) \end{cases} \tag{11.55c}$$

$$K^2-K \text{ terms} \begin{cases} +p(n_1+1,n_2-1,\ldots,n_K;t)q_{1,2}(n_1+1) \\ \vdots \\ +P(n_1,n_2,\ldots,n_{K-1}-1,n_K+1;t)q_{K,K-1}(n_K+1) \end{cases} \tag{11.55d}$$

$$-p(n_1,n_2,\ldots,n_K;t)\sum_{i=1}^{K}(q_{0,i}n_0+q_{K+1,i}n_{K+1}) \tag{11.55e}$$

$$-p(n_1,n_2,\ldots,n_K;t)\sum_{i=1}^{K}(q_{i,0}+q_{i,K+1})n_i \tag{11.55f}$$

$$-p(n_1,n_2,\ldots,n_K;t)\sum_{i=1}^{K}\sum_{j=1,j\neq i}^{K}q_{i,j}n_i. \tag{11.55g}$$

Let us consider that at time $t=0$, the system is assumed to be empty, so

$$p(n_1,\ldots,n_K;0) = \begin{cases} 1 & \text{if } n_1 = \cdots = n_K = 0 \\ 0 & \text{otherwise} \end{cases}. \tag{11.55h}$$

Denoting $(n_1,\cdots,n_K) = \mathbf{n}$, multiplying both sides of the master equation (11.55) by n_i, summing over all $\mathbf{n}$, and identifying the expected values

$$\langle n_i(t)\rangle \equiv \mathbb{E}[\mathbf{N}_i(t)] = \sum_{\mathbf{n}} n_i p(\vec{n};t),$$

we have:

$$\sum_{\mathbf{n}} n_i \frac{\mathrm{d}p(n_1,n_2,\ldots,n_K;t)}{\mathrm{d}t} = \frac{\mathrm{d}\langle n_i(t)\rangle}{\mathrm{d}t}$$

$$= \langle n_i(t)\rangle \sum_{j=1}^{K}(q_{0,j}n_0+q_{K+1,j}n_{K+1}) + (q_{0,i}n_0+q_{N+1,i}n_{N+1})$$

$$+\sum_{j=1}^{K}\left(\langle n_i(t)n_j(t)\rangle(q_{j,0}+q_{j,K+1})\right) - \langle n_i(t)\rangle(q_{i,0}+q_{i,K+1})$$

$$+\sum_{j,k=1,j\neq k}^{K}\left(\langle n_i(t)n_j(t)\rangle q_{j,k}\right) + \sum_{j=1,j\neq i}^{K}\left(\langle n_j(t)\rangle q_{j,i} - \langle n_i(t)\rangle q_{i,j}\right)$$

$$-\langle n_i(t)\rangle \sum_{j=1}^{K}(q_{0,j}n_0+q_{K+1,j}n_{K+1})$$

$$-\sum_{j=1}^{K}\left(\langle n_i(t)n_j(t)\rangle(q_{j,0}+q_{j,K+1})\right) - \sum_{j,k=1,j\neq k}^{K}\left(\langle n_i(t)n_j(t)\rangle q_{j,k}\right). \tag{11.56}$$

Therefore, upon simplification, we see that the expected number of molecules in state i at time t must satisfy

$$\frac{d \langle n_i(t) \rangle}{dt} = \sum_{j=0, j \neq i}^{K+1} \left(\langle n_j(t) \rangle q_{j,i} - \langle n_i(t) \rangle q_{i,j} \right), \tag{11.57}$$

for all $1 \leq i \leq K$, where $\langle n_0(t) \rangle = n_0$ and $\langle n_{K+1}(t) \rangle = n_{K+1}$, $\forall t$. The system is assumed to be empty initially so that the initial conditions (ICs) of (11.57) are

$$\text{ICs: } \langle n_i(0) \rangle = 0, \quad 1 \leq i \leq K. \tag{11.58}$$

For convenience, let $\langle n_i \rangle^{ss}$ be the expected value of the number of molecules in state i when the system is in NESS and define $\Delta n_i \triangleq n_i - \langle n_i \rangle^{ss}$. Then, one can derive the equations that the variances and covariances must satisfy

$$\frac{d \langle (\Delta n_i(t))^2 \rangle}{dt} = \sum_{j=0, j \neq i}^{K+1} \left[\left(\langle n_j(t) \rangle + 2 \langle \Delta n_i(t) \Delta n_j(t) \rangle \right) q_{j,i} \right. \tag{11.59a}$$

$$\left. + \left(\langle n_i(t) \rangle - 2 \langle (\Delta n_i(t))^2 \rangle \right) q_{i,j} \right], \tag{11.59b}$$

and

$$\frac{d \langle \Delta n_i(t) \Delta n_j(t) \rangle}{dt} = \sum_{k=0, k \neq i}^{K+1} \left(\langle \Delta n_j(t) \Delta n_k(t) \rangle q_{k,i} - \langle \Delta n_i(t) \Delta n_j(t) \rangle q_{i,k} \right)$$

$$+ \sum_{k=0, k \neq j}^{K+1} \left(\langle \Delta n_i(t) \Delta n_k(t) \rangle q_{k,j} - \langle \Delta n_i(t) \Delta n_j(t) \rangle q_{j,k} \right)$$

$$- \langle n_i(t) \rangle q_{i,j} - \langle n_j(t) \rangle q_{j,i}. \tag{11.60a}$$

To sustain the boundary condition such that $\langle n_0(t) \rangle \equiv n_0$ and $\langle n_{K+1}(t) \rangle \equiv n_{K+1}$, $\forall t$, the observer must continuously add or remove molecules to states 0 and $K+1$ at rates

$$J_0^{ext}(t) = \sum_{j=1}^{K} \left(n_0 q_{0,j} - \langle n_j(t) \rangle q_{j,0} \right) \tag{11.61a}$$

$$J_{K+1}^{ext}(t) = \sum_{j=1}^{K} \left(n_{K+1} q_{K+1,j} - \langle n_j(t) \rangle q_{j,K+1} \right) \tag{11.61b}$$

respectively. These rates are positive if an experimenter adds and negative if the experimenter removes the respective molecules. When the system is in NESS, J_0^{ext}

and $-J_{K+1}^{ext}$ must be equal, , *i.e.*, , the total number of molecules in the system must be constant.

It can be shown through direct substitution that the joint probability function for the open system is

$$p(n_1,\ldots,n_K;t) = \prod_{i=1}^{K}\left[\frac{\langle n_i(t)\rangle^{n_i}}{n_i!}e^{-\langle n_i(t)\rangle}\right]. \tag{11.62}$$

These random variables are actually independent of each other when the system is subject to open boundaries, which is not the case for the closed system. The Poisson distribution is the major result of the Gibbsian grand partition function. However, the factor $n!$ continues to be a mystery in physics, known as Gibbs' paradox [92].

11.4.3 Correlation Functions of Open Systems

It is straightforward to calculate the correlation and covariance functions of the random variables describing the number of molecules in each state of the open boundary-driven Markov system. In fact, there is a time lag in their correlation functions.

The ODE system in (11.57) can be expressed in terms of a matrix. Substituting $n_i(t) = \Delta n_i(t) + \langle n_i\rangle^{ss}$, we have

$$\frac{d\langle \Delta \mathbf{n}(t)\rangle}{dt} = \mathbf{Q}\langle \Delta \mathbf{n}(t)\rangle \tag{11.63}$$

where the column vector $\langle \Delta \mathbf{n}(t)\rangle \in \mathbb{R}^N$ has i^{th} entry $\langle \Delta n_i(t)\rangle$ and the $K \times K$ matrix $\mathbf{Q}$ has the form

$$\mathbf{Q} = \begin{bmatrix} -\sum\limits_{j=0,j\neq 1}^{K+1} q_{1,j} & q_{2,1} & q_{3,1} & \cdots & q_{K,1} \\ q_{1,2} & -\sum\limits_{j=0,j\neq 2}^{K+1} q_{2,j} & q_{3,2} & \cdots & q_{K,2} \\ \vdots & & \ddots & & \vdots \\ q_{1,K-1} & q_{2,K-1} & \cdots & -\sum\limits_{j=0,j\neq K-1}^{K+1} q_{K-1,j} & q_{K,K-1} \\ q_{1,K} & q_{2,N} & \cdots & q_{K-1,K} & -\sum\limits_{j=0,j\neq K}^{K+1} q_{K,j} \end{bmatrix}. \tag{11.64}$$

The solution of (11.63) with initial conditions $\Delta \mathbf{n}(0) = \Delta \mathbf{n}_0$ is

$$\begin{bmatrix} \langle \Delta n_1(t) | \Delta \mathbf{n}(0) = \Delta \mathbf{n}_0 \rangle \\ \langle \Delta n_2(t) | \Delta \mathbf{n}(0) = \Delta \mathbf{n}_0 \rangle \\ \vdots \\ \langle \Delta n_N(t) | \Delta \mathbf{n}(0) = \Delta \mathbf{n}_0 \rangle \end{bmatrix} = e^{\mathbf{Q}t} \Delta \mathbf{n}_0 = \mathbf{V} e^{\Lambda t} \mathbf{V}^{-1} \Delta \mathbf{n}_0, \qquad (11.65)$$

where the columns of the $K \times K$ matrix $\mathbf{V}$ are the eigenvectors of $\mathbf{Q}$, and $K \times K$ matrix $e^{\Lambda t}$ is a diagonal matrix with i^{th} diagonal entry $e^{\lambda_i t}$, where λ_i is the eigenvalue associated with the ith column of $\mathbf{V}$.

To calculate the autocorrelation function $\langle \Delta n_i(t) \Delta n_i(0) \rangle$, we start with

$$\langle \Delta n_i(t) \Delta n_i(0) \rangle = \langle n_i(t) n_i(0) \rangle - (\langle n_i \rangle^{ss})^2 \qquad (11.66a)$$

$$= \sum_{k=0}^{\infty} k \sum_{j=0}^{\infty} j P\{n_i(t) = j, n_i(0) = k\} - (\langle n_i \rangle^{ss})^2 \qquad (11.66b)$$

$$= \sum_{\mathbf{n}} k \sum_{j=0}^{\infty} j P\{n_i(t) = j, \mathbf{n}(0) = \mathbf{n}\} - (\langle n_i \rangle^{ss})^2, \qquad (11.66c)$$

where $\mathbf{n}$ represents the vector of integers $(n_1, \ldots, n_K)$ with i^{th} entry $n_i = k$. Then, continuing from (11.66c),

$$\langle \Delta n_i(t) \Delta n_i(0) \rangle = \sum_{\mathbf{n}} k \sum_{j=0}^{\infty} j P\{n_i(t) = j | \mathbf{n}(0) = \mathbf{n}\} P\{\mathbf{n}(0) = \mathbf{n}\} - (\langle n_i \rangle^{ss})^2$$

$$= \sum_{\mathbf{n}} k \langle n_i(t) | \mathbf{n}(0) = \mathbf{n} \rangle P\{\mathbf{n}(0) = \mathbf{n}\} - (\langle n_i \rangle^{ss})^2$$

$$= \sum_{\mathbf{n}} k \langle \Delta n_i(t) + \langle n_i \rangle^{ss} | \Delta \mathbf{n}(0) = \mathbf{n} - \langle \mathbf{n} \rangle^{ss} \rangle P\{\mathbf{n}(0) = \mathbf{n}\} - (\langle n_i \rangle^{ss})^2$$

$$= \sum_{\mathbf{n}} k \langle \Delta n_i(t) | \Delta \mathbf{n}(0) = \mathbf{n} - \langle \mathbf{n} \rangle^{ss} \rangle P\{\mathbf{n}(0) = \mathbf{n}\}$$

$$+ \langle n_i \rangle^{ss} \sum_{\mathbf{n}} k P\{\mathbf{n}(0) = \mathbf{n}\} - (\langle n_i \rangle^{ss})^2$$

$$= \sum_{\mathbf{n}} k \langle \Delta n_i(t) | \Delta \mathbf{n}(0) = \mathbf{n} - \langle \mathbf{n} \rangle^{ss} \rangle P\{\mathbf{n}(0) = \mathbf{n}\}. \qquad (11.67)$$

Substituting (11.65) into (11.67) yields

$$\langle \Delta n_i(t) \Delta n_i(0) \rangle = \left[e^{\mathbf{Q}t} \right]_{i,i} \langle n_i \rangle^{ss} = \sum_{k=1}^{K} \left(\mathbf{V}_{i,k} e^{\lambda_k t} \mathbf{V}_{k,i}^{-1} \langle n_i \rangle^{ss} \right), \qquad (11.68a)$$

$$\langle \Delta n_j(t) \Delta n_i(0) \rangle = \left[e^{\mathbf{Q}t} \right]_{j,i} \langle n_i \rangle^{ss} = \sum_{k=1}^{K} \left(\mathbf{V}_{j,k} e^{\lambda_k t} \mathbf{V}_{k,i}^{-1} \langle n_i \rangle^{ss} \right). \qquad (11.68b)$$

Cross-correlation functions can be used to quantify the NESS fluxes of specific reaction pathways. This result can be generalized to any linear reaction network. For example, consider the reaction between states i and j and that the NESS flux of that reaction is related to the initial portion of the cross-correlation functions:

$$\lim_{t \to 0} \frac{\langle \Delta n_j(t) \Delta n_i(0) \rangle - \langle \Delta n_i(t) \Delta n_j(0) \rangle}{t}$$

$$= \lim_{t \to 0} \frac{1}{t} \sum_{k=1}^{K} \left(\mathbf{V}_{j,k} e^{\lambda_k t} \mathbf{V}_{k,i}^{-1} \langle n_i \rangle^{ss} - \mathbf{V}_{i,k} e^{\lambda_k t} \mathbf{V}_{k,j}^{-1} \langle n_j \rangle^{ss} \right)$$

$$= \sum_{k=1}^{K} \left(\mathbf{V}_{j,k} \lambda_k \mathbf{V}_{k,i}^{-1} \langle n_i \rangle^{ss} - \mathbf{V}_{i,k} \lambda_k \mathbf{V}_{k,j}^{-1} \langle n_j \rangle^{ss} \right)$$

$$= \mathbf{Q}_{j,i} \langle n_i \rangle^{ss} - \mathbf{Q}_{i,j} \langle n_j \rangle^{ss}$$

$$= q_{i,j} \langle n_i \rangle^{ss} - q_{j,i} \langle n_j \rangle^{ss}. \tag{11.69}$$

The one-way NESS fluxes can be calculated from the initial slopes of the individual cross-correlation functions. This idea was first suggested by Qian and Elson in [226].

The method for calculating fluxes from cross-correlation functions also allows discerning which states are connected by reactions because the values $q_{i,j}$ and $q_{j,i}$ are nonzero if and only if states i and j are connected by a reaction. This is a potentially powerful method for studying the connectivities of reaction pathways and measuring NESS fluxes.

11.4.4 An Example

We shall use a simple example to illustrate what we have thus far presented. Let us turn our attention to fluctuations in the number of molecules in the kinetic pathway

$$0 \underset{q_{1,0}}{\overset{q_{0,1}}{\rightleftharpoons}} 1 \underset{q_{2,1}}{\overset{q_{1,2}}{\rightleftharpoons}} 2 \overset{q_{2,3}}{\longrightarrow} 3 \tag{11.70}$$

with input at state 0, which is fixed at n_0, and output at state 3, which is fixed at $n_3 = 0$.

The joint probability $p(n_1, n_2; t)$ of having n_1 and n_2 molecules in states 1 and 2, respectively, at time t satisfies the chemical master equation

$$\frac{dp(n_1,n_2;t)}{dt} = p(n_1-1,n_2,t)q_{0,1}n_0 + p(n_1,n_2+1;t)q_{2,3}(n_2+1)$$
$$+ p(n_1+1,n_2;t)q_{1,0}(n_1+1)$$
$$+ p(n_1+1,n_2-1;t)q_{1,2}(n_1+1)$$
$$+ p(n_1-1,n_2+1;t)q_{2,1}(n_2+1) \tag{11.71}$$
$$- p(n_1,n_2;t)\left[q_{0,1}n_0 + (q_{1,0}+q_{1,2})n_1 + (q_{2,1}+q_{2,3})n_2\right].$$

As expected, we can solve for the NESS joint probability function

$$p^{ss}(n_1,n_2) = \frac{(\langle n_1\rangle^{ss})^{n_1}}{n_1!}e^{-\langle n_1\rangle^{ss}}\frac{(\langle n_2\rangle^{ss})^{n_2}}{n_2!}e^{-\langle n_2\rangle^{ss}}. \tag{11.72}$$

We can show that the NESS expectations of the number of molecules in the interior states of this kinetic pathway are

$$\langle n_1\rangle^{ss} = \frac{q_{0,1}(q_{2,1}+q_{2,3})N}{q_{1,0}(q_{2,1}+q_{2,3})+q_{1,2}q_{2,3}} \tag{11.73a}$$

$$\langle n_2\rangle^{ss} = \frac{q_{0,1}q_{1,2}N}{q_{1,0}(q_{2,1}+q_{2,3})+q_{1,2}q_{2,3}} \tag{11.73b}$$

which is the steady-state solution of the ODE system

$$\frac{d}{dt}\begin{pmatrix}\langle n_1(t)\rangle\\\langle n_2(t)\rangle\end{pmatrix} = \mathbf{Q}\begin{pmatrix}\langle n_1(t)\rangle\\\langle n_2(t)\rangle\end{pmatrix} + \begin{pmatrix}q_{0,1}n_0\\0\end{pmatrix} = \begin{pmatrix}0\\0\end{pmatrix} \tag{11.74}$$

where

$$\mathbf{Q} = \begin{pmatrix}-q_{1,0}-q_{1,2} & q_{2,1}\\q_{1,2} & -q_{2,1}-q_{2,3}\end{pmatrix}. \tag{11.75}$$

Additionally, from (11.59) and (11.60), we obtain the NESS variances and co-variance

$$\langle(\Delta n_1)^2\rangle^{ss} = \langle n_1\rangle^{ss}, \quad \langle(\Delta n_2)^2\rangle^{ss} = \langle n_2\rangle^{ss}, \quad \langle\Delta n_1\Delta n_2\rangle^{ss} = 0. \tag{11.76}$$

The autocorrelation and cross-correlation functions

$$\langle\Delta n_1(0)\Delta n_1(t)\rangle = \frac{\langle n_1\rangle^{ss}}{\lambda_1-\lambda_2}\left((\lambda_1+q_{1,0}+q_{1,2})e^{\lambda_2 t} - (\lambda_2+q_{1,0}+q_{1,2})e^{\lambda_1 t}\right)$$
$$\tag{11.77a}$$

$$\langle\Delta n_2(0)\Delta n_2(t)\rangle = \frac{\langle n_2\rangle^{ss}}{\lambda_1-\lambda_2}\left((\lambda_1+q_{2,1}+q_{2,3})e^{\lambda_2 t} - (\lambda_2+q_{2,1}+q_{2,3})e^{\lambda_1 t}\right)$$
$$\tag{11.77b}$$

$$\langle\Delta n_1(0)\Delta n_2(t)\rangle = \frac{q_{1,2}\langle n_1\rangle^{ss}}{\lambda_1-\lambda_2}\left(e^{\lambda_1 t} - e^{\lambda_2 t}\right) \tag{11.77c}$$

$$\langle \Delta n_2(0) \Delta n_1(t) \rangle = \frac{q_{2,1} \langle n_2 \rangle^{ss}}{\lambda_1 - \lambda_2} \left(e^{\lambda_1 t} - e^{\lambda_2 t} \right) \qquad (11.77d)$$

where λ_1 and λ_2 are the two eigenvalues of the matrix $\mathbf{Q}$.

The net flux through that reaction can be calculated from

$$\lim_{t \to 0} \frac{\langle \Delta n_1(0) \Delta n_2(t) \rangle - \langle \Delta n_2(0) \Delta n_1(t) \rangle}{t} = q_{1,2} \langle n_1 \rangle^{ss} - q_{2,1} \langle n_2 \rangle^{ss}. \qquad (11.78)$$

11.5 Grand Canonical Markov Model

11.5.1 Equilibrium Grand Canonical Ensemble

Consider the situation in which the source and the sink are the same. Let state 0 be that state and suppose there are N internal states. In this case, the system is in exchange with a single material reservoir and will tend to reach equilibrium, which fits the classic grand canonical ensemble of Gibbsian statistical mechanics [20].

The total number of particles, n, in the system can take on any value with probability

$$P(n) \propto \frac{Q(\beta, n)}{n!} e^{\beta \mu n} \qquad (11.79)$$

where $\beta = (k_B T)^{-1}$, k_B is the Boltzmann constant, T is absolute temperature, and μ is the constant chemical potential of the system and the material reservoir in equilibrium [20]. $Q(\beta, n)$ is known as the partition function for the canonical ensemble consisting of n particles, which is the normalization constant for the probability of finding a specific state having energy E in the canonical ensemble with fixed β and n. If $n = 1$,

$$Q(\beta, 1) \triangleq \sum_{i=0}^{N} e^{-\beta E_i} \qquad (11.80)$$

where E_i is the energy of state i.

If states i and 0 communicate via the path $i \to i_1 \to i_2 \to \cdots \to i_m \to 0$, the energy of state i is

$$E_i - E_0 = k_B T \ln \left(\frac{q_{i,i_1} q_{i_1,i_2} \cdots q_{i_m,0}}{q_{i_1,i} q_{i_2,i_1} \cdots q_{0,i_m}} \right) = k_B T \ln \left(\frac{\tilde{q}_{i,0}}{\tilde{q}_{0,i}} \right). \qquad (11.81)$$

The ratio $\tilde{q}_{i,0}/\tilde{q}_{0,i}$ is unique regardless of the path taken because of the detailed balance condition in (11.43a). For n molecules,

$$Q(\beta, n) = \sum_{\substack{n_0, n_1, \ldots, n_N \geq 0 \\ n_0 + n_1 + \cdots + n_N = n}} \binom{n}{n_0, n_1, \ldots, n_N} e^{-\beta(n_0 E_0 + n_1 E_1 + \cdots + n_N E_N)} \tag{11.82a}$$

$$= \left(e^{-\beta E_0} + e^{-\beta E_1} + \cdots + e^{-\beta E_N} \right)^n \tag{11.82b}$$

$$= (Q(\beta, 1))^n; \tag{11.82c}$$

hence, the normalizing condition for the probability $P(n)$ is

$$\Xi(\beta, \mu) = \sum_{n=0}^{\infty} \frac{Q(\beta, n)}{n!} e^{\beta \mu n} \tag{11.83a}$$

$$= e^{Q(\beta, 1) e^{\beta \mu}} \tag{11.83b}$$

which is known as the grand canonical partition function.

$\langle n \rangle^{eq}$ is the expected number of molecules in the system at equilibrium; then,

$$\langle n \rangle^{eq} = k_B T \left(\frac{\partial \ln \Xi(\beta, \mu)}{\partial \mu} \right)_{T,V} \tag{11.84a}$$

$$= Q(\beta, 1) e^{\beta \mu}, \tag{11.84b}$$

which yields

$$\mu = k_B T \ln \langle n \rangle^{eq} - k_B T \ln Q(\beta, 1). \tag{11.85}$$

Using (11.79), (11.82), (11.83), and (11.84), the probability that the total number of particles in the system is n can be written as

$$P(n) = \frac{(Q(\beta, 1) e^{\beta \mu})^n}{n!} e^{-Q(\beta, 1) e^{\beta \mu}} \tag{11.86a}$$

$$= \frac{(\langle n \rangle^{eq})^n}{n!} e^{-\langle n \rangle^{eq}} \tag{11.86b}$$

which is a Poisson distribution with mean $\langle n \rangle^{eq}$. In addition, (11.62) becomes

$$P(n_0, n_1, \ldots, n_N) = \prod_{i=0}^{N} \left[\frac{(\langle n_i \rangle^{eq})^{n_i}}{n_i!} e^{-\langle n_i \rangle^{eq}} \right]. \tag{11.87}$$

11.5.2 Open Boundaries, Potential Function and Reaction Conductance

There is a deep mathematical relation between stochastic linear reaction systems with detailed balance but open boundaries and the theory of an electrical circuit with batteries [51]. This motivates the concept of chemical reaction *conductance* through the nonequilibrium chemical potential function.

We begin with Eqn. (11.57). The only nonequilibrium effect comes from the boundary species X_0 and X_{N+1}, and we assume detailed balance among the internal chemical species $X_1, \cdots, X_N$. One notices the analogy between this setup and an electrical circuit that is just an inert material in the absence of a battery. Even though a chemical reaction has $q_{i,j} \neq q_{j,i}$ while an electric resistor has the same conduction in both directions, the concept of conductance is actually related to "traveling time". Eqn. (11.57) can be rewritten as

$$\frac{d \langle n_i(t) \rangle}{dt} = \sum_{\substack{j=0 \\ j \neq i}}^{N+1} \left[\left(\langle n_j(t) \rangle \frac{q_{j,i}}{q_{i,j}} - \langle n_i(t) \rangle \right) q_{i,j} \right], \tag{11.88a}$$

and

$$\frac{1}{n_0} \frac{\pi_0}{\pi_i} \left(\frac{d \langle n_i(t) \rangle}{dt} \right) = \sum_{\substack{j=0 \\ j \neq i}}^{N+1} \left[\left(\frac{\langle n_j(t) \rangle}{n_0} \frac{\pi_0}{\pi_j} - \frac{\langle n_i(t) \rangle}{n_0} \frac{\pi_0}{\pi_i} \right) q_{i,j} \right], \tag{11.88b}$$

in which π_i represents the probability of being in state i at equilibrium, and with the substitution

$$u_i(t) = \frac{\langle n_i(t) \rangle}{n_0} \frac{\pi_0}{\pi_i} \tag{11.89}$$

we obtain

$$\frac{du_i(t)}{dt} = \sum_{\substack{j=0 \\ j \neq i}}^{N+1} [(u_j(t) - u_i(t)) q_{i,j}], \qquad \forall i \in \{1, 2, \ldots, N\}; \tag{11.90a}$$

$$\text{ICs:} \quad u_i(0) = 0, \quad \forall i \in \{1, 2, \ldots, N\}, \tag{11.90b}$$

where $u_0(t) = 1$ and $u_{N+1}(t) = \frac{n_{N+1}}{n_0} \frac{\pi_0}{\pi_{N+1}}$, $\forall t$.

The chemical potential of state i relative to state 0 can be defined as

$$\mu_i(t) \triangleq \ln u_i(t) = -\ln \frac{\pi_i}{\pi_0} + \ln \langle n_i(t) \rangle - \ln n_0, \tag{11.91}$$

in which $\ln \langle n_i(t) \rangle$ is the Boltzmann entropy of state i, and $-\ln(\pi_i/\pi_0)$ is the standard chemical potential of state i relative to that of state 0 [119]. The chemical potential of state 0 is taken to be 0 and is used here as the reference potential for all other chemical potentials.

By defining one-way fluxes as $J_{i,j}(t) \triangleq \langle n_i(t) \rangle q_{i,j}$ for the flux in the $i \to j$ direction, the chemical potential difference for a reaction between states i and j can be written as

$$\Delta \mu_{i,j}(t) = \mu_i(t) - \mu_j(t) = \ln \frac{\langle n_i(t) \rangle q_{i,j}}{\langle n_j(t) \rangle q_{j,i}} = \ln \frac{J_{i,j}(t)}{J_{j,i}(t)} \qquad (11.92)$$

using (11.42) and (11.91). It then immediately follows that

$$(J_{i,j}(t) - J_{j,i}(t)) \Delta \mu_{i,j}(t) = (J_{i,j}(t) - J_{j,i}(t)) \ln \frac{J_{i,j}(t)}{J_{j,i}(t)} \geq 0. \qquad (11.93)$$

The product of the net flux through a reaction and the corresponding potential difference is related to the heat dissipation rate of that reaction, which must be positive unless the reaction is in equilibrium, i.e., $J_{i,j}(t) = J_{j,i}(t)$, in which case it is necessarily equal to 0. This result is equivalent to the second law of thermodynamics, stating that each reaction with nonzero flux must dissipate energy, and the entropy of the system must increase.

In terms of the chemical affinity, $u(t) = e^{\mu(t)}$, one has

$$\Delta u_{i,j}(t) = u_i(t) - u_j(t) = \frac{J_{i,j}(t) - J_{j,i}(t)}{n_0 \pi_i q_{i,j}} \pi_0 \qquad (11.94)$$

in which case

$$(J_{i,j}(t) - J_{j,i}(t)) \Delta u_{i,j}(t) = \frac{(J_{i,j}(t) - J_{j,i}(t))^2}{n_0 \pi_i q_{i,j}} \pi_0 \geq 0. \qquad (11.95)$$

Note that (11.94) bears a likeness to the linear Ohm's law of electrical circuit theory. A reaction conductance can be defined for each reaction as

$$c_{i,j} \triangleq \frac{J_{i,j}(t) - J_{j,i}(t)}{\Delta u_{i,j}(t)} = \frac{n_0 \pi_i q_{i,j}}{\pi_0} = c_{j,i}. \qquad (11.96)$$

The last equality is a result of (11.42) and shows that there is a unique conductance associated with each reaction in a NESS. Defining the reaction conductance in this way does not require the assumption that the system is near equilibrium to justify a linear expansion of the flux-potential relationship, as it must be for Onsager-Hill's theory [119].

Chapter 12
Nonlinear Stochastic Reaction Systems with Simple Examples

12.1 Bimolecular Nonlinear Reaction

The independence between individual molecules in unimolecular reactions does not hold for bimolecular reactions. Therefore, one expects that the treatment of a bimolecular reaction cannot be simplified to a single-molecule model. Thus, we must officially abandon the state-tracking molecular representation and adopt the number-counting representation discussed in Chapter 11.

Let us consider the following second-order reversible association reaction with forward second-order rate constant k_+ and backward first-order rate constant k_-:

$$A + B \underset{k_-}{\overset{k_+}{\rightleftharpoons}} C. \tag{12.1}$$

The probability distribution $p(n;t) = \mathrm{P}\{\# \text{ of } C \text{ molecules at time } t = n\}$ satisfies the birth-death equation

$$\frac{\mathrm{d}p(n;t)}{\mathrm{d}t} = k_1(N_A - n + 1)(N_B - n + 1)p(n - 1) + k_-(n + 1)p(n + 1)$$

$$- \left[k_1(N_A - n)(N_B - n) + k_- n\right]p(n), \tag{12.2}$$

where N_A is the total number of A and $C \equiv AB$ molecules, and N_B is the total number of B and $C \equiv AB$ molecules in (12.1). Without loss of generality, let us assume $N_B \geq N_A$. Note there is a key difference between the traditional, macroscopic reaction rate constant k_+, which has dimensions $[\text{time}]^{-1}[\text{concentration}]^{-1}$, while the Markov transition rate $k_1 = k_+/V$ has dimensions $[\text{time}]^{-1}$, where V is the volume of a mesoscopic reaction tank.

© The Author(s), under exclusive license to Springer Nature Switzerland AG 2021
H. Qian, H. Ge, *Stochastic Chemical Reaction Systems in Biology*,
Lecture Notes on Mathematical Modelling in the Life Sciences,
https://doi.org/10.1007/978-3-030-86252-7_12

The stationary equilibrium distribution can be obtained by setting the rhs of (12.2) to zero:

$$\frac{p^{eq}(n_C+1)}{p^{eq}(n_C)} = \frac{k_1(N_A - n_C)(N_B - n_C)}{k_-(n_C+1)}.$$

Thus,

$$p^{eq}_{n_C}(\ell) = \frac{1}{Z(N_A,N_B)}\left[\frac{N_A!N_B!}{\ell!(N_A-\ell)!(N_B-\ell)!}\left(\frac{k_1}{k_-}\right)^{\ell}\right], \quad 0 \le \ell \le N_A, \qquad (12.3)$$

in which $n_C = \ell$, $n_A = N_A - \ell$, $n_B = N_B - \ell$, and $Z(N_A,N_B)$ should be determined from the normalization of $p^{eq}_{n_C}(\ell)$. The distribution has a maximum at

$$\frac{n_C^*+1}{n_A^* n_B^*} = \frac{k_1}{k_-}, \qquad (12.4)$$

where $n_A^* + n_C^* = N_A$ and $n_B^* + n_C^* = N_B$. This is known as the chemical equilibrium of the association reaction in (12.1).

12.2 One-Dimensional Birth-Death Processes of a Single Species

The simplest deterministic population dynamics model is the nonlinear, autonomous $\frac{dx}{dt} = b(x)$ for a single species, where $x \in \mathbb{R}^+$ is the density of the population. In a similar vein, equation such as (12.2) can be written for a general one-dimensional birth-death process $\mathbf{N}(t)$ with birth rate $u_+(n)$ and death rate $u_-(n)$. Let $p(n;t) = P\{\mathbf{N}(t) = n\}$; then

$$\frac{dp(n,t)}{dt} = u_+(n-1)p(n-1) - \left[u_+(n)+u_-(n)\right]p(n) + u_-(n+1)p(n+1), \quad (12.5)$$

where $0 \le n$, $u_+(n)$ and $u_-(n)$ are usually nonlinear functions of n. The diagram, called the *master equation graph* in [17], in Fig. 12.1 illustrates the master equation in Eqn. (12.5), in which n represents the number of individual molecules in the reaction species in the stochastic kinetics in a small tank.

Noting the steady state $p^{ss}(0)u_+(0) - p^{ss}(1)u_-(1) = 0$ and so on, one can immediately obtain the stationary probability distribution for the stochastic $n(t)$:

$$p^{ss}(n) = C\prod_{j=1}^{n}\frac{u_+(j-1)}{u_-(j)}, \qquad (12.6)$$

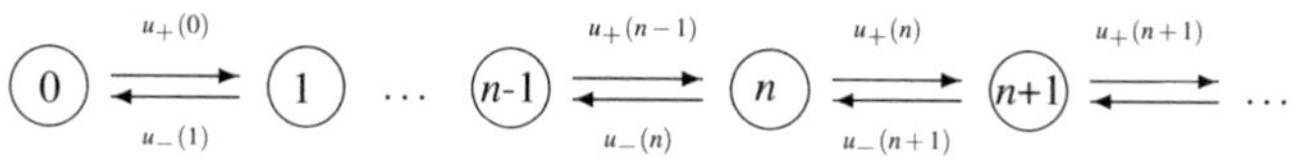

Fig. 12.1 Master equation graph for the Markov process of a single species with birth (synthesis) rate $u_+(n)$ and death (degradation) rate $u_-(n)$. n represents the number of molecules in the population. Both transitions $u_+(n)$ and $u_-(n)$ have dimensions of $[\text{time}]^{-1}$. The continuous-time Markov process $N(t)$ has an exponentially distributed waiting time for the transition, and $u_\pm(n)$ are the rates. They can be understood as the per capita birth and death rates in the traditional, differential equation-based population dynamics.

where C is a normalization constant. $p^{ss}(n)$ reaches a local peak j^* when $u_+(j-1)/u_-(j) \geq 1$ with $j \leq j^*$ followed by $u_+(j-1)/u_-(j) < 1$ with $j > j^*$.

Both $u_+(n,V)$ and $u_-(n,V)$ are also functions of the volume of the reaction vessel V. In the limits as $V \to \infty$ and $N \to \infty$ but fixing $\frac{N}{V} = x$ as the concentration of the chemical species, one expects macroscopic rates

$$\hat{u}_+(x) = \lim_{V \to \infty} \frac{u_+(Vx,V)}{V}, \quad \hat{u}_-(x) = \lim_{V \to \infty} \frac{u_-(Vx,V)}{V}, \tag{12.7}$$

and the deterministic kinetic equation is then

$$\frac{dx(t)}{dt} = \hat{u}_+(x) - \hat{u}_-(x). \tag{12.8}$$

This is a special case of a general result, with explicit computation, where in the limit as $V \to \infty$, the CME and DGP descriptions of the stochastic nonlinear kinetics become the traditional kinetic rate equation. In Section 5.5, we already encountered this problem; a more general discussion will be found in Section 12.7.

12.2.1 Landscape For Stochastic Population Kinetics

As $V \to \infty$, as in Eqn. (11.41) with $N \to \infty$, the stationary probability distribution in (12.6) has an interesting and very important asymptotic expression:

$$- \lim_{V \to \infty} \frac{\ln \prod_{j=1}^{Vx} \frac{u_+(j-1)}{u_-(j)}}{V} = - \lim_{V \to \infty} \frac{1}{V} \sum_{j=1}^{Vx} \ln \left(\frac{u_+(j-1,V)}{u_-(j,V)} \right)$$

$$= - \lim_{V \to \infty} \sum_{j=1}^{x/\Delta z} \ln \left(\frac{u_+(V(j-1)\Delta z, V)}{u_-(Vj\Delta z, V)} \right) \Delta z$$

$$= - \lim_{\Delta z \to 0} \sum_{j=1}^{x/\Delta z} \ln \left(\frac{\hat{u}_+(j\Delta z)}{\hat{u}_-(j\Delta z)} \right) \Delta z$$

$$= \int_0^x \ln \left(\frac{\hat{u}_-(z)}{\hat{u}_+(z)} \right) dz = \varphi^{ss}(x), \qquad (12.9)$$

in which we let $V^{-1} = \Delta z$. The local minima of function $\varphi^{ss}(x)$ correspond to the peaks of the probability distribution $p_n^{ss}(Vx) \simeq A_V e^{-V\varphi^{ss}(x)}$, where constant A_V is a normalization factor.

In connection to the mathematical theory of large deviations,

$$- \lim_{V \to \infty} \frac{\ln p_n^{ss}(Vx)}{V} = - \lim_{V \to \infty} \frac{1}{V} \ln \left(A_V e^{-V\varphi^{ss}(x)} \right) = \varphi^{ss}(x) - \lim_{V \to \infty} \frac{1}{V} \ln \int_{-\infty}^{\infty} e^{-V\varphi^{ss}(x)} dx$$

$$= \varphi^{ss}(x) - \inf_{x \in \mathbb{R}} \varphi^{ss}(x).$$

The last step follows Laplace's method for integrals from Section 4.2. Since the large deviation rate function is strictly defined as the limit above, it always has the essential property of nonnegativity with a zero global minimum. In the present text, however, we shall adopt a less stringent notion and consider the normalization factor to be arbitrary; this translates to the addition of an arbitrary constant to $\varphi^{ss}(x)$. One can obtain a glimpse of a mathematical origin of the partition function in statistical mechanics.

We call the function $\varphi^{ss}(x)$, with the arbitrary additive constant, a "landscape" for the nonlinear dynamics in (12.8) since it is a Lyapunov function: for a solution to the nonlinear differential equation (12.8), $\xi(t)$,

$$\frac{d}{dt} \varphi^{ss}(\xi(t)) = \left(\frac{d\varphi^{ss}(x)}{dx} \right)_{x=\xi(t)} \frac{d\xi(t)}{dt}$$

$$= \left(\hat{u}_+(\xi) - \hat{u}_-(\xi) \right) \ln \left(\frac{\hat{u}_-(\xi)}{\hat{u}_+(\xi)} \right) \leq 0. \qquad (12.10)$$

The equality holds true only at the fixed point of the (12.8).

12.2.2 More Discussions On Diffusion Approximation

We obtained the asymptotic landscape $\varphi^{ss}(x)$ in (12.9) by first solving the stationary probability distribution of the stochastic kinetics, e.g., $t \to \infty$, and then taking the limit as $V \to \infty$. We now demonstrate that the diffusion approximation in Section 11.2.1 can possibly yield an erroneous result that is inconsistent with Eqn. (12.9). This is because the two limits—the limiting processes $V \to \infty$, which is related to the diffusion approximation, and $t \to \infty$—cannot exchange order. The convergence of $p(Vx;t)$ as $V \to \infty$ is not uniform for all $t \in [0,\infty)$.

Following the same steps as in Section 11.2.1, the chemical master equation in (12.5) can be approximated by the PDE

$$\frac{\partial f(x;t)}{\partial t} = \frac{\partial^2}{\partial x^2}\left(\frac{\hat{u}_+(x)+\hat{u}_-(x)}{2V}f(x;t)\right) - \frac{\partial}{\partial x}\left[\left(\hat{u}_+(x)-\hat{u}_-(x)\right)f(x;t)\right]+O\left(V^{-2}\right),$$
(12.11)

which yields the asymptotic stationary distribution

$$f^{ss}(x) \propto \frac{1}{\hat{u}_+(x)+\hat{u}_-(x)}\exp\left(-2V\int_{x_0}^{x}\frac{\hat{u}_-(z)-\hat{u}_+(z)}{\hat{u}_+(z)+\hat{u}_-(z)}dz\right),$$
(12.12)

for any given x_0.

Note that $\frac{\hat{u}_-(x)-\hat{u}_+(x)}{\hat{u}_+(x)+\hat{u}_-(x)}$ equals zero when $\hat{u}_+(x^*)=\hat{u}_-(x^*)$. Hence, the exponential term has an extremum at $x=x^*$. This result is consistent with the $\varphi^{ss}(x)$ in (12.9), which has a very different expression.

Let us denote

$$\tilde{\varphi}(x) = 2\int_{x_0}^{x}\frac{\hat{u}_-(z)-\hat{u}_+(z)}{\hat{u}_+(z)+\hat{u}_-(z)}dz.$$
(12.13)

Both the locations of all the extrema of $\varphi(x)$ and $\tilde{\varphi}(x)$ and the curvatures of the two functions at each extremum agree:

$$\begin{aligned}\left(\frac{d\tilde{\varphi}(x)}{dx^2}\right)_{x=x^*} &= \left[2\frac{d}{dx}\left(\frac{\hat{u}_-(x)-\hat{u}_+(x)}{\hat{u}_+(x)+\hat{u}_-(x)}\right)\right]_{x=x^*}\\ &= \left[4\frac{\frac{d}{dx}\left(\frac{\hat{u}_-(x)}{\hat{u}_+(x)}\right)}{\left(1+\frac{\hat{u}_-(x)}{\hat{u}_+(x)}\right)^2}\right]_{x=x^*} = \left[\frac{d}{dx}\left(\frac{\hat{u}_-(x)}{\hat{u}_+(x)}\right)\right]_{x=x^*}\\ &= \left[\frac{d^2}{dx^2}\varphi^{ss}(x)\right]_{x=x^*}.\end{aligned}$$
(12.14)

The matched curvatures mean that an equivalent Gaussian approximation can be made for the probability distribution near its peak at x^*. Locally, Eqn. (12.11) is consistent with a Gaussian approximation.

For a unimolecular reaction with linear variables $\hat{u}$, we have $\hat{u}_+(x) = k_1 x$, and $\hat{u}_-(x) = k_{-1}(x_{tot} - x)$. Hence, the Gaussian variance is

$$-\left[\frac{d}{dx}\left(\frac{\hat{u}_-(x)}{\hat{u}_+(x)}\right)\right]^{-1}_{x=x^*} = \frac{k_1 k_{-1} x_{tot}}{(k_1 + k_{-1})^2} = \left[\frac{1}{\langle x_A \rangle^{eq}} + \frac{1}{\langle x_B \rangle^{eq}}\right]^{-1},$$

which agrees with (11.16) and (11.17). For the association-dissociation reaction in (12.1) with $\hat{u}_+(x) = k_+ x(b - a + x)$ and $\hat{u}_-(x) = k_-(a - x)$,

$$-\left[\frac{d}{dx}\left(\frac{\hat{u}_-(x)}{\hat{u}_+(x)}\right)\right]^{-1}_{x=x^*} = \left[\frac{1}{\langle x_A \rangle^{eq}} + \frac{1}{\langle x_B \rangle^{eeq}} + \frac{1}{\langle x_C \rangle^{eq}}\right]^{-1}. \tag{12.15}$$

This is a well-known result [217].

The real difference between $\varphi^{ss}(x)$ and $\tilde{\varphi}(x)$, however, occurs when they have multiple minima; in that case, it is possible that the global minimum of $\varphi^{ss}(x)$ corresponds to a nonglobal minimum of $\tilde{\varphi}(x)$. This difference becomes very significant in the limit as $V \to \infty$; the probability distribution $p^{ss}(Vx) = A_V e^{-V\varphi^{ss}(x)}$ converges to the global minimum of $\varphi^{ss}(x)$ with probability one [265]. The origin of the error in the diffusion approximation can be understood as follows: the diffusion approximation, or the linear noise approximation, only preserves terms of order $O(V^{-1})$. This is sufficient for estimating local Gaussian fluctuations. However, the probabilities of transitions between two minima of $\varphi^{ss}(x)$, as Kramers' formula informed us, is of order $\sim (e^{-\alpha V})$ with $\alpha > 0$. This is too small relative to the $\sim O(V^{-1})$ order approximation.

12.3 Bimolecular association-dissociation reaction

We now return to the nonlinear bimolecular reaction in Eqn. (12.1) and study it in greater detail. We are particularly interested in a comparison between the deterministic nonlinear kinetics based on an ordinary differential equation (ODE) and the stochastic reaction kinetics based on a Delbrück–Gillespie process (DGP). Let $c_A(t)$, $c_B(t)$, and $c_C(t)$ be the concentrations of species A, B, and C, respectively; then, according to the law of mass action, we have

$$\frac{dc_A(t)}{dt} = -k_+c_Ac_B + k_-c_C, \tag{12.16a}$$

$$\frac{dc_B(t)}{dt} = -k_+c_Ac_B + k_-c_C, \tag{12.16b}$$

$$\frac{dc_C(t)}{dt} = k_+c_Ac_B - k_-c_C. \tag{12.16c}$$

We note that $c_A(t) + c_C(t) = a$ and $c_B(t) + c_C(t) = b$ are independent of time, they are conserved. Then, in terms of the a and b, and writing $c_C(t) = x(t)$, we have

$$\frac{dx(t)}{dt} = k_+(a-x)(b-x) - k_-x, \tag{12.17}$$

which can be solved using the method of separation of variables:

$$\frac{dx}{(x-a)(x-b) - \kappa^{-1}x} = k_+dt,$$

$$\left(\frac{1}{x-r_1} - \frac{1}{x-r_2}\right)\frac{dx}{r_1-r_2} = k_+dt,$$

$$\ln\left(\frac{x-r_1}{x-r_2}\right) = (r_1-r_2)k_1t + \ln\left(\frac{x(0)-r_1}{x(0)-r_2}\right). \tag{12.18}$$

where r_1 and r_2 are the two roots of the quadratic equation $(a-x)(b-x) - \kappa^{-1}x = 0$, $\kappa = k_+/k_-$,

$$r_{1,2} = \frac{a+b+\kappa^{-1} \pm \sqrt{(a+b+\kappa^{-1})^2 - 4ab}}{2}.$$

They are both positive.

How do the deterministic kinetics in (12.18) compare with the stochastic kinetics shown in Fig. 12.1? In Fig. 12.2, a Monte Carlo simulation of the stochastic kinetics with initial $N_A(0) = N_A = 30$, $N_B(0) = N_B = 40$, and $N_C(0) = 0$ is compared with the deterministic kinetics. To make this comparison, we need to relate the concentrations a, b, and x to the corresponding numbers of molecules used in the simulations, N_A, N_B, and n_C. Let reaction (12.1) occur in a small reaction volume of V, then $a = N_A/V = 30/V$, $b = N_B/V = 40/V$, and $\kappa = 0.1V$. This yields two roots, $r_1 = 60/V$, $r_2 = 20/V$. Then, the solution to the ODE given in Eqn. (12.18) gives $n(t) \equiv Vx(t)$ as a function of t through its inverse function:

$$t = \frac{1}{4}\ln\left(\frac{n-60}{3n-60}\right). \tag{12.19}$$

This is the red curve in Fig. 12.2.

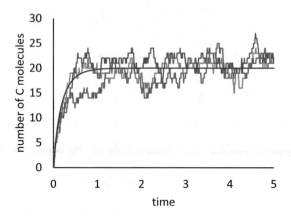

Fig. 12.2 Three sample trajectories for $N_C(t)$, all with $N_A(0) = N_A = 30$, $N_B(0) = N_B = 40$, $N_C(0) = 0$. $k_1 = 0.1$, $k_-/k_1 = \kappa^{-1}V = 10$. They all plateau at approximately $N_C = 20$. The red smooth curve is given by Eqn. (12.19). It agrees well with the simulation results.

Allowing the trajectory simulations to run for a much longer time, Fig. 12.3 shows the stationary statistics for $N_C(t)$. It should be compared with the distribution given in Eqn. (12.3):

$$p_n^{ss} = C \prod_{j=1}^{n} \left\{ \frac{(N_A - j + 1)(N_B - j + 1)\kappa}{j} \right\} = C \left(\frac{N_A! N_B! \kappa^n}{n!(N_A - n)!(N_B - n)!} \right), \quad (12.20)$$

where $N_A = aV = 30$, $N_B = bV = 40$, $\kappa = k_1/k_{-1} = 0.1$, and C is a normalization constant:

$$C^{-1} = \sum_{n=0}^{N_A} \frac{N_A! N_B! \kappa^n}{n!(N_A - n)!(N_B - n)!}.$$

The modal value of the distribution p_n^{ss}, n^*, can be obtained from the equation $p_{n^*}^{ss} = p_{n^*+1}^{ss}$, or equivalently by setting the term inside $\{ \cdots \}$ in Eqn. (12.20) to 1:

$$\left(\frac{n_A n_B}{n_C} \right)_{eq} = \frac{(N_A - n^* + 1)(N_B - n^* + 1)}{n^*} = \frac{k_-}{k_1},$$

which corresponds macroscopically to

$$\left(\frac{c_A c_B}{c_C} \right)_{eq} = \frac{1}{V} \left(\frac{n_A n_B}{n_C} \right)_{eq} \simeq \frac{k_-}{V k_1} = \frac{k_-}{k_+} = \frac{1}{\kappa},$$

in which "$\simeq$" is used due to replacing $n^* - 1$ by n^*. κ is the *equilibrium constant* of the association reaction (12.1).

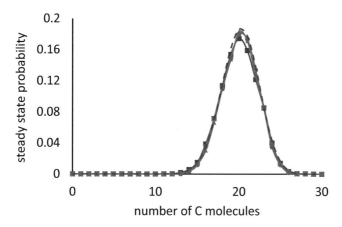

Fig. 12.3 Three simulations for the steady state probability distribution of $\mathbf{N}_C(t)$ using different initial values, $\mathbf{N}_C(0) = 0$ (blue), 20 (green), and 30 (orange), all with $N_A = aV = 30$, $N_B = bV = 40$. They all peak at approximately $\mathbf{N}_C = 20$. In comparison, the smooth dashed red curve is given by Eqn. (12.20) below, with an appropriately chosen C. This agrees well with the Monte Carlo simulation results.

Using $k_1 = 0.1$, $k_- = 1$, $N_A = 30$, and $N_B = 40$, we solve this quadratic equation for the equilibrium number of molecule C, n^*:

$$n^* = \frac{1}{2}\left(N_A + N_B + \tfrac{k_-}{k_1} - \sqrt{\left(N_A + N_B + \tfrac{k_-}{k_1}\right)^2 - 4N_A N_B}\right)$$
$$= \frac{1}{2}\left(80 - \sqrt{6400 - 4800}\right) = 20.$$

The negative sign is chosen since n^* has to be smaller than $\min(N_A, N_B)$. $n^* = 20$ is precisely the peak position of the distributions in Fig. 12.3.

12.4 Simple Bifurcations in One-Dimensional Nonlinear Dynamics

We expect the readers to have a working knowledge of dynamical systems theory and nonlinear bifurcations [255]. Here we give a very brief summary of the key notions and associated equations of the three most widely encountered types of bifurcations in single-variable systems. We encountered saddle-node bifurcation and pitch-fork bifurcation in Chapter 8; the following sections will involve the third type, *transcritical bifurcation*. We focus on the contradistinctions among these three different phenomena.

In terms of a simple, 1-dimensional differential equation $\frac{dx}{dt} = b(x; \lambda)$ with parameter λ these three types of bifurcations have the canonical forms

$$\frac{dx}{dt} = \lambda - x^2, \quad \text{saddle-node;} \tag{12.21}$$

$$\frac{dx}{dt} = \lambda x - x^2, \quad \text{transcritical;} \tag{12.22}$$

$$\frac{dx}{dt} = \lambda x - x^3, \quad \text{pitch-fork.} \tag{12.23}$$

12.4.1 Saddle-node Bifurcation

The equation in (12.21) has no real-valued steady state when $\lambda < 0$; both roots to $b(x; \lambda) = 0$ are complex numbers. However, when $\lambda > 0$, a pair of two steady states appears "out of the blue" at $x_1^* = \sqrt{\lambda}$ and $x_2^* = -\sqrt{\lambda}$. x_1^* is stable while x_2^* is unstable. They start together when $\lambda = 0$ and grow apart as λ increases (see Fig. 12.4A).

12.4.2 Transcritical Bifurcation

The two steady states are $x_1^* = 0$ and $x_2^* = \lambda$. x_1^* is stable when $\lambda < 0$ but becomes unstable when $\lambda > 0$. At the same time, x_2^* is unstable when $\lambda < 0$ and becomes stable when $\lambda > 0$. The stabilities of the two steady states are "exchanged" at $\lambda = 0$ (see Fig. 12.4B). It is always the lower branch that is unstable.

Transcritical bifurcation is not a robust phenomenon; it is called *imperfect* in [255]. To see this, consider

$$\frac{dx}{dt} = \lambda x - x^2 + \varepsilon, \tag{12.24}$$

with $\varepsilon > 0$. In this case, the two branches are

$$x_1^*(\lambda) = \frac{\lambda - \sqrt{\lambda^2 + 4\varepsilon}}{2}, \quad \text{and} \quad x_2^*(\lambda) = \frac{\lambda + \sqrt{\lambda^2 + 4\varepsilon}}{2}. \tag{12.25}$$

They never cross. A bifurcation behavior that cannot persist under an ε-perturbation, such as the systems in (12.22) and (12.24), is called imperfect. In biological terms, this bifurcation phenomenon is not robust.

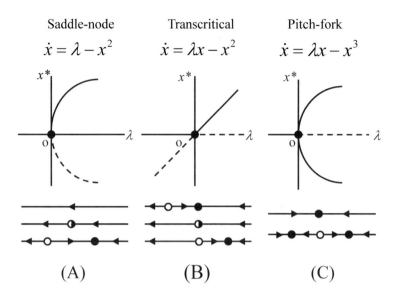

Fig. 12.4 Canonical representation of the three different types of bifurcations in the 1-dimensional nonlinear dynamics $dx/dt = b(x; \lambda)$, shown in Eqns. (12.21), (12.22), and (12.23). An unstable steady state is represented by a dashed line and an open circle, while a stable steady state is represented by a solid line and a filled circle. A half-filled circle represents a *semistable* fixed point, corresponding to ODE $\frac{dx}{dt} = -x^2$.

12.4.3 Pitch-fork Bifurcation

There is one stable steady state $x_1^* = 0$ when $\lambda < 0$. However, when $\lambda > 0$, $x_1^* = 0$ becomes unstable, and two additional stable steady states emerge at $x_{2,3}^* = \pm\sqrt{\lambda}$ (see Fig. 12.4C).

The system in Eqn. (12.23) is also imperfect. By employing the triple-root condition for a cubic equation, one sees that

$$\frac{dx}{dt} = \lambda x - x^3 - \varepsilon(x^2 - 1), \tag{12.26}$$

if $\varepsilon \neq 0$, will no longer have a pitch-fork bifurcation, but only a saddle-node bifurcation, with respect to changing λ.

Therefore, both imperfect bifurcations in Eqns. (12.22) and (12.23) are reduced to saddle-node bifurcations. The loss of the pair of roots in (12.21) is preserved under perturbation by a generic $\varepsilon(x)$ term on its right-hand side as long as $\varepsilon(x)$ is sufficiently small. The theory of saddle-node bifurcation has a topological origin in terms of the *catastrophe theory*. It is a more fundamental phenomenon.

12.5 Keizer's Paradox and Time Scale Separation

We now consider the nonlinear reaction kinetics

$$A + X \underset{k_{-1}}{\overset{k_1}{\rightleftharpoons}} 2X, \quad X \xrightarrow{k_2} B, \tag{12.27}$$

in which A and B are chemical species with fixed concentrations, and the kinetic rate equation according to the law of mass action yields the ODE

$$\frac{dx(t)}{dt} = (k_1 a - k_2)x - k_{-1}x^2. \tag{12.28}$$

The system has two steady states, $x_1^* = 0$ and $x_2^* = \frac{k_1 a - k_2}{k_{-1}}$. It is easy to see that when $k_1 a < k_2$, state x_1^* is stable, while x_2^* is unstable, negative, and thus not physically meaningful. When $k_1 a > k_2$, however, x_2^* becomes positive and stable. There is a transcritical bifurcation when $k_1 a = k_2$.

In the latter case, the deterministic kinetics clearly show that, for any initial concentration $x(0) > 0$, as $t \to \infty$, $x(t) \to x_2^*$. However, the stochastic kinetics tells us a rather different story.

The stochastic DGP $N(t)$ has a chemical master equation (CME) for the probability of n X molecules in a reaction volume of V:

$$\frac{dp_n(t)}{dt} = u(n-1)p_{n-1} - \Big(u(n) + w(n)\Big)p_n + w(n+1)p_{n+1}. \tag{12.29a}$$

in which the state-dependent birth and death rates are

$$u(n) = k_1 a n, \quad w(n) = k_2 n + \frac{k_{-1}n(n-1)}{V}. \tag{12.29b}$$

We note that the death rate $w(1) = k_2 > 0$ but the birth rate $u(0) = 0$. Therefore, state 0 is an absorbing state for the Markov process: one can verify that $p_0^{ss} = 1$ and $p_n^{ss} = 0$, $n \geq 1$ is a stationary solution to Eqn. (12.29). In other words, according to the stochastic model, the stationary distribution is "population extinction with probability 1".

There is a clear contradiction between the deterministic kinetics, which state that if $x(0) > 0$, $x(t) \to \frac{k_1 a - k_2}{k_{-1}}$ as $t \to \infty$, and the stochastic kinetics, which state that $\mathbf{N}(t) \to 0$ as $t \to \infty$. Actually, while the ODE can have a different long-time "fate" depending upon its initial condition, the DGP has a unique, stationary distribution that is independent of its initial condition. This discrepancy had been noted by J. Keizer [145] and is known as *Keizer's paradox* in [264].

The resolution to the paradox involves a *time scale separation* of the kinetic problem. It turns out that the time scale for $x(t)$ to reach x_2^* and for $\mathbf{N}(t)$ to reach the corresponding

$$n_2^* = \frac{(k_1 a - k_2)V}{k_{-1}} + 1$$

is on the order of $\sim O(V)$. In contrast, moving toward 0 is very unlikely since $u(n) > w(n)$ for all $n < n_2^*$. This translates into a time scale of $\sim O(e^{\alpha V})$ with $\alpha > 0$ arriving at 0!

The DGP $\mathbf{N}(t)$, starting with any nonzero value, quickly reaches n_2^* and then fluctuates around that value for a long time. The time for $\mathbf{N}(t)$ to actually reach 0 the very first time is on the order of $e^{\alpha V}$, where $\alpha > 0$. Nevertheless, as long as V is finite, this will occur with certainty. However, if one takes the limit $V \to \infty$ before $t \to \infty$, then this possibility is lost: deterministic dynamics cannot move against the drift direction. Therefore, the resolution of Keizer's paradox ultimately resides in understanding the separation of time scales in the phase transition problem, discussed in Chapter 8. Even though the present problem does not have bistability, the barrier crossing time is essentially the time required to reach the saddle point; e.g., extinction in the current setting.

12.5.1 Mean Time to Extinction

We now formulate a problem to solve the mean time to extinction for the discrete birth-death process in (12.29).

Let us assume $\tilde{n} \gg n_2^*$ is a reflect boundary and $\hat{n} = 0$ is the absorbing state for the DGP $\mathbf{N}(t)$. We shall denote the random first-arriving time by $T(n_0)$, where n_0 is the initial value of process $\mathbf{N}(t)$, and its expected value $\mathbb{E}[T(n_0)] = \tau_{n_0}$. Then, τ_n satisfies the equation

$$\tau_n = \frac{1}{u(n)+w(n)} + \frac{u(n)\tau_{n+1}+w(n)\tau_{n-1}}{u(n)+w(n)}, \tag{12.30}$$

which can be rearranged into

$$\mathscr{L}^*\big[\{\tau_n\}\big] \triangleq u(n)\tau_{n+1} - \big(u(n)+w(n)\big)\tau_n + w(n)\tau_{n-1} = -1. \tag{12.31}$$

We encountered an equation like this in Section 10.2.1 for the Michaelis–Menten kinetics. One should observe the close similarity between this equation and the right-hand side of Eqn. (12.29a). In terms of matrix representation, the middle part of (12.31), which defines $\mathscr{L}^*$, is the transpose of

$$\mathscr{L}\big[\{p_n\}\big] \triangleq p_{n-1}u(n-1) - p_n\big(u(n)+w(n)\big) + p_{n+1}w(n+1). \tag{12.32}$$

To solve Eqn. (12.31), by introducing $s_n = \tau_{n+1} - \tau_n$, we can express Eqn. (12.31) in terms of a first-order difference equation for s_n:

$$u(n)s_n - w(n)s_{n-1} = -1, \tag{12.33}$$

with boundary condition at $\tilde{n}$: $s(\tilde{n}) = 0$.

The solution to (12.33) can be obtained using the method of variation of parameters for inhomogeneous linear equations:

$$s_n = \left[\sum_{k=n}^{\tilde{n}} \frac{1}{w(k)} \prod_{j=1}^{k-1} \frac{u(j)}{w(j)}\right] \prod_{i=1}^{n} \frac{w(i)}{u(i)}. \tag{12.34}$$

Thus,

$$\tau_n = \sum_{\ell=\hat{n}+1}^{n} s_\ell = \sum_{\ell=\hat{n}+1}^{n} \left[\sum_{k=\ell}^{\tilde{n}} \frac{1}{w(k)} \prod_{j=1}^{k-1} \frac{u(j)}{w(j)}\right] \prod_{i=1}^{n} \frac{w(i)}{u(i)}, \tag{12.35}$$

recalling that $\hat{n} = 0$.

Interestingly, the τ_n in (12.35) can be expressed in terms of the stationary solution to (12.32), given in (12.6):

$$\tau_n = \sum_{m=\hat{n}}^{n-1} \sum_{\ell=m+1}^{\tilde{n}} \frac{p_\ell^{ss}}{w(m)p_m^{ss}}. \tag{12.36}$$

Recognizing the V-dependence of the stationary distribution p_n^{ss} and $w(n)$, we have an asymptotic approximation of (12.36) in terms of the landscape function $\varphi^{ss}(x)$ in Eqn. (12.9):

$$\tau_{xV} \simeq V \int_{\hat{x}}^{x} \frac{dz}{\widehat{w}(z)} \int_{z}^{\tilde{x}} e^{V[\varphi^{ss}(z) - \varphi^{ss}(y)]} dy, \tag{12.37}$$

in which $x = n/V$ and $\hat{x} = \hat{n}/V$. When a particular form of the function $\varphi^{ss}(x)$ is given, with local minima and maxima, this integral can be evaluated using Laplace's method asymptotically as $V \to \infty$. One then obtains an explicit result that is analogous to Kramers' rate formula, as in Section 4.6. This is the rate of crossing a saddle point in a stochastic nonlinear chemical reaction system, in contrast to Kramers' rate for crossing a saddle point in a fluctuating single molecule.

12.6 Schlögl's Model and Nonlinear Bistability

We now discuss a nonlinear reaction system that actually has bistability:

$$A + 2X \underset{k_{-1}}{\overset{k_{+1}}{\rightleftharpoons}} 3X, \quad X \underset{k_{-2}}{\overset{k_{+2}}{\rightleftharpoons}} B, \tag{12.38}$$

which is widely known as the Schlögl model. Again, A and B in (12.38) are chemical species with fixed concentrations, and the kinetic rate equation according to the law of mass action yields the ODE

$$\frac{dx(t)}{dt} = k_{+1} a x^2 - k_{-1} x^3 - k_{+2} x + k_{-2} b = b(x), \tag{12.39}$$

in which $b(x)$ is a third-order polynomial. The system can exhibit bistability, saddle-node bifurcation phenomena, and cusp catastrophe. This, however, only occurs under driven conditions, when chemical potentials $\mu_A \neq \mu_B$. Note that in the chemical equilibrium: $\mu_A = \mu_A^o + k_B T \ln a = \mu_B^o + k_B T \ln b$, and

$$\left(\frac{b}{a}\right)^{eq} = \frac{k_{+1} k_{+2}}{k_{-1} k_{-2}}. \tag{12.40}$$

ODE (12.39) under the condition $a k_{+1} k_{+2} = b k_{-1} k_{-2}$ has the right-hand side

$$\begin{aligned}
b(x) &= k_{+1} a x^2 - k_{-1} x^3 - k_{+2} x + k_{-2} b \\
&= k_{+1} a x^2 - k_{-1} x^3 - k_{+2} x + \frac{a k_{+1} k_{+2}}{k_{-1}} \\
&= \left(x^2 + \frac{k_{+2}}{k_{-1}}\right)\left(a k_{+1} - k_{-1} x\right).
\end{aligned} \tag{12.41}$$

Therefore, $b(x)$ has only a single real-valued fixed point at $x = x^* \equiv ak_{+1}/k_{-1}$, the chemical equilibrium. In general, system (12.39) can exhibit chemical bistability; but this is only possible when A and B have a sufficiently large chemical potential difference, e.g., a sustained chemostat.

The stochastic DGP $\mathbf{N}(t)$ again has a CME (12.29a) in which the state-dependent birth and death rates are

$$u(n) = \frac{k_{+1}an(n-1)}{V} + k_{-2}bV, \tag{12.42a}$$

$$w(n) = \frac{k_{-1}n(n-1)(n-2)}{V^2} + k_{+2}n. \tag{12.42b}$$

When $ak_{+1}k_{+2} = bk_{-1}k_{-2}$ holds, these two rates can be further simplified

$$u(n) = \frac{k_{+1}a}{V}\left(n(n-1) + \frac{k_{+2}V^2}{k_{-1}}\right),$$

$$w(n) = \frac{k_{-1}n}{V^2}\left((n-1)(n-2) + \frac{k_{+2}V^2}{k_{-1}}\right).$$

Then, the stationary distribution, according to Eqn. (12.6),

$$p_n^{ss} = C\prod_{\ell=0}^{n-1}\frac{k_{+1}a/V}{k_{-1}(\ell+1)/V^2} = \frac{\lambda^n}{n!}e^{-\lambda}, \quad \lambda = \left(\frac{k_{+1}aV}{k_{-1}}\right). \tag{12.43}$$

This is a Poisson distribution, with expected value $\mathbb{E}[\mathbf{N}^{eq}] = \lambda$, corresponding to the expected concentration $x^* = k_{+1}a/k_{-1}$.

More generally, even when $a_{+1}k_{+2} \neq bk_{-1}k_{-2}$, one still has a stationary distribution by (12.6)

$$p_n^{ss} = C\prod_{j=1}^{n}\frac{u(j-1)}{w(j)} = C\prod_{j=1}^{n}\frac{k_{+1}Va(j-1)(j-2)+k_{-2}bV^3}{k_{-1}j(j-1)(j-2)+k_{+2}V^2j}, \tag{12.44}$$

which yields a continuous landscape on $\mathbb{R}^+$ as in Eqn. (12.9):

$$\begin{aligned}
\varphi^{ss}(x) &= \int_0^x \ln\left(\frac{\hat{w}(z)}{\hat{u}(z)}\right)dz \\
&= \int_0^x \ln\left(\frac{k_{-1}z^3 + k_{+2}z}{k_{+1}az^2 + k_{-2}b}\right)dz \\
&= A\int_0^x \ln\left(\frac{\alpha v(1+v^2)+v}{\alpha\beta(1+v^2)}\right)dv \\
&= A\int_0^x \left[\ln(\sigma^2+v^2)+\ln v - \ln(1+v^2) - \ln\beta\right]dv,
\end{aligned}$$

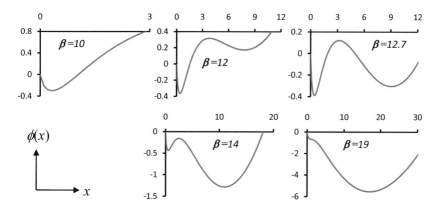

Fig. 12.5 Landscapes $\varphi^{ss}(x)$ for the Schlögl nonlinear reaction system given by Eqn. (12.45), $A = 1$, with different values of β and $\alpha = 0.03$, corresponding to $\sigma = \sqrt{(1+\alpha)/\alpha} = 5.86$.

in which

$$\alpha = \frac{k_{-1}k_{-2}b}{k_{+1}k_{+2}a - k_{-1}k_{-2}b}, \quad \beta = \frac{k_{+1}a}{k_{-1}}\sqrt{\frac{k_{+1}a}{k_{-2}b}}, \quad A = \sqrt{\frac{k_{-2}b}{k_{+1}a}}, \quad \sigma^2 = \frac{k_{+1}k_{+2}a}{k_{-1}k_{-2}b}.$$

Thus,

$$\varphi^{ss}(x) = A\left[x\ln\left(\frac{\sigma^2 + x^2}{1 + x^2}\right) + 2\sigma\arctan\left(\frac{x}{\sigma}\right) + x\ln\left(\frac{x}{\beta}\right) - 2\arctan x - x\right].$$
$$(12.45)$$

A few examples of $\varphi^{ss}(x)$ with bistability are shown in Fig. 12.5.

12.7 General Theory of Nonlinear Reaction Systems

We now present the general theory of a system of many nonlinear stochastic elementary chemical reactions. In particular, we introduce the concept of a *stochastic elementary reaction*, which implies that the numbers of all chemical species in the system follow discrete, integer-valued Markov processes in continuous time, i.e., a DGP.

Some of the materials in the present section have already appeared in Chapter 5 and Chapter 7. With the specific examples we studied in detail, it will be pedagogically useful to have the general theory again discussed, this time more systematically. For the general discussion, we again assume that in a CSTR, there are N chemical species, $A_1, \cdots, A_N$, and M chemical reactions:

$$v_{j1}A_1 + v_{j2}A_2 + \cdots + v_{jN}A_N \xrightarrow{r_j(\mathbf{n})} \kappa_{j1}A_1 + \kappa_{j2}A_2 + \cdots + \kappa_{jN}A_N, \qquad (12.46)$$

$1 \leq j \leq M$. The nonnegative integers v_{ji} and κ_{ji} are called *stoichiometric coefficients*. They relate the ith species to the jth reaction. The state of the chemical reaction system is represented by $\mathbf{N}(t) = (N_1, \cdots, N_N)(t)$, the nonnegative integers counting the number of molecules in species $A_1, \cdots, A_N$ at time t. In probabilistic terms, a "reaction" is a random event that changes $\mathbf{n}$ instantaneously. By *stochastic elementary reaction*, we assume that all events have exponentially distributed waiting times, with rate r_j for the jth reaction. Note that $r_j(\mathbf{n}, V)$ is actually a function of $\mathbf{n}$ as well as the volume of the reaction vessel V; it has physical dimensions $[\text{time}]^{-1}$, and the function $r_j(\mathbf{n}, V)$ is called a *rate law*. Mass action kinetics assumes a very special rate law; the stochastic chemical kinetics can be much more general.

12.7.1 Deterministic Chemical Kinetics

In a macroscopic-sized vessel, $\mathbf{n}$ is very large: A micromolar is 10^{17} per liter. One introduces the concentration for i^{th} species, $x_i = \frac{n_i}{V}$, and the macroscopic reaction flux:

$$R_j(\mathbf{x}) = \lim_{V \to \infty} \frac{r_j(V\mathbf{x}, V)}{V}, \qquad (12.47)$$

in which $\mathbf{x} = (x_1, \cdots, x_N)$. Then, one expects from the conservation of molecules:

$$\frac{dx_i}{dt} = \sum_{j=1}^{M} \left(\kappa_{ji} - v_{ji} \right) R_j(\mathbf{x}). \qquad (12.48)$$

The law of mass actions further assumes that

$$R_j(\mathbf{x}) = k_j x_1^{v_{j1}} x_2^{v_{j2}} \cdots x_n^{v_{jn}}. \qquad (12.49)$$

12.7.2 Chemical Master Equation (CME)

The stochastic elementary reactions with exponential waiting times naturally define a continuous-time, integer-valued Markov process on $\mathbb{Z}^N$, with each grid point having no more than $2M$ connections. Stochastic chemical kinetics therefore are a Markov jump process $\mathbf{N}(t)$, whose Kolmogorov forward equation, also called its *chemical master equation*, takes the form:

$$\frac{dp(\mathbf{n};t)}{dt} = \sum_{j=1}^{M} p(\mathbf{n}-\mathbf{v}_j;t)r_j(\mathbf{n}-\mathbf{v}_j) - p(\mathbf{n};t)r_j(\mathbf{n}), \qquad (12.50)$$

in which $\mathbf{v}_j = (\kappa_{j1} - v_{j1}, \cdots, \kappa_{jN} - v_{jN})$.

At any state $\mathbf{n}$, *e.g.*, the grid point $\mathbf{n}$, there are at most M possible reactions that can occur. The jth reaction moves $(\mathbf{n}-\mathbf{v}_j) \to \mathbf{n}$ with rate $r_j(\mathbf{n}-\mathbf{v}_j,V)$, and $\mathbf{n} \to (\mathbf{n}+\mathbf{v}_j)$ with rate $r_j(\mathbf{n},V)$. The first reaction that occurs at $\mathbf{n}$ also follows exponential time, with the rate being the sum of the M reactions:

$$\sum_{j=1}^{M} r_j(\mathbf{n},V). \qquad (12.51)$$

The probability of choosing reaction k is equal to $\frac{r_k(\mathbf{n},V)}{\sum_{j=1}^{M} r_j(\mathbf{n},V)}$. Furthermore, the waiting time and the reaction to be chosen are statistically independent.

To be consistent with the mass-action kinetics, the rate law can be rewritten as

$$r_j(\mathbf{n},V) = k_j V \prod_{\ell=1}^{N} \left(\frac{n_\ell!}{(n_\ell - v_{j\ell})! V^{v_{j\ell}}} \right). \qquad (12.52)$$

However, the general theory discussed thus far can be applied to arbitrary cases.

12.7.3 Integral Representation With Random Time Change

Our readers might have learned about the Fokker–Planck equations and stochastic differential equations (SDE) as two complementary descriptions of a stochastic process with continuous time in continuous space, the former in terms of a probability density function $f_{\mathbf{X}}(\mathbf{x};t)$ and the latter in terms of a stochastic trajectory $\mathbf{X}(t)$. They are related via

$$f_{\mathbf{X}}(\mathbf{x};t)d\mathbf{x} = P\{\mathbf{x} < \mathbf{X}(t) \leq \mathbf{x}+d\mathbf{x}\}.$$

For stochastic chemical kinetics, the chemical master equation corresponds to a Fokker–Planck equation. One therefore naturally asks, "where is the SDE-like representation for stochastic $\mathbf{n}(t)$ with continuous time in $\mathbb{Z}^N$?"

Recall that a multidimensional SDE is defined based on a set of i.i.d. standard Brownian motions. Similarly, $\mathbf{N}(t)$ can be represented based on M i.i.d. *standard Poisson processes* with unit rate, $\mathbf{Y}_j(t)$, each of which follows

$$P\{\mathbf{Y}(t) = k\} = \frac{t^k e^{-t}}{k!}. \tag{12.53}$$

Each reaction in (12.46) is represented by a Poisson process. Then, $\mathbf{n}(t)$ is represented through an integral equation:

$$\mathbf{N}_i(t) = n_i(0) + \sum_{j=1}^{M} \left(\kappa_{ji} - \nu_{ji} \right) \mathbf{Y}_j \left(\int_0^t r_j(\mathbf{N}(t)) dt \right). \tag{12.54}$$

$r_j(\mathbf{n})$ is called the *propensity* of reaction j. The time argument in $Y_j(t)$, and thus the rate of the Poisson process, is continuously updated according to the propensity at time t.

We see that in the limit of $\mathbf{n}, V \to \infty$ and $\mathbf{x} = \frac{\mathbf{n}}{V}$, $r_j(\mathbf{n}) \simeq VR(\mathbf{x})$, which is very large. For large r_j, $\mathbf{Y}_j(r_j) \simeq r_j$. Therefore, Eqn. (12.54) becomes

$$x_i(t) = x_i(0) + \sum_{j=1}^{M} \left(\kappa_{ji} - \nu_{ji} \right) \int_0^t R_j(\mathbf{x}(t)) dt, \tag{12.55}$$

which implies Equation 12.48. This little exercise shows the analytic power of the integral representation given in (12.54).

12.8 Stochastic Bistability

The bistability in Section 12.6 shows a correspondence between the deterministic nonlinear fixed points and the two basins of attraction in Fig. 12.5, which imply bimodality in the steady state probability distribution $p^{ss}(n) \propto e^{-V \varphi^{ss}(n/V)}$ for sufficiently large V. However, for small V, it is possible to have a bimodal $p^{ss}(n)$ while the corresponding deterministic kinetics have only a single stable fixed point. We have termed this phenomenon *stochastic bistability*, or noise-induced bistability [27].

Consider the following kinetics,

$$A + X \underset{k_{-1}}{\overset{k_1}{\rightleftharpoons}} 2X, \ X \underset{k_{-2}}{\overset{k_2}{\rightleftharpoons}} A. \tag{12.56}$$

It is nearly identical to the reaction system in (12.27) with an addition step of k_{-2}. It is inspired by a phosphorylation-dephosphorylation cycle $A \rightarrow X \rightarrow A$ if one identifies A and X as the unphosphorylated and phosphorylated states of a protein, respectively. $A + X \longrightarrow 2X$ implies that the phosphorylation reaction is autocatalytic.

The kinetic rate equation according to the law of mass action yields the ODE for $x(t)$, the concentration of X:

$$\frac{dx(t)}{dt} = k_1 a x - k_{-1} x^2 - k_2 x + k_{-2} a$$
$$= -(k_1 + k_{-1}) x^2 + (k_1 x_{tot} - k_2 - k_{-2}) x + k_{-2} x_{tot} = b(x), \quad (12.57)$$

in which, since $[A] + [X] = x_{tot}$ is conserved, $a = x_{tot} - x$. $b(x)$ is a second-order polynomial. Since the coefficients of x^2 and x^0 have opposite signs, $b(x) = 0$ has only one positive root. The ODE is monostable at

$$x_2^* = \frac{(k_2 + k_{-2} - k_1 x_{tot}) + \sqrt{(k_2 + k_{-2} - k_1 x_{tot})^2 + 4(k_1 + k_{-1})k_{-2} x_{tot}}}{2(k_1 + k_{-1})}. \quad (12.58)$$

However, the stochastic DGP $\mathbf{N}(t)$ has a CME (12.29a), in which the state-dependent birth and death rates are

$$u(n) = \left(\frac{k_1 n}{V} + k_{-2}\right)(n_{tot} - n), \quad w(n) = \frac{k_{-1} n(n-1)}{V} + k_2 n. \quad (12.59)$$

When $k_1 k_2 = k_{-1} k_{-2}$ holds, the stationary distribution according to Eqn. (12.6) is:

$$p_n^{ss} = C \prod_{\ell=1}^n \frac{k_1(n_{tot} - \ell + 1)}{k_{-1}\ell} = \frac{n_{tot}! \kappa^n}{n!(n_{tot} - n)!}, \quad \kappa = \left(\frac{k_1}{k_1 + k_{-1}}\right), \quad (12.60)$$

which is a binomial distribution for n_{tot} i.i.d. molecules, each with two states, A and X. More generally, when $k_1 k_2 \neq k_{-1} k_{-2}$,

$$p_n^{ss} = C \prod_{\ell=1}^n \frac{(k_1(\ell-1) + k_{-2}V)(n_{tot} - \ell + 1)}{k_{-1}\ell(\ell-1) + k_2 V \ell}. \quad (12.61)$$

Because of the presence of nonzero k_{-2}, p_n^{ss} is nonzero for all $n \geq 0$, in contrast to the case in Keizer's paradox,, where $p^{ss}(n) = \delta_{n0}$.

It is expected, therefore, that for small k_{-2}, there will be two modal values at $n_1^* = 0$ and n_2^* near $V x_2^*$. This is the phenomenon of *stochastic bistability*. We can determine the modal values and the location of the local minimum of distribution p_n^{ss}:

$$\left(k_1 n + k_{-2} V\right)\left(n_{tot} - n\right) - \left(k_{-1} n + k_2 V\right)\left(n + 1\right) = 0,$$

$$\Rightarrow \left(k_1 + k_{-1}\right) n^2 - \left(k_1 n_{tot} - k_{-2} V - k_2 V - k_{-1}\right) n + k_2 V - k_{-2} V n_{tot} = 0,$$

$$\Rightarrow \left(k_1 + k_{-1}\right) c^2 - \left(k_1 c_{tot} - k_{-2} - k_2 - \frac{k_{-1}}{V}\right) c + \frac{k_2}{V} - k_{-2} c_{tot} = 0,$$

where concentration $c = \frac{n}{V}$. As long as $k_2 > k_{-2} n_{tot}$, there is a local minimum that separates the two modal values in p_n^{ss}. This feature, termed the *extinction effect at zero*, is a recurring feature in several stochastic effects [217].

Chapter 13
Kinetics of the Central Dogma of Molecular Cell Biology

In a living cell, there is usually only one copy of a particular gene. The genetic information encoded in DNA needs to be "expressed" in terms of biochemically functioning proteins and enzymes, and their numbers, as well as their locations, inside a cell matter. The permanent gene residing in the DNA double helix is first transcribed to the more dynamic and labile messenger RNA (mRNA) based on a "templet copying" mechanism through base paring.The mRNA then is translated to polypeptides again based on a templet mechanism using transfer RNA (tRNA) as converters. The biochemical processes that carry out this information flow, referred to as the *central dogma of molecular biology*, are RNA polymerization and polypeptide biosynthesis, respectively; they involve the machinery of RNA polymerase and ribosomal complexes together with a myriad of molecular regulators. These biochemical processes and their players are described in every modern biology textbook. The molecular structural determinations of RNA polymerase (~ 5000 atoms) and ribosomes (~ 50000 atoms) at the atomic level are major accomplishments of modern science.

The atomic structures give a rather static portrayal of these macromolecules. In a living cell, however, the biochemical processes of gene expression and regulation performed by these molecular assemblies are highly stochastic. The stochasticity in each and every individual process leads to *stochastic gene expression*; this becomes particularly relevant for proteins with very low copy numbers in a single cell [166]. Traditional biochemical studies with bulk quantities or large populations of cells have obscured this aspect of single living cells. Thanks to advanced fluorescence microscopy with single-molecule sensitivity and the method of RNA sequencing, experimentalists can capture real-time temporal fluctuations in the numbers of mRNA molecules and proteins in a single cell and sample the presence and even the copy numbers of mRNA in thousands of individual cells [257, 59, 246, 256]. These observations have firmly established the cell-to-cell variability in the copy numbers of mRNA and proteins as gene products.

Measurement of stochastic processes leads to the acquisition of "noisy data" that are different each time; only the statistical characteristics are truly reproducible in measurements from single molecules and single cells.Note that the differences here are not due to measurement errors; they are fundamental properties of macromolecules at finite temperature and living cells with stochastic gene expressions. Stochastic kinetics descriptions are based on a given mechanism and are able to provide both predictions for the statistical characteristics of dynamic behavior and simulations of noisy data. They are the link between a biochemical mechanism and experimental observables, especially when one records the real-time trajectories of gene transcription and protein translation. In this chapter, we will discuss the stochastic biochemical kinetic portrayal of the central dogma in living cells, leading to quantitative descriptions of mRNA and protein synthesis and gene regulation.

13.1 The Simplest Model of the Central Dogma

Fig. 13.1A is the simplest kinetic representation of the central dogma. It is assumed that the synthesis of both mRNA and protein are zeroth-order processes with constant rates due to the steady level of source materials, e.g., nucleotides and amino acids. The degradation of mRNA and protein is a first-order process with single rate-limiting steps. Examples of this ideal case are the consecutive promoter or the *Lac* operon under repressed conditions [32]. The state of a single cell is denoted by the copy numbers (m,n) of an mRNA and its corresponding protein molecules, and $p(m,n;t)$ is the joint probability of a cell having m copies of the mRNA and n copies of the protein at time t. The chemical master equation (CME) for the system (see Fig. 13.2A) is

$$
\begin{aligned}
\frac{dp(m,n;t)}{dt} = &\, k_1 p(m-1,n;t) - k_1 p(m,n;t) + d_1(m+1)p(m+1,n;t) \\
&- d_1 m p(m,n;t) + k_2 m p(m,n-1;t) - k_2 m p(m,n;t) \\
&+ d_2(n+1)p(m,n+1;t) - d_2 n p(m,n;t).
\end{aligned} \tag{13.1}
$$

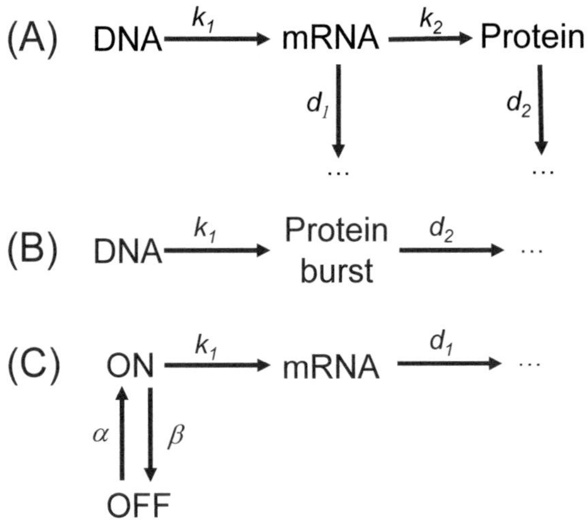

Fig. 13.1 Kinetic representation of the central dogma of molecular biology with transcription and translation without feedback. (A) Simplest model. (B) Translational burst model. (C) Transcriptional burst model.

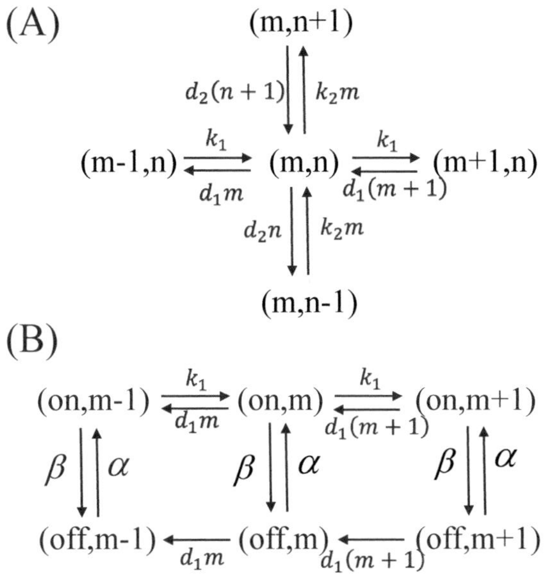

Fig. 13.2 Chemical master equation for the central dogma without feedback. (A) Simplest model. (B) Two-state transcriptional burst model.

Denoting

$$\langle m \rangle (t) = \sum_{m,n=0}^{\infty} m p(m,n;t)$$

as the mean copy number of mRNA at time t, it follows from the CME that

$$\frac{d}{dt}\langle m \rangle(t) = k_1 - d_1\langle m \rangle(t).$$

Denoting

$$\langle n \rangle(t) = \sum_{m,n=0}^{\infty} n p(m,n;t)$$

as the mean copy number of the protein at time t, it follows from the CME that

$$\frac{d}{dt}\langle n \rangle(t) = k_2\langle m \rangle(t) - d_2\langle n \rangle(t).$$

Similarly, denoting the expected value of the number of protein molecules as

$$\langle m^2 \rangle(t) = \sum_{m,n=0}^{\infty} m^2 p(m,n;t)$$

$$\langle n^2 \rangle(t) = \sum_{m,n=0}^{\infty} n^2 p(m,n;t),$$

and the cross-moment as

$$\langle mn \rangle(t) = \sum_{m,n=0}^{\infty} mn p(m,n;t),$$

then

$$\frac{d}{dt}\langle m^2 \rangle(t) = k_1\left[2\langle m \rangle_t + 1\right] - d_1\left[2\langle m^2 \rangle(t) - \langle m \rangle(t)\right];$$

$$\frac{d}{dt}\langle n^2 \rangle(t) = k_2\left[2\langle mn \rangle(t) + \langle m \rangle(t)\right] - d_2\left[2\langle n^2 \rangle(t) - \langle n \rangle(t)\right];$$

$$\frac{d}{dt}\langle mn \rangle(t) = k_1\langle n \rangle(t) - d_1\langle mn \rangle(t) + k_2\langle m^2 \rangle(t) - d_2\langle mn \rangle(t).$$

In the limit as $t \to \infty$, the mean and variance of the stationary distributions of mRNA and protein are [200]

$$\langle m \rangle_{ss} = \frac{k_1}{d_1}, \quad \langle n \rangle_{ss} = \frac{k_1 k_2}{d_1 d_2}, \tag{13.2}$$

$$\text{var}[m]_{ss} = \langle m \rangle = \frac{k_1}{d_1}, \quad \text{var}[n]_{ss} = \frac{k_1 k_2}{d_1 d_2}\left(1 + \frac{k_2}{d_1 + d_2}\right), \tag{13.3}$$

$$\text{cov}(m,n)_{ss} = \frac{k_1 k_2}{d_1(d_1 + d_2)}. \tag{13.4}$$

More importantly, their correlation coefficient

$$\rho = \frac{\langle (m - \langle m \rangle_{ss})(n - \langle n \rangle_{ss}) \rangle_{ss}}{\sqrt{\mathrm{var}[m]_{ss}} \cdot \sqrt{\mathrm{var}[n]_{ss}}} = \sqrt{\frac{d_2}{d_1 + d_2} \cdot \frac{1}{1 + \dfrac{d_1 + d_2}{k_2}}}.$$

We see that if d_1 is large, $\gg d_2$, this correlation coefficient is nearly zero, which has been experimentally validated in E. coli [257].

In this simplest model, the dynamics of mRNA do not depend on the corresponding "downstream" protein concentrations. There is no feedback; the stochastic mRNA kinetics are a Poisson process, with a constant influx and first-order efflux. The stationary distribution is then a Poisson distribution with equal mean and variance. Using a Poisson distribution as a reference, the concept of the *Fano factor*, which is simply the variance divided by the mean, has been introduced as a measure of the noise. The Fano factor is truly meaningful only for counting statistics. When a measurement signal is proportional to the number of molecules, the Fano factor is not scale free. In this case, the coefficient of variation (CV), which is the variance divided by the square of the mean, or its square is used to measure the noise.

Although the Fano factor is not universally suitable for quantifying counting noise, it is still useful for comparing the measured distribution with a Poisson distribution, which can shed light on the underlying molecular reaction mechanisms. For instance, experimentally measured mRNA distributions inside living bacterial cells are actually super-Poisson, i.e., the Fano factor is greater than 1 [257]. One possible reason is transcriptional bursts, which will be discussed next.

13.2 Translational Burst

Although the simplest model is linear, its exact analytic solution is complicated in terms of hypergeometric functions [247]. Solvable approximations to the model based on additional assumptions, often motivated by experimental observations, are a highly desirable part of the mathematical approach to scientific problems. Fig. 13.1B shows one simplification with only the copy-number dynamics of protein molecules, based on the observation of translational bursts. Translational burst refers to the observation of bursts of protein synthesis in a single cell, which has been shown to be ubiquitous not only in bacteria but also in eukaryotic cells [32, 256]. These bursts are due to the intermittent transcription of a single mRNA molecule.

The underlying mechanism may be complicated for eukaryotic cells, but it turns out to be quite simple for a bacterium. Inside a typical living bacterial cell, the lifetime of an mRNA molecule is much shorter than that of a protein molecule, as is the time between two consecutive DNA transcriptional events [32]. The initiation of each transcriptional event is controlled by the binding or unbinding of specific transcription factor molecules as a rate-limiting step.

Based on these observations, one can formulate a simple, mathematical representation that describes translational burst events. As in the simplest kinetic model previously described, we assume that the rate of transcription is k_1, i.e., waiting times are exponentially distributed with mean $1/k_1$ between consecutive transcriptions of mRNA molecules. Therefore, the number of protein bursts per cell cycle follows a Poisson distribution with parameter $k_1 T_{cycle}$, in which T_{cycle} is the time of each cell cycle.

Once an mRNA molecule is synthesized, it will be available for producing protein molecules through a Poisson process with parameter k_2. It will then be degraded after an exponentially distributed distribution with parameter d_1. The consequence of this mechanism is a burst in protein production. Since d_1 is very large, we regard the lifetime of an mRNA molecule as instantaneous. However, we still need to know the distribution of the number of protein molecules being synthesized during each burst. As first theoretically predicted in the 1970s [24], the synthesized protein number during each burst is geometrically distributed with parameter $q = k_2/d_1$, i.e.,

$$G_n \triangleq p(n) = \int_0^{+\infty} d_1 e^{-d_1 t} \cdot e^{-k_2 t} \left(\frac{(k_2 t)^n}{n!} \right) dt = (1-q)q^n, \qquad (13.5)$$

which is just the discrete version of the exponential distribution. Denote $b = k_2/d_1 = q/(1-q)$ as the mean of this geometric distribution.

In the case of bursty translation, the chemical master equation could be simplified to

$$\frac{dp(n;t)}{dt} = k_1 \left(\sum_{j=1}^n G_j p(n-j,t) - qp(n;t) \right) + d_2 \left[(n+1)p(n+1;t) - np(n;t) \right],$$
$$(13.6)$$

where d_2 is the degradation/dilution rate of protein.

The stationary distribution $p^{ss}(n)$ of the protein copy number satisfies

$$k_1 \left(\sum_{j=1}^{n} G_j p^{ss}(n-j) - q p^{ss}(n) \right) + d_2 \left[(n+1) p^{ss}(n+1) - n p^{ss}(n) \right] = 0.$$

Solving the equation, we arrive at the stationary distribution, which is a negative binomial distribution [201]

$$p^{ss}(n) = \frac{b^n}{(1+b)^{a+n}} \frac{\Gamma(a+n)}{\Gamma(a)n!}, \tag{13.7}$$

in which $a = k_1/d_2$. It is the discrete version of the continuous gamma distribution [247].

Typically, the decrease in protein levels is mainly due to cell division, and thus $e^{-d_2 T_{cycle}} = 1/2$. Hence, the period of the cell cycle $T_{cycle} = (1/d_2)\ln 2 \approx 1/d_2$, and $a = k_1/d_2$ could be realized as the burst frequency, i.e., the mean number of bursts per cell cycle. Experimentally, a and b can be determined by single-cell trajectories as well as fitting the cell-to-cell variation in the copy numbers [32].

13.3 Transcriptional Burst and a Two-state Model

Single-molecule studies on counting mRNA molecules in single cells have shown that transcription can also exhibit burst-like behavior itself [105]. Accordingly, the copy-number distribution is super-Poissonian, implying that the cell-to-cell variation is significantly greater than previously expected from a single, rate-limiting process [257]. The simplest two-state model for transcriptional burst is shown in Fig. 13.1C. The state of the system is characterized by two possible states of the gene, either "on" or "off", and the copy number of the mRNA. Let $p_{on}(m;t)$ and $p_{off}(m;t)$ be the probability of the cell state at time t with the gene state on or off and m copies of mRNA molecules. The corresponding chemical master equation is

$$\frac{dp_{on}(m;t)}{dt} = k_1 p_{on}(m-1) + d_1(m+1)p_{on}(m+1) + \alpha p_{off}(m)$$
$$- (k_1 + d_1 m + \beta) p_{on}(m;t),$$
$$\frac{dp_{off}(m)}{dt} = d_1(m+1)p_{off}(m+1) + \beta p_{on}(m) - (d_1 m + \alpha) p_{off}(m).$$

The stationary distribution of the mRNA copy number again cannot be analytically solved with simple expressions, but its mean and variance could be exactly derived due to the linearity of the model:

$$\langle m \rangle_{ss} = \frac{\beta k_1}{(\alpha + \beta)d_1}, \tag{13.8a}$$

$$\text{var}_{ss}[m] = \frac{\beta k_1}{(\alpha + \beta)d_1} + \frac{\alpha \beta k_1^2}{(\alpha + \beta)^2(\alpha + \beta + d_1)d_1}, \tag{13.8b}$$

with Fano factor

$$F = 1 + \frac{\alpha \beta k_1}{(\alpha + \beta)(\alpha + \beta + d_1)\beta}. \tag{13.8c}$$

Hence, the stationary distribution of mRNA would be super-Poissonian. If $d_1 \gg \alpha + \beta$, then $\langle m \rangle + F \approx 1 + k_1/d_1$, independent of α and β.

When α and β are both sufficiently large, we can merge the (on, m) and (off, m) states into one state; that is, the rate at which the gene switches between the on and off states is too fast, so it cannot be distinguished during transcription and translation. This is equivalent to the simplest mechanism shown in Fig. 13.1A, with effective transcription rate $\alpha k_1/(\alpha + \beta)$, so the final stationary distribution of mRNA becomes Poissonian.

When both α and β are very small, the connection between the two gene states can almost be considered broken. Therefore, under each gene state, the mRNA distribution forms a Poisson distribution, with mean values of $\lambda_1 = k_1/d_1$ and $\lambda_2 = 0$, respectively. The system reaches each of the two Poisson distributions well before the switching. Therefore if the peaks for the two distribution are far enough apart, then the peaks of the two distributions are thus far enough apart that the total stationary distribution can be approximated as

$$p(0) = \frac{\alpha}{\alpha + \beta}e^{-\lambda_1} + \frac{\beta}{\alpha + \beta}, \quad p(m) = \frac{\alpha}{\alpha + \beta}\frac{\lambda_1^m}{m!}e^{-\lambda_1} \quad (m \geq 1). \tag{13.9}$$

Such a distribution is called a "zero-inflated Poisson distribution". It actually has two modal values! This is also known as "stochastic bistability", because if the deterministic equation is built according to Fig. 13.1C, only one stable fixed point exists.

While mathematical models alone without real experiments cannot determine a molecular mechanism, they significantly help in uncovering the molecular mechanisms of cellular processes. Furthermore, one can even measure all the biochemical activities inside a cell *in vivo*, which does not mean one can unequivocally deduce a mechanism from data. The mechanism of transcriptional burst in terms of the existence of the gene on/off states and transitions between them is a good example.

In 2014, Chong et al. [37] showed that the transcription burst in bacteria results from the stochastic association and dissociation of gyrase molecules, which are responsible for unwinding twisted DNA, with the DNA molecule. The authors proposed a hypothesis for the mechanism of transcription burst based on the following observations. They realized that even in bacteria, DNA molecules are divided into segments and anchored with nucleoid-associated proteins such as H-NS and Fis. During the process of DNA transcription, the DNA strands in front of the RNA polymerase undergo so-called positive supercoiling, i.e., the DNA molecule becomes increasingly tighter, and the number of base pairs in each turn decreases. In contrast, the DNA strand behind the RNA polymerase undergoes negative supercoilings. These two types of DNA deformations are unable to cancel each other out because the DNA is anchored to macromolecules, resulting in a topological constraint. Therefore, there are two kinds of enzymes in the cell: topoisomerase, which is responsible for releasing the negative supercoiling behind RNA polymerase, and gyrase, which is responsible for releasing the positive supercoiling in front of RNA polymerase. While the topoisomerase activity is very high, there is only approximately one gyrase molecule for each anchored DNA fragment.

To verify the mechanism they proposed, the authors performed high-throughput *in vitro* single-molecule experiments, which revealed that if the gyrase molecules are dissociated from the DNA molecule, positive supercoiling continuously occurs in the front part of the RNA polymerase during the transcription process. This gradually slows the elongation of the RNA polymerase and eventually completely halts the initiation of transcription.

Unfortunately, the same experiment with similar resolution cannot be carried out in living cells thus far. The two-state model therefore played a crucial role in bridging the gap between the processes inside living cells and the measurements. The single-cell mRNA fluorescence in situ hybridization (FISH) technique allows one to measure the stationary distribution of the number of mRNA molecules in living cells; furthermore, it is already known that the association and dissociation rates for gyrase and DNA are quite slow. Hence, one can use the zero-inflated Poisson distribution in Eqn. (13.9) to fit the measured FISH data and obtain an inference on the duty cycle (β/α) of the transcription burst. The ratio depends on the concentration and activity of gyrase in the cell, and the variation in the distribution is consistent with the model for all the different measured operons along the bacterial genome. Combining single-molecule experiments with stochastic kinetic models has uncovered, although indirectly but rigorously, the mechanism of cellular transcriptional burst *in vivo* [37].

13.4 Self-Regulating Genes in a Cell: the Simplest Model

When a transcription factor (TF) protein regulates its own gene expression, a feedback loop is formed. For example, we assume here that the single DNA molecule inside a cell undergoes the biochemical reaction

$$\text{DNA} + \text{TF} \underset{f}{\overset{h}{\rightleftharpoons}} \text{DNA·TF}. \tag{13.10}$$

We shall call the bare DNA the off-state and the DNA bound with the TF the on-state. Fig. 13.3 gives the state transition diagram for a stochastic DGP. This scenario describes weak and linear feedback. The following section discusses models with a strong feedback.

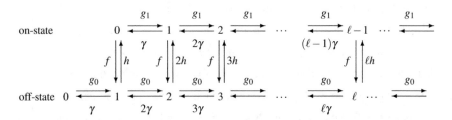

Fig. 13.3 The off-state and on-state represent the unbound and bound states, respectively, of a single copy of DNA (gene) with a TF. ℓ denotes the copy number of free TF protein. Monomeric TF binds DNA with on-rate constant h and off-rate constant f. Binding reduces the copy number of free TF by 1. The TF biosynthesis rate is g_1 and g_0 when the gene is bound and unbound, respectively. The TF degradation rate is γ.

When $f = h = 0$, the steady state distribution for the protein copy numbers in both the on- and off-states are Poissonian:

$$p_P(n|i) = \frac{(g_i/\gamma)^n}{n!} e^{-g_i/\gamma}, \tag{13.11}$$

where $i = 0$ for the off-state and $i = 1$ for the on-state. The modal value of a Poisson distribution is its mean value g_i/γ. Therefore, for very different g_i, the two modal values will be very different. For very small f and h, i.e., the TF binding-unbinding to DNA are much slower than protein biosynthesis and degradation, the probability distribution for the number of proteins is simply

$$p_P(n) = \frac{\gamma f p_P(n|0) + g_0 h p_P(n|1)}{\gamma f + g_0 h}$$

$$= \left(\frac{\gamma^{-n}}{n!}\right) \frac{\gamma f g_0^n e^{-g_0/\gamma} + g_0 h g_1^n e^{-g_1/\gamma}}{\gamma f + g_0 h}, \tag{13.12}$$

which has two modal values approximately at g_0/γ and g_1/γ.

At the other extreme, if f and h are very large, that is, the TF binding-unbinding to DNA are much faster than the protein biosynthesis and degradation, then there is a mean g value at each n, the total number of protein copies:

$$\bar{g}(n) = \frac{f g_0 + n h g_1}{f + nh}, \quad \bar{\gamma}(n) = n\left(\frac{f + h(n-1)}{f + nh}\right)\gamma, \tag{13.13}$$

then the protein copy number distribution is

$$p_P(n) = C \prod_{i=1}^{n} \frac{\bar{g}(i-1)}{\bar{\gamma}(i)} = \frac{C}{n!}\left(\frac{g_1}{\gamma}\right)^n \prod_{i=1}^{n} \frac{[f g_0/(h g_1)] + (i-1)}{(f/h) + (i-1)}, \tag{13.14}$$

where C is a normalization factor. This is a unimodal distribution with an intermediate modal value.

Note that the macroscopic ODE kinetics in a test tube containing a large number of DNA molecules with concentration $x(t)$ at time t ($0 \le x \le 1$) is

$$\begin{cases} \dfrac{dx(t)}{dt} = hy(1-x) - fx, \\[2mm] \dfrac{dy(t)}{dt} = \left(g_0(1-x) + g_1 x\right) - \gamma y, \end{cases} \tag{13.15}$$

in which $y(t)$ is the concentration of protein at time t. $g_1 < g_0$ indicates that the TF is its own gene expression repressor and that the feedback is negative. $g_1 > g_0$ indicates that the TF is its own gene expression enhancer and that the feedback is positive. In the latter case, the two nullclines for the ODE system are

$$y_1(x) = \frac{fx}{h(1-x)}, \quad y_2(x) = \gamma^{-1}\left(g_0(1-x) + g_1 x\right). \tag{13.16}$$

Both $y_1(x)$ and $y_2(x)$ are increasing functions of x; $y_1(x)$ is a convex function and $y_2(x)$ is linear. There is only a single intersection, $x^* \in (0,1]$. The macroscopic fixed point is located at (x^*, y^*), in which y^* is the positive root of

$$y^2 + \left(\frac{f}{h} - \frac{g_1}{\gamma}\right) y - \frac{f g_0}{h\gamma} = 0.$$

We note this is the same equation as $\bar{g}(n) = \bar{\gamma}(n)$, both given in (13.13). Therefore, the macroscopic kinetics are closer to the rapid TF-binding scenario, with the term $(g_0(1-x) + g_1 x)$ in Eqn. (13.15) representing the law of large numbers. Therefore, there is no bistability according to the ODE model. The bimodality in the stochastic model is a consequence of the single copy of DNA with slow gene-state switching rates. This phenomenon is another example of *stochastic bistability*, in contrast to the *nonlinear bistability* discussed in the following section.

13.5 Fluctuating-Rate Model and Landscape Function of Self-Regulating Genes

A single cell behaves stochastically with time as a consequence of gene expression and biochemical regulation. The intrinsic stochasticity of cellular kinetics has two major origins: stochastic gene-state switching and copy-number fluctuations of proteins. The former is pertinent to the fact that there is only a single copy of DNA inside a typical cell that leads to stochastic production of mRNA and protein, while the latter results from the low copy numbers of certain proteins. According to the stochastic chemical kinetics, a biochemical reaction system in a living cell under a nonequilibrium chemostat with continuous exchange of materials and energies with its surroundings can have multiple modal values that define the phenotypic states of the cell, corresponding to different modalities of the copy-number distribution. It has been suggested that the coexistence of multiple phenotypic states and transitions among them provides one possible mechanism for the survival of unicellular organisms in unpredictable environments.

In the present section, we consider a simple kinetic theory that consists of a self-regulating gene and its protein, multiple gene states associated with different rates of protein synthesis, and positive feedback, as illustrated in Fig. 13.4A. We assume that the protein functions as a dimer that binds and activates its own gene. At any

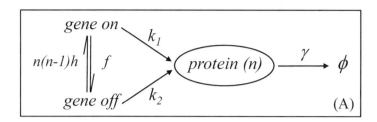

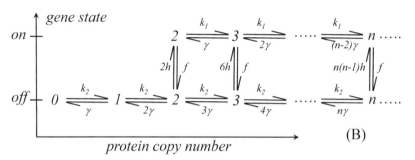

Fig. 13.4 (A) A simple gene network, with two different gene states, "on" and "off", for a transcriptional activator protein. The transcriptional activator forms positive feedback, in which the protein catalyzes the "on-rate" due to cooperative binding of the protein to the DNA: $k_1 \gg k_2$. (B) The master equation graph with the state transitions for the full stochastic biochemical kinetics. The corresponding CME is given in Eqn. (13.17). The model assumes that the activator molecule, as a dimer, binds cooperatively to DNA and turns the gene state "on". This leaves two copies less for the unbound activators.

given moment, the biochemical state of the system is described by both the gene state $\{1 : \text{on}, 2 : \text{off}\}$ and the protein copy number $\{n = 0, 1, 2, \cdots\}$, which includes the two proteins bound with the DNA. In terms of chemical kinetics, as illustrated in Fig. 13.4B, the time evolution of the probability distribution of the biochemical state (i, n) is governed by a CME,

$$\frac{\partial p_1(n,t)}{\partial t} = k_1 p_1(n-1,t) - k_1 p_1(n,t)$$
$$+ \gamma(n-1)p_1(n+1,t) - \gamma(n-2)p_1(n,t)$$
$$+ hn(n-1)p_2(n,t) - f p_1(n,t); \tag{13.17a}$$
$$\frac{\partial p_2(n,t)}{\partial t} = k_2 p_2(n-1,t) - k_2 p_2(n,t)$$
$$+ \gamma(n+1)p_2(n+1,t) - \gamma n p_2(n,t)$$
$$- hn(n-1)p_2(n,t) + f p_1(n,t), \tag{13.17b}$$

in which k_1 and k_2 are the protein synthesis rates corresponding to gene states 1 and 2, respectively, γ is the decay rate of the protein copy number that consists of protein degradation as well as cell division, and the switching rates between the two gene states are f and $hn(n-1)$.

We assume that the protein synthesis rate k_1 corresponding to the activated gene state 1 is much greater than the rate k_2 that is associated with the inactive gene state 2. Consequently, there are three time scales within this simple gene network model:

(i) the decay rate γ of the protein copy number;

(ii) the switching rates f and $hn(n-1)$ between the gene states;

(iii) the larger protein synthesis rate k_1.

Normally, the typical copy number of a protein is quite high in a cell when the gene is fully activated: e.g., $k_1/\gamma \gg 1$. This implies that the time scale (iii) is usually much faster than (i).

There are two limiting scenarios that we mentioned in the previous section:

(a) When (ii) is much slower than both (i) and (iii), a bimodal distribution of the protein copy number can occur without the need for positive feedback; and

(b) When (ii) is much faster than (iii), the gene states are in rapid pre-equilibrium. Often a diffusion approximation of the CME is applied to this case.

We now focus on a third, intermediate scenario in which the gene-state switching rates, f and $n(n-1)h$, are in between γ and the much faster k_1. It turns out that this scenario is most relevant for real laboratory systems, such as the *Lac* operon [101, 102]. The ubiquitous transcriptional and translational bursts observed in living cells, ranging from bacterial to mammalian cells, indicate the relatively slow switching between the ON and OFF states of genes.

For this third, intermediate regime, one can develop a simplification of the full CME system in Fig. 13.4B and obtain a much simpler stochastic model [101, 102]. This will allow us to further propose a barrier-crossing rate formula for the transition rate between the two phenotypic states as well as an emerging nonequilibrium landscape function. The resulting behaviors are very different from those of the other two limiting cases.

13.5.1 Kinetics In the Intermediate Regime

Since the mean number of proteins at the active gene state, $\bar{n} = k_1/\gamma$, is sufficiently large, the dynamics of the protein copy number can be represented by a continuous variable $x(t) = n(t)/\bar{n} \in \mathbb{R}$. If, additionally, the switching between the two gene states is rapid, then $x(t)$ follows a deterministic rate equation:

$$\frac{dx}{dt} = g(x) - \gamma x, \qquad (13.18a)$$

in which the dynamically averaged protein synthesis rate

$$g(x) = \frac{\bar{h}x^2 k_1 + fk_2}{\bar{n}(\bar{h}x^2 + f)} = \frac{\gamma\left(x^2 + K_{eq}k_2/k_1\right)}{x^2 + K_{eq}}, \qquad (13.18b)$$

in which $K_{eq} = f/\bar{h}$ and $\bar{h} = h \cdot \bar{n}^2$. This result is widely called a *mean field approximation*. Note the $x(t)$ is not the expected value of the stochastic $n(t)$; rather, since $\bar{n}$ is large, $x(t)$ behaves macroscopically with negligible variance.

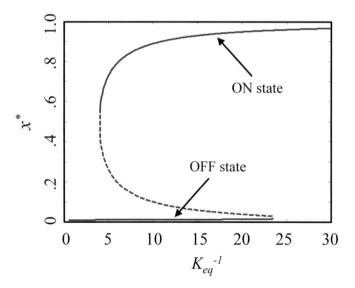

Fig. 13.5 Bistability and bifurcation due to positive feedback in the deterministic nonlinear ODE system in Eqn. (13.18). Parameters used: $k_1 = 10\ min^{-1}$, $k_2 = 0.1\ min^{-1}$ and $\gamma = 0.02\ min^{-1}$. These values are actually taken from reasonable ranges for bacteria. Figure redrawn based on [101].

In the presence of positive feedback, the nonlinear dynamical system in Eqn. (13.18), for a certain range of the parameters, has two stable fixed points, the OFF and ON states, separated by an unstable state, as shown in the bifurcation diagram in Fig. 13.5. With multiple parameters, one can further draw a phase diagram, as shown in Fig. 8.1b, in the space of the parameters with regions indicating different numbers of stable fixed points. Our present system undergoes both saddle-node bifurcation and cusp catastrophe.

We now focus on the case in which gene-state switching is much slower than the activated protein synthesis rate k_1 but much faster than γ, a realistic situation for many operons in bacteria. Because of the large $\bar{n}$, the protein copy-number fluctuations are negligible, but the stochastic switches between the two gene states still have to be addressed. The full CME in Eqn. (13.17) can be reduced to a simpler model in which the protein copy number changes deterministically for each gene state, but stochastic gene state switching leads to stochastic fluctuations in the synthesis rate. This is depicted in Fig. 13.6A. This type of model is also applicable to a single-molecule conformational transition with fluctuating rates.

The precise mathematical condition allowing the validity of this fluctuating rate model being valid is $k_1 \to \infty$. This leads to $\bar{n} = k_1/\gamma \gg 1$, which guarantees the existence of the real-valued $x_i = n_i/\bar{n}$, and all kinetic processes are slow with respect to the dimensionless time scale $(k_1 t)$. Thus, it actually covers both cases of gene-state switching being slower and faster than γ.

One can compute the stochastic trajectories of this fluctuating-rate model using the method of Doob, Boltz–Kalos–Lebowitz, and Gillespie (see Section 3.3.3 and [101]) combined with an ODE solver. The time evolution of the probability $p_i(x)$ of state $\{(i,x)|i \in \mathbb{Z}, x \in \mathbb{R}\}$ is described by

$$\frac{\partial p_1}{\partial t} = -fp_1 + \bar{h}x^2 p_2 - \frac{\partial}{\partial x}\left[\left(\frac{k_1}{\bar{n}} - \gamma x\right)p_1\right];$$

$$\frac{\partial p_2}{\partial t} = fp_1 - \bar{h}x^2 p_2 - \frac{\partial}{\partial x}\left[\left(\frac{k_2}{\bar{n}} - \gamma x\right)p_2\right]. \tag{13.19}$$

This system is known as a piecewise deterministic Markov process (PDMP), random evolution, and Markov switching ordinary differential equations, among many other names.

At the end of this section, we mention another case that we have already encountered in the previous section, in which the gene-state switching is much more rapid than both synthesis and degradation (see Fig. 13.6C). Following a similar method, we can obtain a different simplification of the full CME in Eqn. (13.17) in terms of a single birth-and-death process $\mathbf{N}(t)$, with its CME

$$\frac{dp(n,t)}{dt} = k(n-1)p(n-1,t) - k(n)p(n,t)$$
$$+ \tilde{\gamma}(n+1)p(n+1,t) - \tilde{\gamma}(n)p(n,t), \qquad (13.20a)$$

in which

$$k(n) = \frac{k_1 hn(n-1) + k_2 f}{hn(n-1) + f}, \quad \tilde{\gamma}(n) = \frac{hn(n-1)(n-2) + fn}{hn(n-1) + f}\gamma, \qquad (13.20b)$$

are the rapidly equilibrated protein synthesis (birth) and degradation (death) rates.

13.5.2 Global Landscapes and Phenotypic Switching

For each of the two simplified models in Eqns. (13.19) and (13.20), one can obtain a nonequilibrium landscape function of x using the method presented in Chapters 4 and 7, assuming the WKB ansatz as they approach the mean-field dynamics in (13.18).More precisely, for the fluctuating-rate model in Eqns. (13.19), we let the switching rates between the gene states go to infinity but maintain their ratio and choose f as the scaling parameter; hence, the WKB ansatz is $p_n^{ss} \sim e^{-f\tilde{\Phi}_0(x)}$ as $f \to \infty$ for the single birth and death process in Eqns. (13.20). We first replace n with $x = n/\bar{n}$, and then let $\bar{n}$ go to infinity, or equivalently let k_1 go to infinity; hence, the WKB ansatz here is $p_n^{ss} \sim e^{-k_1\tilde{\Phi}_1(x)}$. In statistical physics, such a landscape has been called a nonequilibrium potential [107, 153, 189, 268].

The landscape function $\tilde{\Phi}_0(x)$ associated with the fluctuating-rate model (Fig. 13.6A) satisfies

$$\frac{d\tilde{\Phi}_0(x)}{dx} = \frac{1}{\gamma(1-x)} + \frac{K_{eq}^{-1}x^2}{\gamma(\sigma-x)}, \qquad (13.21)$$

in which $\sigma = k_2/k_1$ and $K_{eq} = \frac{f}{h}$. The landscape function $\tilde{\Phi}_1(x)$ for the reduced CME model (Fig. 13.6C) satisfies

$$\frac{d\tilde{\Phi}_1(x)}{dx} = -\frac{1}{\gamma}\ln\left(\frac{x^2 + \sigma K_{eq}}{x(x^2 + K_{eq})}\right). \qquad (13.22)$$

See [101] for a detailed derivation. The term "nonequilibrium" here signifies that the landscape functions are for a nonlinear stochastic dynamical system, as will be discussed in Section 15.1.

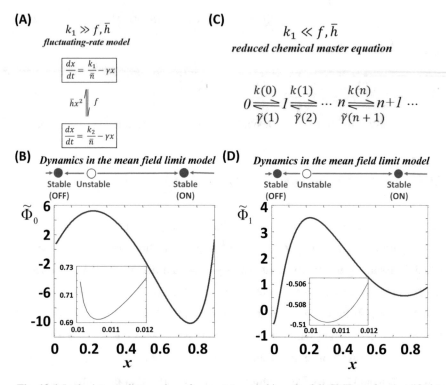

Fig. 13.6 In the intermediate region of gene-state switching, the full CME can be simplified to a fluctuating-rate model (A), the steady-state distribution of which corresponds to a normalized landscape function $\tilde{\Phi}_0(x)$ (B). If gene-state switching is extremely rapid, a reduced CME (C) and a different landscape function $\tilde{\Phi}_1(x)$ (D) can also be derived. The insets in (B) and (D) are magnified views of the functions near $x = 0$. The parameters in (B) and (D) are the same as those in Fig. 13.5 with $K_{eq} = 1/5.5$. Figure adapted from [101] with permission.

The deterministic dynamics in (13.18) depend on three independent parameters, one with dimensions of $[\text{time}]^{-1}$, γ, and two dimensionless parameters that represent relative time scales, K_{eq} and σ, as do the landscape functions.

As we have already discussed several times throughout the previous chapters, one of the most important features of the nonequilibrium, global landscape function is the fact that the corresponding deterministic, macroscopic dynamics in (13.18) are always downhill. This implies that local minima, maxima, and saddle points of the landscape function correspond to stable steady states, unstable steady states, and the transition (saddle) state of the deterministic dynamics, respectively. This is illustrated in Fig. 13.6B and Fig. 13.6D. Note that both $\tilde{\Phi}_0(x)$ and $\tilde{\Phi}_1(x)$ are Lyapunov functions of the same ODE (13.18); this implies that the parameter region for double-well shaped landscape functions are the same for the two models. The variance of the local fluctuations near each stable steady state x^*, which represents a phenotypic state of a cell, can be approximated by the curvature $\left[d^2\Phi_i(x)/dx^2\right]_{x=x^*}$. One can clearly see from Fig. 13.6B and Fig. 13.6D that the local fluctuations in the intermediate case can be very different from those in the case of rapid gene-state switching, even though their corresponding macroscopic limits, in Eqn. (13.18), are the same.

13.5.3 Transition Rate Formula For Phenotypic Switching

The transition rates between two phenotypic states of a cell, as a stochastic reaction kinetic system, can usually be defined as the reciprocal of the mean first-passage time from one steady state to another. They are a well defined concept because the mean first-passage time is nearly independent of the initial state within the same basin of attraction as long as there is a time-scale separation between the intrabasin fluctuations and the interbasin transitions.

In the case where the gene-state switching is very slow, it becomes the rate-liming step for the transition between two phenotypic states. However, the rate formulae for the phenotypic switching have much more complicated expressions for the intermediate regime and for the case of rapid gene-state switching. In the latter two cases, the transition rates from one phenotypic state A to another B can be expressed as

$$k_{AB} \simeq k_{AB}^0 \exp\left(-f\Delta\Phi_{AB}^{\ddagger}\right),\tag{13.23}$$

and

$$k_{AB} \simeq k_{AB}^0 \exp\left(-k_1\Delta\Phi_{AB}^{\ddagger}\right)\tag{13.24}$$

respectively, where $\Delta\Phi_{AB}^{\ddagger}$ is called the "activation barrier height" and k_{AB}^0 is a prefactor with dimensions of $[\text{time}]^{-1}$. The barrier of the landscape function is $\Delta\Phi_{AB}^{\ddagger} = \Phi^{\ddagger} - \Phi_A$, where $\Phi^{\ddagger}$ is the Φ value of the unstable fixed point, , e.g., the local maximum of Φ along a transition path from one stable fixed point A to another stable fixed point B, and Φ_A is the Φ-value of the local minimum A. This is illustrated in Fig. 13.7a. This result is analogous to Kramers' formula; the exponential dependence of the rate on the barrier height guarantees the stability and robustness of a phenotypic state against various perturbations. It provides the "identity" of an individual, as will be discussed in Section 15.2.

Since the expressions of the two landscape functions $\tilde{\Phi}_0$ and $\tilde{\Phi}_1$ are given explicitly in (13.21) and (13.22), respectively, the corresponding rates can be straightforwardly computed.

Setting $d\tilde{\Phi}_0(x)/dx = 0$,

$$\frac{K_{eq}}{1-x} + \frac{x^2}{\sigma - x} = 0,$$

and $d\tilde{\Phi}_1(x)/dx = 0$

$$\frac{x^2 + \sigma K_{eq}}{(x^2 + K_{eq})x} = 1,$$

we obtain the same polynomial Eqn. (13.25) for the extrema of the two landscapes:

$$x^3 - x^2 + K_{eq}x - \sigma K_{eq} = 0. \tag{13.25}$$

Under the condition in which it has three positive roots x_1^*, $x^\ddagger$, and x_2^*, we obtain

$$\Delta\tilde{\Phi}_{0,12}^\ddagger = \int_{x_1^*}^{x^\ddagger} \left(\frac{1}{\gamma(1-z)} + \frac{K_{eq}^{-1}z^2}{\gamma(\sigma - z)} \right) dz$$

$$= \left[-\frac{K_{eq}^{-1}}{2\gamma}z^2 - \frac{K_{eq}^{-1}}{\gamma}\sigma z + \frac{1}{\gamma}\ln(1+z) - \frac{K_{eq}^{-1}\sigma^2}{\gamma}\ln(\sigma - z) \right]\Bigg|_{x_1^*}^{x^\ddagger}$$

and

$$\Delta\tilde{\Phi}_{1,12}^\ddagger = \int_{x_1^*}^{x^\ddagger} \frac{1}{\gamma}\ln\left(\frac{z(z^2 + K_{eq})}{z^2 + \sigma K_{eq}} \right) dz$$

$$= \left[\frac{1}{\gamma}\left(z\ln z - z + 2\sqrt{K_{eq}}\arctan\frac{z}{\sqrt{K_{eq}}} - 2\sqrt{\sigma K_{eq}}\arctan\frac{z}{\sqrt{\sigma K_{eq}}} \right) \right]\Bigg|_{x_1^*}^{x^\ddagger}.$$

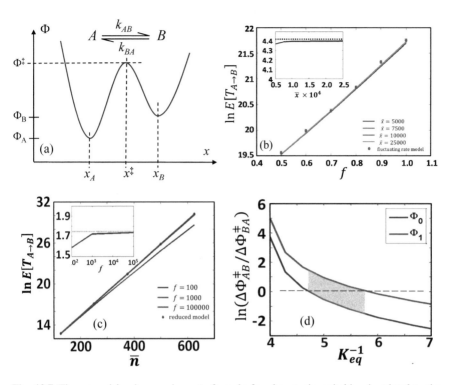

Fig. 13.7 The general barrier-crossing rate formula for phenotypic switching is related to the nonequilibrium landscape $\Phi(x)$ as illustrated in (a). The mean transition times from the A (OFF) state to the B (ON) state are obtained as either, from the full CME mode and two simplified models (with different colors), the rate of gene-state switching in the intermediate regime (b) and being very fast (c). All results are obtained by computing the mean first-passage time, except for the fluctuating-rate model, which is obtained through stochastic simulations of the trajectories. (d) The relative stability of the two phenotypic states with respect to Φ_0 (blue) and Φ_1 (green) as a function of K_{eq}^{-1}. The different colored lines in (b) and (c) are nearly indistinguishable. The insets in (b) and (c) compare the numerically determined normalized barrier heights from the full CME (solid blue lines) and the two simplified models (dashed black lines). Parameters used: $\sigma^{-1} = 3000$ in (b), 50 in (c), and 100 in (d); $K_{eq}^{-1} = 6$ in both (b) and (c); and $\gamma = 0.02 \ min^{-1}$ in all. Figure adapted from [101] with permission.

fluctuating-rate model	landscape function & associated rate formula	reduced CME model	landscape function & associated rate formula
$k_1 \gg f, \bar{h}, \gamma$	$k_1 \gg f, \bar{h} \gg \gamma$	$f, \bar{h} \gg k_1, \gamma$	$f, \bar{h} \gg k_1 \gg \gamma$

Table 13.1 Summary of the conditions for the validity of the simplified models and associated landscape functions with barrier-crossing rate formulae. Table reproduced from [101] with permission.

Fig. 13.7 summarizes the results from numerical computations that validate the rate formula in Eqn. (13.23) and (13.24) by either Monte Carlo simulations of the stochastic trajectories or by numerically solving the equations of the mean first-passage time and obtaining $\mathbb{E}\left[\mathbf{T}_{\text{off}\to\text{on}}\right]$ from the OFF state to the ON state. We fix the parameters K_{eq}, k_1/k_2 and γ to keep the ODE unchanged, thus preserving the position of the steady states, and vary f and $\bar{n}$. The results show that $\mathbb{E}\left[\mathbf{T}_{\text{off}\to\text{on}}\right]$ in the full CME model is well approximated by the two simplified models in their separate regions of gene-state switching (Fig. 13.7b, c), and the barriers $\varDelta\tilde{\varPhi}^{\ddagger}_{\text{off}\to\text{on}}$ with respect to f or $\bar{n}$ can be determined from such Arrhenius-like plots (Fig. 13.7b, c, inset), which matches those predicted from the normalized landscape functions $\tilde{\varPhi}_0(x)$ or $\tilde{\varPhi}_1(x)$. Fig. 13.7b and c can be experimentally observed once we can simultaneously tune at least two of the parameters and keep the mean-field model unchanged [1]. In such experiments, we will see that in the intermediate region, $\mathbb{E}\left[\mathbf{T}_{\text{off}\to\text{on}}\right]$ can be insensitive to the total number of protein molecules (Fig. 13.7b) but varies dramatically in terms of the gene-state switching rates, while in the extremely rapid region, the situation is reversed (Fig. 13.7c). The conditions under which the simplified models and associated landscape functions, as well as saddle-crossing rate formulas, are valid are summarized in Table 13.1.

The complex parameter-dependence relation implies that the relative stability between the two phenotypic states, defined as the ratio of the two "activation energies" $\varDelta\varPhi^{\ddagger}_{AB}$ and $\varDelta\varPhi^{\ddagger}_{BA}$, might be reversed, as the gene-state switching rates are in the intermediate region or the extremely rapid region, even keeping the same equilibrium constants among all the gene states (see the shaded region in Fig. 13.7d and compare Fig. 13.6B and Fig. 13.6D).

Chapter 14
Stochastic Macromolecular Mechanics and Mechanochemistry

14.1 Mechanics and Chemistry: Two Complementary Representations

Newtonian mechanics and biochemistry ultimately meet in the motions of atoms in a macromolecule, such as a protein. As a collection of point masses in a viscous environment, such as an aqueous solvent, macromolecules have been viewed through two equivalent but different lenses: in terms of mechanics, one is interested in the *force* involved in molecular events and the positions of atoms, but in terms of biochemistry, one is interested in the *free energy* of a molecule and its conformational states. Of course, the force is simply the directional derivative of a free energy function, and a conformational change is just an essentialized, emergent description of the collective change in the positions of the atoms with cooperativity.

A stochastic mathematical description of a macromolecule, its states, free energy, dynamics, and functions, is an integrative lens that provides a unifying perspective on macromolecular systems at a mesoscopic level [212]. Nowhere is this powerful new view more clearly demonstrated than in the studies of biological macromolecules that function through generating forces.

Although the present chapter exclusively focuses on mechanochemistry at the level of single macromolecules, the material presented is highly relevant to the growing, exciting research in the new field of *mechanobiology* at the cellular and tissue levels, pioneered by, among many others, E. L. Elson [60] and M. P. Sheetz [131], two life-long mentors of the first author.

© The Author(s), under exclusive license to Springer Nature Switzerland AG 2021
H. Qian, H. Ge, *Stochastic Chemical Reaction Systems in Biology*,
Lecture Notes on Mathematical Modelling in the Life Sciences,
https://doi.org/10.1007/978-3-030-86252-7_14

14.2 Potential of Mean Force

This concept was first developed by J. G. Kirkwood in 1935 [148]. The fundamental postulate of the statistical mechanics of a collection of atoms is as follows. Let x be the center of mass of a macromolecule that consists of many atoms as point masses and $\bar{\mathbf{y}}$ be all the internal degrees of freedom w.r.t. x. We further assume that all the degrees of freedom are immersed in a continuous medium at temperature T. Then, the total free energy of the mechanical system is

$$-k_B T \ln \int \int e^{-U(x,\bar{\mathbf{y}})/k_B T}\,d\mathbf{y}dx = -k_B T \ln \int e^{-A(x;T)/k_B T}\,dx, \qquad (14.1)$$

in which $U(x,\mathbf{y})$ is the complete, temperature-independent mechanical potential energy function for all the point masses. The right-hand side of Eqn. (14.1) defines a function of x, $A(x;T)$, based on $U(x,\mathbf{y})$:

$$A(x;T) = -k_B T \ln \int e^{-U(x,\bar{\mathbf{y}})/k_B T}\,d\mathbf{y}, \qquad (14.2)$$

and it has the clear meaning of a *potential of mean force* [148]. This is because

$$-\frac{\partial U(x,\mathbf{y})}{\partial x}$$

is the force acting on x when all the point masses are located at $\mathbf{y}$, where

$$\frac{\int \left(-\dfrac{\partial U(x,\mathbf{y})}{\partial x}\right) e^{-U(x,\bar{\mathbf{y}})/k_B T}\,d\mathbf{y}}{\int \int e^{-U(x,\bar{\mathbf{y}})/k_B T}\,d\mathbf{y}}$$

is the mean force averaged over all the degrees of freedom of $\mathbf{y}$. Since

$$-\frac{\partial A(x;T)}{\partial x} = \frac{\int \left(-\dfrac{\partial U(x,\mathbf{y})}{\partial x}\right) e^{-U(x,\bar{\mathbf{y}})/k_B T}\,d\mathbf{y}}{\int \int e^{-U(x,\bar{\mathbf{y}})/k_B T}\,d\mathbf{y}},$$

$A(x;T)$ is the potential of the mean force acting on x.

   If an experimenter applies an external force to x that is independent of $\mathbf{y}$, then that force simply adds to the potential of the mean force,

$$-\frac{\partial A(x;T)}{\partial x} + F_{ext}(x). \qquad (14.3)$$

If the external force is a Hookean spring, then one has a total "mechanical potential" of $A(x;T) + \frac{1}{2}k(x - x_0)^2$.

In a more abstract setting, x can represent a "reaction coordinate" along which two basins of attraction define two conformational states of a macromolecule. Then, one can apply an external force $F_{ext}(x)$ to modify $A(x;T)$, thus mechanically manipulating the macromolecular conformational change at the single-molecule level [212].

14.3 Chemical Model of Molecular Motors

Muscles in a human body generate force by using the free energy from the hydrolysis of cellular adenosine triphosphate (ATP) to adenosine diphosphate (ADP) and inorganic phosphate (Pi):

$$ATP \; \rightleftharpoons \; ADP + Pi. \tag{14.4}$$

The normal cellular concentrations of ATP, ADP, and Pi are such that the

$$\Delta G = \Delta G^o + RT \ln \left(\frac{[ATP]}{[ADP][Pi]} \right)_{cellular}$$

of the reaction is approximately 12 kcal per mol. In the 1980s, experimental studies of single myosin molecular motion along a one-dimensional actin filament, with nanometer movements and piconewton forces, provided fertile ground for stochastic single motor protein chemomechanics, or mechanochemistry.

14.3.1 Chemical Model of Single Molecular Motors

Although a myosin molecule moves as a molecular motor, it is actually an ATPase, the enzyme that catalyzes the hydrolysis reaction in (14.4). The ingredients for establishing a kinetic model for a myosin motor are (i) a linear track with discrete, periodic sites, e.g., the actin filament, and (ii) the motor protein, which simply goes through an internal kinetic cycle while its center of mass steps along the track:

$$M_n \cdot ADP + ATP \underset{k_{-1}^o}{\overset{k_1^o}{\rightleftharpoons}} M_n \cdot ATP + ADP, \tag{14.5a}$$

$$M_n \cdot ATP \underset{k_{-2}^o}{\overset{k_2}{\rightleftharpoons}} M_n^* \cdot ADP + Pi, \tag{14.5b}$$

$$M_n^* \cdot ADP \underset{k_{-3}^o}{\overset{k_3^o}{\rightleftharpoons}} M_{n+1} \cdot ADP. \tag{14.5c}$$

Here M_n denotes the motor located at the nth position along the linear track. Hence, the third step involves a translocation of the motor. If there is an external resistant force F that is independent of n, then in general, $k_3(F)$ and $k_{-3}(F)$ are functions of the external force:

$$\frac{k_{-3}(F)}{k_3(F)} = \frac{k_{-3}^o}{k_3^o} e^{Fd/k_B T} \tag{14.6}$$

where d is the size of the step. Eqn. (14.6) is the mechanochemical analog of the Nernst–Planck equation for electrochemical system.

14.3.2 Chemomechanical Energy Transduction

We set constant concentrations for ATP, ADP and Pi, and denote $k_1 = k_1^o[\text{ATP}]$, $k_{-1} = k_{-1}^o[\text{ADP}]$ and $k_{-2} = k_{-2}^o[\text{Pi}]$. Then, the amount of energy from the hydrolysis of a single ATP molecule is

$$\Delta\mu = k_B T \ln \frac{k_1 k_2 k_3^o}{k_{-1} k_{-2} k_{-3}^o}. \tag{14.7}$$

We note that the system of rate equations for the reaction scheme in (14.5) is a set of infinite equations for $[M_n \cdot ADP]$, $[M_n \cdot ATP]$, and $[M_n^* \cdot ADP]$, with all integer n. To obtain the steady state of the kinetic system, we use a periodic boundary by setting

$$[M_n \cdot ADP] = [M_{n+1} \cdot ADP]. \tag{14.8}$$

This is equivalent to computing all the motor proteins in state $M \cdot ADP$ irrespective of their position on the track. We thus can obtain the steady-state probabilities and steady-state velocity of the motor:

$$V = dJ_{ss} \tag{14.9}$$

where J_{ss} is the number of cycles per unit time, which is given in Eqn. (2.28):

$$J_{ss} = \frac{k_1 k_2 k_3 - k_{-1} k_{-2} k_{-3}}{\left\{ \begin{array}{l} k_2 k_3 + k_3 k_{-1} + k_{-1} k_{-2} + k_3 k_1 + k_1 k_{-2} + k_{-2} k_{-3} \\ + k_1 k_2 + k_2 k_{-3} + k_{-3} k_{-1} \end{array} \right\}}. \tag{14.10}$$

In the presence of resistant force F, k_3 and k_{-3} are functions of F; hence, we obtain $V(F) = d J_{ss}(F)$, which is a function of the force.

One value of the force is particularly interesting and important: the value F^* at which $V(F^*) = 0$. This is known as the stalling force. To obtain it, we note that $V = 0$ when and only when

$$\frac{k_1 k_2 k_3(F^*)}{k_{-1} k_{-2} k_{-3}(F^*)} = \frac{k_1 k_2 k_3^o}{k_{-1} k_{-2} k_{-3}^o} e^{-F^* d / k_B T} = 1. \tag{14.11}$$

That is,

$$F^* = \frac{k_B T}{d} \ln \frac{k_1 k_2 k_3^o}{k_{-1} k_{-2} k_{-3}^o} = \frac{\Delta \mu}{d}. \tag{14.12}$$

This is precisely the conservation of mechanochemical energy: at F^*, the work done by the external mechanical force is exactly balanced by the chemical energy from ATP hydrolysis. Because of fluctuations, chemical energy is required to even just maintaining a static force at the macromolecular level.

14.3.3 Efficiency of a Molecular Motor

Let us now see the efficiency of the mechanochemical motor. Each step, the work done by the motor against the resistant force is Fd, where $F \leq F^*$. Note that when $F > F^*$, the motor is in fact doing negative work. The amount of chemical energy input for each step is $\Delta \mu$. Hence, the efficiency of the motor is

$$\eta = \frac{Fd}{\Delta \mu} = \frac{F}{F^*}. \tag{14.13}$$

This means that when the $F = F^*$, the efficiency is maximal. However, under this condition, the motor is in fact not moving at all! Thus, although the efficiency is the greatest, the power output of the motor, i.e., the amount of work done per unit time, is zero.

The simple "one-cycle" motor protein model in (14.5) has a *tight coupling*, e.g., each kinetic cycle and mechanical stepping are precisely linked. In general, a protein as an ATPase might, every so often, catalyze a hydrolysis reaction without stepping.

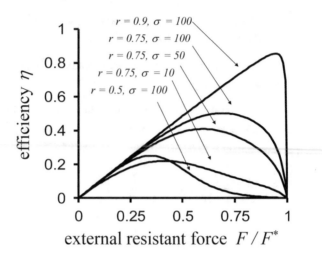

Fig. 14.1 The chemomechanical efficiency η of a motor protein with slippage. $e^{\Delta\mu/k_BT} = 10^{10}$, $F^*d = \Delta\mu/d = 21$. For both small and large F, η diminishes, corresponding respectively to too light a load and too slow a motion. Figure adapted from [215] with permission.

If this occurs, then the efficiency is determined by the ratio of the steady-state fluxes for the stepping cycle to those of the hydrolysis cycle, which will be strictly less than 1.

For example, if

$$k_3(F) = k_3^o \left(1 + \sigma e^{-(1-r)Fd/k_BT}\right) \text{ and } k_{-3}(F) = \sigma k_{-3}^o e^{rFd/k_BT}, \qquad (14.14)$$

where $\frac{1}{1+\sigma}$ represents the "slippage" of the motor stepping, $\sigma = \infty$ corresponds to the tight coupling, and $\sigma = 0$ represents a "futile cycle", we then have [215]

$$\eta = \frac{Fd}{\left(1 + \dfrac{\left(e^{\Delta\mu/k_BT} - 1\right)e^{-rFd/k_BT}}{\sigma\left(e^{(\Delta\mu - Fd)/k_BT} - 1\right)}\right)\Delta\mu}. \qquad (14.15)$$

Fig. 14.1 shows that with decreasing slippage, the maximal efficiency approaches 100% while the force F approaches the stalling force F^*.

14.4 Polymerization Motors

The motor protein in the previous section can be thought of as a train moving along its track. Another force generating device inside a living cell is like the plumber's "snake", which generate force by elongation.

We are all very familiar with polyesters, which are large molecules made of a string of ester groups. Similarly, inside living cells, proteins themselves are globular-shaped molecules, which can be strung together to form a long filament. Actin in muscle is an example of such a polymer made of globular protein monomers. Another example is microtubules, which are made of tubulins as monomers.

If we use P to denote a polymer of any length and M to denote the monomer, then the simplest chemical reaction scheme we can have is

$$P + M \underset{k_-}{\overset{k_+}{\rightleftarrows}} P. \tag{14.16}$$

In the equilibrium steady state, we have

$$k_+ [P]^{eq} [M]^{eq} = k_- [P]^{eq}. \tag{14.17}$$

Hence, we have $[M]^{eq} = \frac{k_-}{k_+}$. This is known as the *critical concentration*, c_c. It indicates that when $[M] < c_c$, there is no polymer; and when $[M] > c_c$, all the excess monomers will be in the polymer.

There are hidden assumptions in the reaction scheme shown in Eqn. (14.16). It neglects the possibility that new, additional polymer is formed. This, of course, can only be true if the initial "nucleation" of the polymer is much more difficult than the elongation.

14.4.1 Polymerization With Nucleation

We now consider a nucleation-elongation model:

$$M_1 + M_1 \underset{k_-^*}{\overset{k_+^*}{\rightleftarrows}} M_2, \quad M_1 + M_k \underset{k_-}{\overset{k_+}{\rightleftarrows}} M_{k+1}, \quad (k = 2, 3, \cdots) \tag{14.18}$$

in which $\frac{k_+^*}{k_-^*} \ll \frac{k_+}{k_-}$. All the M_k, $k \geq 2$, are considered polymers.

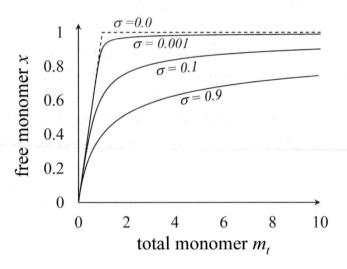

Fig. 14.2 Polymerization with a nucleation step. Decreasing σ means increasing cooperativity. The notion of critical concentration c_c arises in the limit of $\sigma \to 0$.

Consider a closed reaction system in which the total amount of monomer units is M_t:

$$[M_1] + 2[M_2] + \cdots + k[M_k] + \cdots = M_t. \tag{14.19}$$

If we assume all the reactions in (14.18) are in equilibrium, then we have

$$\frac{[M_2]}{[M_1^2]} = \frac{k_+^*}{k_-^*}, \quad \frac{[M_{k+1}]}{[M_k][M_1]} = \frac{k_+}{k_-}, \quad (k = 2, 3, \cdots)$$

and

$$
\begin{aligned}
M_t &= [M_1] + 2\frac{k_+^*}{k_-^*}[M_1]^2 + 3\frac{k_+^* k_+}{k_-^* k_-}[M_1]^3 + 4\frac{k_+^*}{k_-^*}\left(\frac{k_+}{k_-}\right)^2[M_1]^4 + \cdots \\
&= \frac{k_-}{k_+}\left[x + \sigma\left(2x^2 + 3x^3 + 4x^4 + \cdots\right)\right]
\end{aligned}
$$

where $x = \frac{k_+}{k_-}[M_1]$, $\sigma = \frac{k_+^* k_-}{k_-^* k_+}$, and $m_t = \frac{k_+}{k_-}M_t$. x must be less than 1 because otherwise, M_t cannot be conserved. Hence,

$$m_t = \frac{x - 2(1-\sigma)x^2 + (1-\sigma)x^3}{(1-x)^2}. \quad (x < 1) \tag{14.20}$$

Again, $x < 1$ implies that the equilibrium concentration of $[M_1] < \frac{k_-}{k_+} = c_c$, the critical concentration.

Fig. 14.2 shows that the free monomer concentration $[M_1]/c_c$ is a function of total monomer concentration M_t/c_c with a different value of σ. If $\sigma = 1$, then we have $m_t = x/(1-x)^2$. That is,

$$[M_1] = c_c \left\{ 1 + \frac{c_c}{2M_t} - \sqrt{\frac{c_c}{M_t} + \frac{c_c^2}{4M_t^2}} \right\}. \tag{14.21}$$

If $\sigma \ll 1$, then we have

$$[M_1] = M_t \text{ when } M_t \leq c_c, \quad \text{and} \quad [M_1] = c_c \text{ when } M_t \geq c_c, \tag{14.22}$$

as shown by the dashed line in Fig. 14.2. This validates the simplest model in the previous section.

A polymerization process with a very small value of σ, the nucleation parameter, is said to be highly *cooperative*: forming a nucleus is difficult, but as soon as it is formed, all monomers are quickly added to the polymer.

The case of $\sigma = 0$, shown by the dashed curve in Fig. 14.2, corresponds to the kinetic system $M_1 + P \rightleftharpoons P$, with rate equation

$$\frac{d[M_1]}{dt} = \begin{cases} -k_+[M_1][P] + k_-[P] & [P] > 0 \\ 0 & [P] = 0 \end{cases} \tag{14.23}$$

in which M_1 and P represent the monomers and the polymer, respectively. In this case, the creation of a new polymer is impossible. The monomers are continuously added into polymers when $k_+[M_1] > k_-$, but when the total monomer $M_t < k_-/k_+$, there will be no polymer at all, $[P] = 0$ and $[M_1] = M_t$.

14.4.2 Polymerization Against a Force

To model polymerization against a resistant force F, we again introduce a mechano-chemical coupling as in Eqn. (14.6). This means that the polymerization reaction rate constant, k_+, and the depolymerization rate constant, k_-, are both functions of the force and satisfy

$$\frac{k_-(F)}{k_+(F)} = \frac{k_-^o}{k_+^o} e^{F\delta/k_B T}, \tag{14.24}$$

where δ is the size of the monomer in the direction of polymer growth. Again, $F\delta$ is the mechanical work done against the external force required for a polymer to grow one more subunit.

For a given concentration of the monomer, $[M]$, the stalling force F^* at which the polymer neither grows nor shrinks is

$$k_+(F^*)[M] = k_-(F^*). \tag{14.25}$$

That is,

$$F^* = \frac{k_B T}{\delta} \ln \left(\frac{k_+^o}{k_-^o}[M] \right). \tag{14.26}$$

Of course, when $[M] = c_c = \frac{k_{-1}^o}{k_1^o}$, the critical concentration, then $F^* = 0$.

We now study the kinetics of polymerization against the force. To do that, we need to know how $k_+(F)$ and $k_-(F)$ individually depend upon F while their ratio satisfies the thermodynamic relation in Eqn. (14.24). Without more knowledge on the nature of k_- and k_+ based on macromolecular structures and energy landscapes, there is truly not much one can say. Nevertheless, it is not unreasonable to assume, for small F, that

$$k_+(F) = k_+^o e^{-\frac{(\ell-\delta)F+\theta F^2}{k_B T}}, \quad k_-(F) = k_-^o e^{-\frac{\ell F+\theta F^2}{k_B T}}, \tag{14.27}$$

in which ℓ and θ are parameters. Normally, $0 \le \ell \le \delta$ and $\theta \ge 0$. However, this is not necessary.

The polymerization velocity is then

$$V(F) = \delta \left[k_+(F)[M] - k_-(F) \right]$$
$$= k_-^o \delta e^{-\frac{\ell F+\theta F^2}{k_B T}} \left[\frac{[M]}{c_c} e^{-\frac{F\delta}{k_B T}} - 1 \right]. \tag{14.28}$$

To see the shape of the V versus F curve given above, we normalize it:

$$\frac{V(F)}{V(0)} = e^{-\frac{\ell F+\theta F^2}{k_B T}} \left[\frac{e^{-\frac{F\delta}{k_B T}} - \frac{c_c}{[M]}}{1 - \frac{c_c}{[M]}} \right]. \tag{14.29}$$

We note that if

$$\frac{\ell}{\delta} < \frac{[M]}{c_c - [M]}, \tag{14.30}$$

then $V(F)$ can be increasing near $F = 0$. This implies some kind of "instability" since with increasing resistant force, the velocity in fact increases!

14.5 Energy, Force and Stochastic Macromolecular Mechanics

Thus far, all of our discussions have been based on discrete conformational states and the corresponding chemical kinetics by counting. We then used the relation in Eqn. (14.6) to couple the chemical reaction rates with an external mechanical force. As we noted in Sec. 14.2, at a more detailed molecular level, the chemical reaction are just mechanics: a molecule is nothing but a collection of atoms held together by forces, and macromolecular conformational transformations are just a rearrangement of the atoms within a molecule according to the forces among them.

Covalent chemical bonds break and form based on quantum mechanics and quantum chemistry. However, many biochemical reactions involve not chemical bonds but weak molecular forces, such as van der Waals forces and electrostatic forces. The noncovalent bonds can be quantitatively modeled in terms of classical mechanics.

14.5.1 Chemical Transformation As Motion In a Potential Force

In terms of classical mechanics and treating atoms in a molecule as Newtonian point masses, a molecular structure is described by the coordinates of all its atoms, or chemical motifs. There are different energies associated with different configurations of the atoms. We use the term "configuration" for the arrangement of atoms and reserve the term "conformations" for the discrete states emerging from the configurational space: the conformations are the wells in the energy surface.

In the absence of any temperature-causing thermal aggregations, a molecule stays at one of its energy minima. However, in the presence of a thermal environment, the molecule is able to move from one energy well to another by passing although a series of saddle points. Most of the time, a molecule will just stay in a well, fluctuating. This is the physical picture, provided by H. A. Kramers, for the reaction $A \xrightarrow{k} B$. The waiting time is exponentially distributed with mean $1/k$. $-\ln k$ is proportional to the height of the "energy barrier" the system has to cross. See Section 4.6 for more discussion.

14.5.2 Single Receptor-ligand Dissociation Under Force

A protein and a ligand can form a complex; this is usually represented by an association reaction. We now consider the problem of applying a force to a protein-ligand complex and investigate how the external force F_{ext} affects the dissociation process. In a laboratory, this type of experiment can be accomplished by using a device called atomic force microscope (AFM). The phenomenon of forced biomolecular 'bond' rupture was first observed by Florin *et al.* in 1994 [75].

For simplicity, let us consider the centers of mass of the protein and the ligand and assume their interaction force has the standard Lennard–Jones "6-12" potential

$$U_{int}(x) = -\frac{a}{x^6} + \frac{b}{x^{12}} = -U_0 \left[2 \left(\frac{x_0}{x} \right)^6 - \left(\frac{x_0}{x} \right)^{12} \right], \qquad (14.31)$$

where x is the separation distance between the two centers of masses. This potential function has a minimum $-U_0 = -\frac{a^2}{4b}$ at $x_0 = \sqrt[6]{2b/a}$. When $x \to +\infty$, $E(x) \to 0$, and when $x \to 0$, $E(x) \to +\infty$.

The AFM uses an elastic cantilever, a beam supported on only one end. Combining the force from the cantilever and the force between the protein and the ligand, we have the total force on the ligand:

$$F_{tot} = \eta(d - x) - \frac{dU_{int}(x)}{dx}. \qquad (14.32)$$

where η is the stiffness of the cantilever, and d is the base position of the piezoelectrical motor on which the cantilever is mounted. The corresponding total potential for the ligand is then

$$U_{tot}(x) = \frac{\eta}{2}(x - d)^2 + U_{int}(x). \qquad (14.33)$$

The equilibrium point(s) are the minima and maxima of $U_{tot}(x)$, at which the two forces balance:

$$\frac{U_0}{x_0} \left[12 \left(\frac{x_0}{x} \right)^7 - 12 \left(\frac{x_0}{x} \right)^{13} \right] = \eta(d - x). \qquad (14.34)$$

In a laboratory, d as a function of time is controlled by an experimenter. Increasing d from $d = 0$ constitutes a "pulling" of the ligand away from the protein. If one slowly increases d, always letting the system equilibrate at each d value and attain an "equilibrium", then x is the solution to the Eqn. (14.34) as a function of d. There are many parameters in the equation. We shall introduce the nondimensionalized form,

$$z = x/x_0, \quad \delta = d/x_0, \quad \text{and} \quad \alpha = kx_0^2/(12U_0).$$

Then, Eqn. (14.34) becomes

$$z^{-7} - z^{-13} = \alpha(\delta - z). \tag{14.35}$$

There is a cusp catastrophe for the roots of this equation as functions of α and δ. Even though the nonlinear equation cannot be solved in a closed form for z, one can obtain a parametric equation for the root(s) in terms of ξ:

$$z = \left(\frac{1 \pm \sqrt{1 - 4\xi}}{2}\right)^{-\frac{1}{6}}, \quad \delta = z + \frac{\xi}{\alpha z}, \quad \xi \in \left[-\infty, \frac{1}{4}\right]. \tag{14.36}$$

Fig. 14.3A shows several z as functions of δ with different values of α. We see with increasing α, i.e., the spring stiffness, that the "sluggish" behavior disappears.

The mechanical thinking in terms of force can be equally understood in terms of the total energy function $U_{tot}(x)$:

$$\frac{U_{tot}(z)}{U_0} = -\left[2\left(\frac{1}{z}\right)^6 - \left(\frac{1}{z}\right)^{12}\right] + 6\alpha(z - \delta)^2. \tag{14.37}$$

Fig. 14.3B shows the total potential energy function $U_{tot}(z)$ for three different values of δ. It can be clearly seen that with pulling, e.g., increasing δ, the the bound state of the ligand is destabilized while the free state of the ligand is stabilized.

In a real single-molecule AFM forced dissociation experiment, increasing $\delta(t)$ constitutes "pulling", while decreasing $\delta(t)$ toward 0 constitutes "pushing" a ligand into the protein. The speed of $\delta(t)$ matters. The landscapes shown in Fig. 14.3B indicate that if the velocity is too high for an equilibration, then $z(t)$ changes with $\delta(t)$ by following the red curves in Fig. 14.4: *ABCED* for increasing $\delta(t)$, with a "bond rupture" event at C when $\delta = \delta_3$, and *DEFBA* for decreasing $\delta(t)$ with a "snap-on" event at F when $\delta = \delta_1$. In this type of experiment, the thermal fluctuations play a relatively minor role.

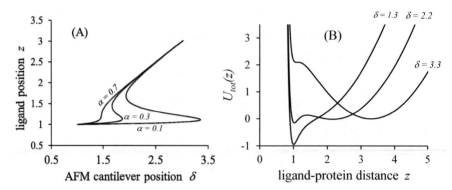

Fig. 14.3 (A) Mechanical equilibrium position of the ligand, z, as a function of δ, the position of the base of the cantilever, with several different values of α, the stiffness of the cantilever, according to Eqn. (14.36). $z = 1$ is the equilibrium position of the ligand in the absent of the AFM force. (B) Total mechanical energy, $U_{tot}(z)$, as a function of the ligand-protein distance z for several different values of δ, as indicated. For all δ, $\alpha = 0.1$.

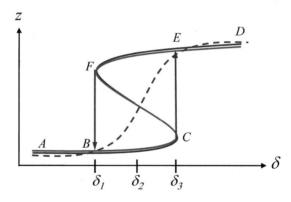

Fig. 14.4 When the value of δ changes, the energy landscape changes as shown in Fig. 14.3B. The velocity of $\delta(t)$ matters regarding how the system responds, *e.g.*, $z(t)$. For fast $\delta(t)$, $z(t)$ follows the red curves $ABCEF$ as $\delta(t)$ increases, and $DEFBA$ as $\delta(t)$ decreases. For slow $\delta(t)$, $z(t)$ has large fluctuations but $\langle z(t) \rangle$ follows the blue dashed line.

On the other hand, if velocity of the $\delta(t)$ is sufficiently low, then $z(t)$ fluctuates and its mean value $\langle z(t) \rangle$ will follow the blue dashed line. Note that in this case, $\delta(t)$ has to be slow enough to allow an equilibration between the two energy wells. In the middle of the curve, when $\delta = \delta_2 \simeq 2.2$, an equilibration means an equal probability between the wells: there will be a large variance in the measurements on z when $\delta = \delta_2$.

Part IV
Epilogue: Beyond Chemical Reaction Kinetics

Chapter 15
Nonequilibrium Landscape Theory, Attractor-State Switching, and Autonomous Differentiation

Living organisms are complex; they are collections of many interacting individuals, each with a large number of internal degrees of freedom of their own. The dynamics of such systems are complicated; they are dominated by collective behaviors due to the interactions and exhibit nonlinearity and stochasticity. Because of this, to an observer, even the simple task of merely "describing" a biological individual requires a great number of "bits" in information-theoretic terms. J. J. Hopfield pinpointed the natural complexity in biology squarely to this effect [123]. The vast internal degrees of freedom necessitate a stochastic representation for a complex individual. In fact, the very notion of "free will" can be described as a useful predictive myth. Hopfield further observed that "actions which are determined by noise are not what we mean by free will, for the entire idea of responsibility is lost in actions resulting from noise." Interestingly, the molecular cell biology one learns from textbook is remarkably definitive and, as a matter of fact, often narrates what happens as a well-defined sequence of biochemical events: "this follows that", "this happens followed by that happening", etc. In the present chapter, using a single cell as an archetype of a biological organism, we shall reconcile this view of molecular cell biology with the nonlinear stochastic kinetic description we have extensively developed thus far. The answer is a more refined, biochemically and kinetically grounded understanding of C. H. Waddington's epigenetic landscape, originally proposed for cell differentiation [267, 269, 70].

We have already firmly established the existence of an emergent landscape for living cells as a driven, open biochemical reaction system under a sustained chemostat. In this chapter, we shall use this newfound analytic concept to help orga-

H. Qian, H. Ge, *Stochastic Chemical Reaction Systems in Biology*,
Lecture Notes on Mathematical Modelling in the Life Sciences,
https://doi.org/10.1007/978-3-030-86252-7_15

nize narratives in molecular cell biology. If one can fully understand Waddington's metaphor in terms of the intracellular nonlinear stochastic chemical kinetics and intercellular nonlinear stochastic cell-population dynamics, then one will be able to reconcile *chance* and *necessity* [182, 111]. On a more conceptual level, the answer is actually hidden in the relationship between the abstract concept of a *random event* and its *realizations* formulated in the mathematical theory of probability: statistics are phenomena to be explained by science, and responsibility is a concept based on mechanical causality within individuals.

At the level of an individual cell, the aforementioned individuals are the major biochemical players, e.g., polymerases, transcription factors, signaling kinases and GTPases. Through biochemical reactions, they form an intracellular *biochemical reaction network*. While individual macromolecules behave stochastically in biochemical reactions, the collective dynamics of a population of "identical" protein molecules can be accurately characterized statistically in terms of a few key kinetic parameters: a diffusion constant in an aqueous solution at room temperature, a few rate constants for conformational transitions, and a pair of rate constants for association and dissociation with a specific partner, *e.g.*, the substrate. We note that all these parameters are characterizations of the highly uncertain timings for the occurrence of various events, called reactions. This observation will be further explored below.

As we discussed in Chapter 8, there is a very meaningful isomorphism between the kinetic pictures for (i) a macromolecule that consists of interacting atoms, (ii) a cell, at a level higher, that consists of interacting proteins, DNA, and biochemicals, and (iii) a tumor tissue, yet another level higher, that consists of distinct interacting cell phenotypes that exist within an isogenic clonal cell population.

Given what is now known as the "nongenetic heterogeneity" of cell phenotypes [34, 168], the cells are the individuals in a tissue: phenotypic switching of each individual cell is kinetically isomorphic to unimolecular conformational transition. The cell birth and death processes can be mapped to the synthesis and degradation of a biochemical species, etc. Then, at the whole tumor tissue level, there are interactive cell population dynamics that involve autocrine and paracrine cell-to-cell communication that influence the rates of the above processes, cell migration, predation and prey, etc. Classifying isogenic individual cells according to biomarkers, gene expression patterns, and biological functions is one of the most important issues in the current study of nongenetic heterogeneity using single-cell technologies. These tasks are not very different from identifying conformational states of biomacro-molecules as a key practice of physical biochemistry using spectroscopies. Indeed, experiments on extracting the cells in the tail portion of a population distribution and observing the re-population kinetics share the same idea as laser ablation of protein conformations followed by observation of relaxation kinetics [34], as well as the method of fluorescence photobleaching recovery (FPR), also known as fluorescence recovery after photobleaching (FRAP), in cell biophysics [11, 12].

Both the kinetics of a biochemical reaction network and the dynamics of a het-erogeneous cell population can be quantitatively represented in terms of a nonlinear stochastic dynamical system (NSDS) [220, 258, 237]. Nonlinear stochastic dynamic theory depicts the behavior of all complex individuals as motions in their state space on a rugged landscape [79]. One of the essential new insights from NSDS theory is the *existence* of an emergent, nonequilibrium, global landscape that represents the dynamics. This is a generalization of the fundamental concept of thermody-namic potential in classical physics and an application of the essential notion of large deviations in modern mathematics. The landscape is not the "mechanism" of the dynamics *per se*; rather, it represents the thermodynamic force that is responsi-ble for the nonequilibrium kinetic transients in the system. In addition to the land-scape, nonequilibrium dynamics are also characterized by a complementary emer-gent quantity, the transport flux.

We would like to note that the entire NSDS perspective agrees with what the prominent mathematical biologist K. Pearson once said, "[a]ll laws must ultimately be merged into laws of motion," and fits P. W. Anderson's theory on the emergent phenomenon, wherein "each level can require a whole new conceptual structure." [203, 7]. The mathematical theory of NSDS provides a rigorous foundation for the landscape as an emergent object: in principle, if one can measure the ergodic proba-bility distribution $p_V^{ss}(n_1, n_2, \cdots, n_N)$ for the nonequilibrium steady state (NESS) of a dynamical system, where n_k is the population size of the kth species and V is the geometrical size of the reaction system, then the landscape

$$\varphi(\mathbf{x}) = -\lim_{V \to \infty} \frac{\ln p_V^{ss}(Vx_1, Vx_2, \cdots, Vx_N)}{V}, \tag{15.1}$$

in which $\mathbf{x} = (x_1, x_2, \cdots, x_N)$, $x_k = \frac{n_k}{V}$ is the number density of the kth subpopulation. Of course, in reality, this is not feasible since a biological organism has only finite n_k, and its environment is constantly changing. Nevertheless, the mathematical object in (15.1) provides our theory with a rigorous foundation, just as the notion of internal energy serves as the foundation of classical thermodynamics.

The landscape provides the notion of *attractor states* as basins of attraction [127]. This is a statistical concept with a precise meaning; it captures a global perspective of a *life history* of an autonomous system, or organism, under a stationary environment. By life history, we mean "the sequences of events undergone by an individual organism during its lifetime": the story of an individual's life. Note that one obtains such a landscape from the statistics of an NSDS, and a "life history" is a nonstationary stochastic process, usually with a well-defined initial state. For each individual system, as a realization of the stochastic process, the overall trend is a downhill motion on the "rugged hillside", hopping from one basin to another. This type of movement is expected to be nearly deterministic since different saddle points that are connected to a given basin have wide disparity in heights in general; the saddle point with the lowest barrier thus dominates the exit probability [112]. In this sense, the life history of a cell is *programmed* in the genome, expressed in terms of a system of biochemical reactions, with the rate determined by the enzymes involved. The milestone events, which can be experimentally observed, are those local minima on the hillside that are connected via the lowest saddle barriers. The fate of an attractor state on the hillside has little uncertainty; the variability mainly comes from the timing.

15.1 Emergent Global Landscape for Nonequilibrium Steady State

There are essential differences between the emergent landscape for a living biochemical system and the free energy landscape, e.g., J. G. Kirkwood's potential of mean force [148], of a nondriven chemical system such as a protein [79]. Let us use a single enzyme cycle, as presented in Chapter 10, to illustrate a key difference between landscapes for a NESS and for an equilibrium steady state, *e.g.*, the Gibbs function in classical equilibrium chemical thermodynamics, as shown in Fig. 15.1.

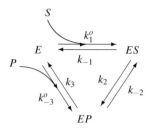

Fig. 15.1 Kinetic scheme for a reversible enzymatic reaction with substrate S and product P. k_{-1}, k_2, k_{-2} and k_3 are first-order rate constants, and k_1^o and k_{-3}^o are second-order rate constants. The system of biochemical reactions is in equilibrium if and only if the concentrations of S and P, $[S]$ and $[P]$, satisfy $k_1 k_2 k_3 = k_{-3} k_{-2} k_{-1}$, in which pseudofirst-order rate constants $k_1 = k_1^o [S]$ and $k_{-3} = k_{-3}^o [P]$.

Recall that the probability of the enzyme being in any one of the three states at time t follows the CME,

$$\frac{dp_E(t)}{dt} = k_3 p_{EP} - (k_{-3} + k_1) p_E + k_{-1} p_{ES},$$

$$\frac{dp_{ES}(t)}{dt} = k_1 p_E - (k_{-1} + k_2) p_{ES} + k_{-2} p_{EP}, \tag{15.2}$$

$$\frac{dp_{EP}(t)}{dt} = k_2 p_{ES} - (k_{-2} + k_3) p_{EP} + k_{-3} p_E.$$

When all k are nonzero, this set of equations has a unique steady state probability distribution $\pi = (\pi_E, \pi_{ES}, \pi_{EP})$.

Now consider a small system that consists of N such enzyme molecules in a reaction volume V. Each undergoes the reactions in Fig. 15.1, among which $\mathbf{N}_E$, $\mathbf{N}_{ES}$ and $\mathbf{N}_{EP} \equiv N - \mathbf{N}_E - \mathbf{N}_{ES}$ are the molecular numbers in states E, ES, and EP, respectively. The $\mathbf{N}$'s are integer-valued random variables that fluctuate, no matter how large N is, as long as it is not infinite. Then, the steady state probability for the fluctuating numbers,

$$P\left\{\mathbf{N}_E = v_1, \mathbf{N}_{ES} = v_2, \mathbf{N}_{ES} = v_3\right\} = \frac{N!}{v_1! v_2! v_3!} \pi_E^{v_1} \pi_{ES}^{v_2} \pi_{EP}^{v_3}$$

$$\simeq Z^{-1}(V) e^{-V \varphi(x_1, x_2, x_3)}, \tag{15.3}$$

in which $x_1 = v_E/V, x_2 = v_{ES}/V, x_3 = v_{EP}/V$ are the *concentrations* of E, ES, and EP, respectively; and,

$$\varphi(x_1,x_2,x_3) = x_1\ln\left(\frac{x_1}{\pi_E}\right) + x_2\ln\left(\frac{x_2}{\pi_{ES}}\right) + x_3\ln\left(\frac{x_3}{\pi_{EP}}\right), \qquad (15.4)$$

$$Z(V) = \int_{x_1+x_2+x_3=x_{tot}} e^{-V\varphi(x_1,x_2,x_3)}\,dx_1\,dx_2\,dx_3. \qquad (15.5)$$

In the state space of the concentrations of chemical species $\mathbf{x} = (x_1,x_2,x_3)$, $\varphi(\mathbf{x})$ in (15.4) is the global, kinetic or nonequilibrium, *landscape*. One can introduce three partial derivatives

$$\begin{aligned}
\frac{\partial \varphi}{\partial x_1} &= \ln x_1 - \ln \pi_E + 1, \\
\frac{\partial \varphi}{\partial x_2} &= \ln x_2 - \ln \pi_{ES} + 1, \\
\frac{\partial \varphi}{\partial x_2} &= \ln x_3 - \ln \pi_{EP} + 1.
\end{aligned} \qquad (15.6)$$

We now observe a very important mathematical result when the six k's satisfy *detailed balance*: $\pi_E k_1 = \pi_{ES} k_{-1}$, $\pi_{ES} k_2 = \pi_{EP} k_{-2}$, and $\pi_{EP} k_3 = \pi_E k_{-3}$. In this case,

$$\frac{\partial \varphi}{\partial x_2} - \frac{\partial \varphi}{\partial x_3} = \ln\left(\frac{k_2}{k_{-2}}\right) + \ln\left(\frac{x_2}{x_3}\right), \qquad (15.7)$$

in which k_2/k_{-2} is the equilibrium constant between *ES* and *EP*. Eqn. (15.7) multiplied by $k_B T$ is the standard chemical potential difference $\mu_{ES} - \mu_{EP}$ for the ideal solution of the enzyme molecules. This leads to

$$\frac{\partial \varphi}{\partial x_1} - \mu_E = \frac{\partial \varphi}{\partial x_2} - \mu_{ES} = \frac{\partial \varphi}{\partial x_3} - \mu_{EP} = \text{constant}.$$

Hence, $\partial \varphi/\partial x_k$ is the chemical potential of species k up to a constant.

$$\varphi(x_1,x_2,x_3) = \sum_{i=1}^{3} x_i \frac{\partial \varphi}{\partial x_i} - \sum_{i=1}^{3} x_i,$$

is equal to the Gibbs' potential function

$$\sum_{i=E,ES,EP} \mu_i x_i$$

of the equilibrium chemical reaction system, up to a constant.

15.1.1 The Emergent Global Nonequilibrium Landscape

When the enzyme reactions in Fig. 15.1 are driven by $S \to P$ with a sustained nonzero chemical potential difference, there will be no detailed balance. Open biochemical reaction "networks", such as the one in Fig. 15.2, are the rule rather than the exception inside living cells.

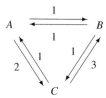

Fig. 15.2 A particular example of the enzymatic reactions in Fig. 15.1. Since $k_1k_2k_3 = 1 \neq k_{-1}k_{-2}k_{-3} = 6$, the system is *driven* under a chemostatic chemical potential $\Delta\mu_{PS} = k_BT\ln\gamma$, where $\gamma = k_1k_2k_3/(k_{-1}k_{-2}k_{-3})$.

In Fig. 15.2, we locally see that the probability of $A \to C$ is twice as likely as $A \to B$. However, the steady state probability distribution for the 3-state kinetic cycle is

$$\pi_A = \tfrac{1}{4}, \quad \pi_B = \tfrac{1}{2}, \quad \pi_C = \tfrac{1}{4}. \tag{15.8}$$

Note that this set of distributions has the following important characteristics:

$$\pi_A k_{AB} - \pi_B k_{BA} = \pi_B k_{BC} - \pi_C k_{CB} = \pi_C k_{CA} - \pi_A k_{AC} = -0.25, \tag{15.9}$$

in which $k_{\xi\eta}$ is the Markov transition rate from state ξ to state η. There is a sustained net cycle flux going from $B \to A \to C \to B$, one quarter of a round per unit time, driven by the external $\Delta\mu = -k_BT\ln3 < 0$.

One notices several very telling features of this kinetic system that has no detailed balance: First, $\pi_B/\pi_C = 2 \neq k_{CB}/k_{BC} = 3$. Second, the NESS state probabilities for B and C, $\pi_B : \pi_C = 2$, are very different from $k_{AB} : k_{AC} = 1/2$. This observation indicates that the local dynamics among the connected states can be completely different from what the global landscape $\pi = (\pi_A, \pi_B, \pi_C)$ tells us. The global steady-state probability π should not be used as the "energy function" for computing local transitions. This is in stark contrast to an equilibrium landscape with detailed balance, which simultaneously serves two well-known functions: (*i*) it gives the equilibrium distribution; and (*ii*) it also provides the information on local transition rates.

Nevertheless, similar to the Gibbs function of equilibrium chemical thermodynamics, π—which is an invariant probability with permanence—has a fundamental role to play in the macroscopic dynamics of the nonequilibrium system in Fig. 15.2. Following the steps in Eqns. (15.3)-(15.5), in the space of the concentrations of chemical species $\mathbf{x} = (x_A, x_B, x_C)$, we have φ in (15.5) again. This time, the global landscape cannot be constructed by local information, e.g., transitions, in the state space; it is a consequence of infinitely long-term dynamics that reaches a NESS. In other words, it is an *emergent global landscape*. The term "emergent" here literally means it takes time to appear.

Consider now the corresponding macroscopic reaction system with concentrations $x_A(t)$, $x_B(t)$ and $x_C(t)$ for molecules in the three states. Then, they follow the deterministic kinetic equation

$$\frac{dx_A}{dt} = x_C - 3x_A + x_B, \tag{15.10a}$$

$$\frac{dx_B}{dt} = x_A - 2x_B + 3x_C, \tag{15.10b}$$

$$\frac{dx_C}{dt} = x_B - 4x_C + 2x_A. \tag{15.10c}$$

We have the important result (see Section 7.1) that

$$\frac{d}{dt}\varphi\left(x_A(t), x_B(t), x_C(t)\right) \le 0. \tag{15.11}$$

The emergent φ has a similar property to the Gibbs function for a closed chemical reaction system. Inequality (15.11) is one origin of the organizational power of the global landscape.

Different from the equilibrium free energy landscape $\varphi^{eq}(\mathbf{x})$, which predicts equilibrium probability distribution as well as the ratio of the transition rates of a reversible reaction, nonequilibrium theory has a *global* potential and a *local* potential, and they are different. This is a key insight from the modern theory of large deviations [80, 94]. The global potential is related to the NESS probability distribution, while the local potential is related to the transitions between two adjacent attractor states. The relationship between the global and local potentials is very analogous to the π's in Eqn. (15.8) and the ratios of rate constants in Fig. 15.2. Their difference is due to nonequilibrium conditions that result in NESS cycle fluxes.

15.2 Bifurcation and the Far-from-Equilibrium State of a System

L. Onsager (1903–1976) pioneered the theory of nonequilibrium thermodynamics [46], but his theory is only applicable to systems near an equilibrium. Since then, extending nonequilibrium thermodynamics to a regime that is far from equilibrium has been one of the greatest challenges in science. However, is there a more objective, even qualitative, distinction between nonequilibrium states that are near and far from an equilibrium state? In [190], Nicolis and Prigogine provided an answer to this question in terms of nonlinear bifurcation.

In addition to the large information content, nonlinearity is another essential feature of a complex system. Multistability is a key characteristic of biochemical reactions networks with positive feedback regulations. Saddle-node (or blue-sky) bifurcation is one of the most widely observed dynamic "mechanisms" that underlies how a system shifts from having a unique steady state to have multiple stable steady states, *e.g.*, attractors.

Let us use the nonlinear Schlögl model studied in Section 12.6 to illustrate the essential features. Gene regulatory networks of transcription factors, phosphorylation-dephosphorylation signaling networks with feedback regulation, and many other biochemical processes can all be conceptually mapped to this simple nonlinear chemical kinetic system as a "model of models".

Fig. 15.3 recaps the crucial points we have learned thus far; it combines the salient features from Figs. 8.1 and 14.4. With decreasing randomness in a system, *e.g.* when the system tends to a deterministic limit, the blue dashed line moves toward the discontinuous orange curve and becomes steeper. The fluctuations in $x(t)$ for each given μ diminish except when $\mu = \mu_2$, the critical value for phase transition.

When parameter μ in a dynamical system changes, one of course expects the corresponding location of its steady state $x^*(\mu)$ to change accordingly. Translating this to the general discussion on NSDSs and taking the mean value of a stationary probability distribution as the most easy observable, this indicates that the mean

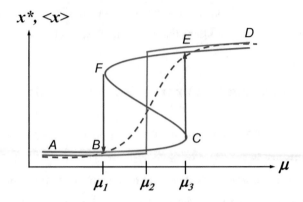

Fig. 15.3 Steady states as functions of parameter μ in a nonlinear dynamical system that undergoes a saddle-node bifurcation. The bifurcation is also called a *blue-sky* bifurcation, since a pair of stable and unstable steady states appears "out of the blue" at F and C. The green S-shaped curve shows three steady states when μ is in the range of (μ_1, μ_3). The middle branch (FC) represents an unstable steady state. For a rapidly decreasing (increasing) μ, x follows the upper (lower) branch and then the red $F \to B$ ($C \to E$) transition to the lower (upper) branch. In the presence of fluctuations and with a slowly changing μ, the average value of x, $\langle x \rangle$, changes with μ, as shown by the blue dashed curve. The dashed blue curve becomes increasingly steeper with decreasing noise (not shown). In the limit of zero noise, it becomes the orange, discontinuous curve, with a critical parameter value at μ_2. In the zero-noise limit, the system behaves differently from that predicated by the deterministic dynamics. Figure redrawn based on [128].

value $\langle x \rangle$ moves with μ accordingly. The blue dashed line in Fig. 15.3 is an example of this phenomenon. When observing a response curve such as this from an NSDS, however, there are actually two fundamentally different scenarios, as illustrated in Fig. 15.4. In the macroscopic limit, the two types of systems behave very differently: a monostable system has a continuous and smooth change, while a bistable system exhibits discontinuous behavior.

This clearly defined mathematical distinction between the two types of behavior, it seems, is at the heart of several important biological debates. In molecular biophysics and in connection to the possible mechanism of cooperative binding of oxygen to hemoglobin, the *conformational selection* versus *induced fit*, proposed by Monod, Wyman, and Changeux, and by Koshland, Nemethy, and Filmer, is one example [71]. In adaptation and evolution of bacteria as a unicellular organism, the Darwinian versus Lamarckian mechanism is another. In both cases, the key issue is whether the "two states" coexist through the transition. In essence, the two mechanisms are "pre-existing bistability with changing relative weights" and "unistability with programmed or induced adaptation".

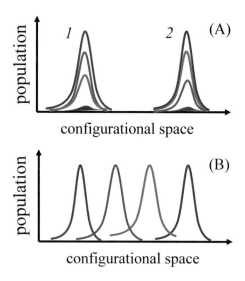

Fig. 15.4 Upon a change in parameter μ, the mean value of a stationary NSDS changes, $\langle x \rangle(\mu)$. However, at the level of the stationary probability distribution, two fundamentally different mechanisms are possible: if the system is unistable with a single deterministic fixed point, then the modal value of the distribution moves with μ, as shown in (B), from red $\rightarrow$ orange $\rightarrow$ green $\rightarrow$ blue. In the deterministic limit, this becomes $x^*(\mu)$. However, systems with bistability exhibits a rather different, "two-state" behavior as shown in (A). There are two modal values; usually, the position of the two model values changes little, and the more pronounced change is in the relative heights of the two peaks. Through the transition region, the identity of the two states, 1 and 2, are preserved; it is their relative stability that changes, from red $\rightarrow$ orange $\rightarrow$ green $\rightarrow$ blue. There are two deterministic limits: on a short time scale, this becomes the red transitions in Fig. 15.3, but on a long time scale, this becomes the discontinuous orange curve.

15.2.1 From Near-equilibrium to Far-from-equilibrium Through Bifurcation

As we have stated, saddle-node bifurcation is one of the "mechanisms" by which a system shifts from having a unique steady state to having multiple stable steady states. In this section, we are particularly interested in how the $\Delta\mu$ from the chemostat induces bistability in an open chemical reaction system. We again employ the Schlögl model

$$A + 2X \underset{k_{-1}}{\overset{k_{+1}}{\rightleftharpoons}} 3X, \quad X \underset{k_{-2}}{\overset{k_{+2}}{\rightleftharpoons}} B. \tag{15.12}$$

As an example, we shall establish a connection between the nonequilibrium thermodynamics on the one hand with nonlinear bifurcation and the emergence of a new "state" on the other [229].

Let the concentrations for X, A, and B be x, a, and b, respectively, with a and b being sustained by a chemostat while x varies dynamically. The reaction, and thus the $x(t)$, eventually reaches the chemical equilibrium if the environmental concentrations a and b satisfy

$$\gamma \equiv \frac{ak_{+1}k_{+2}}{bk_{-2}k_{-1}} = 1. \tag{15.13}$$

Note that the equilibrium constant for the "overall reaction"

$$A \underset{}{\overset{K_{AB}}{\rightleftharpoons}} B$$

is simply $K_{AB} = k_{+1}k_{+2}/(k_{-1}k_{-2})$. Therefore

$$k_B T \ln \gamma = k_B T \ln K_{AB} + k_B T \ln \left(\frac{a}{b}\right) = \Delta \mu_{BA}, \tag{15.14}$$

which is the chemostatic chemical potential difference from the environment upon the system. If $\Delta \mu_{BA} \neq 0$, the reaction reaches a nonequilibrium steady state. Assuming the nonlinear kinetic rates follow the law of mass action, then

$$\frac{dx(t)}{dt} = -k_{-1}x^3 + k_{+1}ax^2 - k_{+2}x + k_{-2}b. \tag{15.15}$$

Introducing nondimensionalized $\tau = k_{+2}t$ and $u = (k_{-1}/k_{+2})^{1/2}x$, then

$$\frac{du}{d\tau} = -u^3 + \gamma \alpha u^2 - u + \alpha = (u^2 + 1)(\alpha - u) + (\gamma - 1)\mu u^2, \tag{15.16}$$

in which

$$\alpha = \left(\frac{k_{-1}}{k_{+2}}\right)^{\frac{1}{2}} \frac{k_{-2}b}{k_{+2}}.$$

When $\gamma = 1$, e.g., $\Delta \mu_{BA} = 0$, the system has a unique equilibrium steady state $u^{eq} = \alpha$. Fig. 15.5 shows chemical steady state(s) u^* as a function, or three functions, of chemical driving force $\ln \gamma = \frac{\Delta \mu_{BA}}{k_B T}$. In a nonequilibrium steady state, there is sustained net transport flux in the reaction system, $J^{ness} = k_{+1}ax^{*2} - k_{-1}x^{*3} = k_2 x^* - k_{-2}b$, from A to B when $\mu_A > \mu_A$. In nondimensionalized variables, u^* is related to γ through $\gamma = \frac{u^*}{\alpha} + \frac{1}{\alpha u^*} - \frac{1}{u^{*2}}$.

With increasing γ, the system is driven ever stronger, and it is farther away from its chemical equilibrium, if measured by the flux. The u^{ss} represented by the blue branch in Fig. 15.5, however, changes very little. One can use

$$\frac{du^{ss}}{d\gamma} = \left(\frac{d\gamma}{du^{ss}}\right)^{-1} = \frac{\alpha u^{*3}}{u^{*3} - u^* + 2\alpha} \tag{15.17}$$

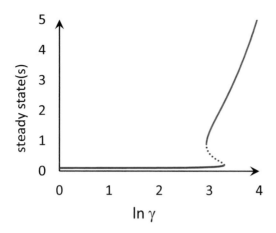

Fig. 15.5 Bifurcation diagram for nonlinear dynamics (15.16) with $\alpha = 0.1$. The steady state(s) is u^*, and the steady-state flux is $(u^* - \alpha)$. When $\ln\gamma = 0$, the chemical steady state is unique, an equilibrium state with $u^* = \alpha$ and zero flux. The blue branch, therefore, is a continuation of the equilibrium state, which we shall call the *near-equilibrium branch*. A saddle-node bifurcation occurs at $\ln\gamma = 2.938$, when a pair of stable (orange) and unstable (dotted red) states emerge. The orange *far-from-equilibrium branch* is separated, and thus protected, from the near-equilibrium branch by the red dotted line. When $\ln\gamma > 3.30$, no spontaneous rare event can occur that endangers the far-from-equilibrium state. A distinct state of *active matter* arises. At a macroscopic scale and in an ergodic limit, the two branches coexist only at a single first-order phase transition critical value of $\gamma^* = 20.36$, $\ln\gamma^* = 3.014 \in (2.938, 3.30)$. Figure reproduced from [128] with permission.

as a measure for characterizing how responsive u^* is with increasing γ. Eqn. (15.17) indicates that when $u^* \ll 1$, the response is nearly zero; however, if $u^* \gg 1$, then the response is approximately linear with slope α. When $\gamma > 2.938$, a new (orange) stable steady state appears, "out of the blue", far from the equilibrium. The bifurcation event, therefore, provides a very natural concept for identifying near- and far-from-equilibrium steady states, which are separated by an unstable steady state, represented by the red dotted line.

The far-from-equilibrium (orange) steady state is characterized by several fundamental features:

(*a*) It is robust against internal and external perturbations due to the attractorial structure. In condensed matter physics, "robustness" has been called "rigidity" and "protected behavior", among many other terms [7, 160];

(*b*) Starting from the near-equilibrium branch, reaching the far-from-equilibrium state spontaneously is an **exponentially** rare event, which takes an exponentially long time. This furnishes another meaning of "being far", in a kinetic sense;

(c) The appearance of the far-from-equilibrium branch is an emergent phenomenon with dynamic symmetry breaking [123, 221];

(d) Last but not least, with a sufficient energy supply $\gamma \gg 1$, the far-from-equilibrium state reaches a complete "safety" without the possibility of spontaneous deterioration to the blue branch: the near-equilibrium branch eventually disappears completely from the state space due to a "saddle-node collision" event.

One can thus legitimately identify the far-from-equilibrium orange branch as a "new form of matter". As we learned from Chapter 8, in the macroscopic limit, there is a discontinuous phase transition; the near- and far-from-equilibrium branches only coexist at a critical condition of γ^*.

The critical γ^*, at which phase transition occurs as introduced in Chapter 8, is defined from the emergent landscape $\varphi(\mathbf{x}; \gamma)$, when it possesses two local minima $\mathbf{x}_1^*$ and $\mathbf{x}_2^* \neq \mathbf{x}_1^*$ with equal height: $\varphi(\mathbf{x}_1^*; \gamma^*) = \varphi(\mathbf{x}_2^*; \gamma^*)$. The landscape of the system (15.12) can then be computed:

$$\varphi(u) = \int \ln\left(\frac{u+u^3}{\alpha+\alpha\gamma u^2}\right) du \tag{15.18}$$

$$= u\ln\left(\frac{u(1+u^2)}{\alpha(1+\gamma u^2)}\right) - u + 2\arctan(u) - \frac{2}{\sqrt{\gamma}}\arctan\left(\sqrt{\gamma}u\right).$$

$\varphi(u)$ is a Lyapunov function for the solution of the differential equation (15.16), $u(t)$:

$$\frac{d}{dt}\varphi[u(t)] = \frac{du(t)}{dt}\ln\left(\frac{u+u^3}{\alpha+\alpha\gamma u^2}\right) \leq 0. \tag{15.19}$$

Fig. 15.6 shows the $\varphi(u)$ for $\alpha = 0.1$ and several different values of γ. The value of γ^* is found to be 20.36.

15.2.2 Thermodynamic vs. Kinetic Branches of Nonequilibrium States of Matter

A piece of inert material under a nonequilibrium condition will have transport flux. Even though the flux might be too small to measure, the condition is nevertheless not at equilibrium. Still, most people would not call such a system "active". Furthermore, when the fluxes are truly small, they follow Onsager's linear theory of irreversibility.

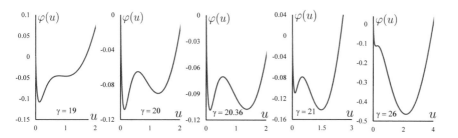

Fig. 15.6 Emergent global landscape $\varphi(u)$ for the reaction kinetic system (15.12), given by (15.18) with $\alpha = 0.1$ and several values of γ.: from left to right, $\ln\gamma = 2.044, 3.00, 3.014, 3.045, 3.258$. The macroscopic phase transition occurs at $\gamma^* = 20.36$, when the two local minima are of equal height. The minima (maxima) correspond to the stable (unstable) steady states in Fig. 15.5. Figure reproduced from [128] with permission.

Is there a qualitative difference between these nonequilibrium systems and systems that are "far from equilibrium", or more intuitively, "active"?

To distinguish the former from the latter, Nicolis and Prigogine [190] articulated the notion of an "equilibrium branch" in phase space and the emergence of a "kinetic branch", or dissipative structure, through a transcritical bifurcation, as shown in Fig. 15.7. The bifurcation parameter here represents the distance from equilibrium.

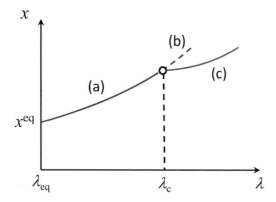

Fig. 15.7 Branching of states as the distance from equilibrium increases. (a) Stable part of the thermodynamic branch; (b) branch becoming unstable; (c) new stable state (dissipative structure) emerging beyond the instability of the thermodynamic branch. Figure adapted from [190] and reproduced from [128] with permission.

As we learned in Section 12.4, transcritical bifurcation is not a robust phenomenon, while in contrast, saddle-node bifurcation is. Therefore, the bifurcation in Fig. 15.5 provides a more reasonable mechanism for the genesis of the dissipative structure through energy (γ) driven saddle-node bifurcation. In the transition region, there is a clear identification of the "thermodynamic branch" and the "kinetic branch", one near and one far from equilibrium. With increasing energy pumping, the inanimate near-equilibrium branch ultimately gives way to the "far-from-equilibrium branch", with all the signatures of a phase transition. This mechanism is robust.

We suggest using the term *inanimate branch* to describe the nonequilibrium continuation ($\gamma > 1$) of the thermodynamic equilibrium, the steady-state branch that passes through the equilibrium $u^{eq} = \alpha$ when $\gamma = 1$. Then, the *far-from-equilibrium branch* has a "barrier" that separates itself from the inanimate branch. The notions of "near" and "far-from" in this sense are separated by the insurmountable barrier in the macroscopic limit. They are qualitatively different.

15.2.3 The Biochemistry of Energy-driven Bifurcation

Chemical energy-dependent saddle-node bifurcation in biology has been demonstrated in a laboratory. The biochemical network and its regulation involved in the yeast cell cycle, e.g., the process of cell division, is one of the best understood cellular systems in terms of biochemical kinetics [266, 282]. In a recent investigation, the role of cellular phosphorylation potential, e.g., the γ for ATP hydrolysis, on cell cycle progression was carefully studied [271]. Cyclin-dependent kinase 1 (Cdk1), also known as cell division cycle protein 2 homolog (Cdc2), is a highly conserved kinase in cell cycle regulation (Fig. 15.8). In fission yeast *S. pombe* and in humans, it is encoded by the cdc2 gene, and in the budding yeast *S. cerevisiae*, by the cdc28 gene. Cdk1 kinase, together with its protein substrate and a cyclin, form a tertiary complex within which phosphorylation can occur. Phosphorylation of the various protein substrates leads to cell cycle progression.

The kinase activity of the Cdc2/Cdc13 complex in fission yeast, where Cdc protein 13 is a B-type cyclin, is itself regulated through a phosphorylation-dephosphorylation cycle (PdPC), with the kinase Wee1 and the phosphatase Cdc25: the dephosphorylation activates the kinase. The enzymatic activities of both Wee1 and Cdc25 themselves are regulated by two respective types of PdPCs, and there are feedback

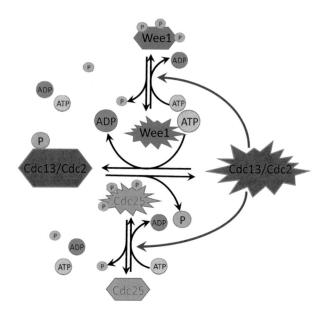

Fig. 15.8 The transition from G2 phase to M phase in fission yeast and human cell cycle progression is controlled by three types of PdPCs, with an intricate interplay of kinase and phosphatase activations [271]. Such a regulatory network will not function when the free energy associated with ATP hydrolysis, a necessary ingredient for each and every PdPC, is low. This has been experimentally demonstrated recently in the nucleoplasmic extract from fission yeast *S. pombe*. Figure reproduced from [128] with permission.

controls: active Cdc12/Cdc13 is actually the kinase for both the Wee1 and CdC25 phosphorylations!

The transition from G2 phase to M phase in a yeast cell cycle is considered to follow the dynamics of a bistable system. The γ-driven saddle-node bifurcation with a bifurcation diagram is remarkably similar to our Fig. 15.5 and has been experimentally observed in the nucleoplasmic extract of fission yeast *S. pombe* [271]. The theory of NSDS and the notion of a landscape provide insights and aid in the understanding of a complex biological process, and laboratory experiments, in turn, support the feasibility of energy-driven bifurcation.

Very recently, it was shown that the behavior of yeast cells exhibited a bimodal distribution in the activation of the S-phase checkpoint, which is a stochastic barrier-crossing process in a double-well system, where the barrier height is determined by both DNA replication stress and autophosphorylation of the key effector kinase Rad53 [285].

15.3 Physics of Complexity Meets Waddington

With the concept of near- vs. far-from-equilibrium states of a system established as above, one naturally asks whether and how a chemical reaction system under a suitable external chemostat can spontaneously develop itself from a near-equilibrium state into a far-from-equilibrium state, not by chance but by necessity [182]. More precisely, how can an autonomous chemical reaction system start in a near-equilibrium state and eventually reach a steady state that is far from equilibrium?This is a key question in the understanding of *self-organization*. The term "self" here is understood as an autonomous dynamical system. In biology, the same question is related to the living process of *organism development and differentiation*.

The solution to this problem might be related to the bifurcation diagram shown in Fig. 15.5. One can simply "embed" the abscissa axis into an extended dynamical system with a slow kinetic transient that corresponds to changes along the abscissa. More specifically, we envision that the γ axis in Fig. 15.5 actually represents a slowly varying dynamic variable. With such multiscale evolving dynamics, a system with any initial condition, originating at $\ln \gamma = 2$, will rapidly settle into the blue, near-equilibrium branch but ultimately in time be in a far-from-equilibrium state on the orange branch when $\ln \gamma = 4$.

15.3.1 Self-organization and Differentiation

In terms of landscapes, let us now embed the saddle-node bifurcation structure, such as that in Fig. 15.5, into autonomous dynamics, with the bifurcation parameter being a slowly changing dynamic variable. In other words, we "stitch" the one-dimensional landscapes in Fig. 15.6 into a single two-dimensional landscape with γ as the second dimension, which contains the interval $(18.9, 27.1)$ and decreases with increasing γ.

Let us consider the following kinetics scheme:

$$A + 2X \underset{k_{-1}}{\overset{k_{+1}y}{\rightleftharpoons}} 3X, \quad X \underset{k_{-2}}{\overset{k_{+2}}{\rightleftharpoons}} B, \quad Y \underset{k_{-3}}{\overset{k_{+3}}{\rightleftharpoons}} C, \tag{15.20}$$

in which now a chemical species Y serves as a catalyst for the reaction $A + 2X \longrightarrow 3X$. (We assume the reaction $3X \longrightarrow A + 2X$ is not the reverse reaction of the former.

In a full thermodynamic analysis, therefore, each requires a reverse reaction of its own.) The kinetic system, according to the law of mass action, is then

$$\frac{dx}{dt} = -k_{-1}x^3 + k_{+1}ayx^2 - k_{+2}x + k_{-2}b,$$
$$\frac{dy}{dt} = -k_{+3}y + k_{-3}c. \tag{15.21}$$

If the rate constant $k_{-3} \ll k_{\pm 2}$, then one can simply use the value y as a bifurcation parameter for the fast dynamics of $x(t)$, while it itself changes with time very slowly, $y(t) = y^* + \left[y(0) - y^*\right]e^{-k_{+3}t}$, where $y^* = k_{-3}c/k_{+3}$, $\gamma = \frac{k_{+1}k_{+2}a}{k_{-1}k_{-2}b}$, $\gamma y(0) < 18.9$ and $\gamma y^* > 27.1$. Note that γy is the free energy of (15.12) for fixed y.

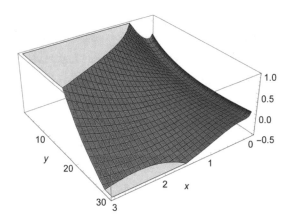

Fig. 15.9 Global, emergent kinetic landscape for the open chemical reaction system in (15.20) according to (15.22). For small y, the trough along $x = 0$ should be identified as the "near-equilibrium" thermodynamic branch; the "lower land" near $x = 3, y = 30$ is a far-from-equilibrium state. One can see a curve like that in Fig. 15.5, and the set of cross-sections like those in Fig. 15.6, embedded in this landscape, which captures the self-organization into the far-from-equilibrium state. Figure reproduced from [128] with permission.

Again, to obtain the emergent global landscape, one computes the NESS probability distribution according to the NSDS in terms of the chemical master equation. When there is a separation of time scales, the steady state probability distribution $p_{XY}^{ss}(\ell, n)$ can actually be obtained in two separated steps. First, one solves the steady-state distribution for n_X with $n_Y = n$ fixed. This yields the so-called conditional probability $p_{X|Y}^{ss}(\ell|n)$. Then, one notices that the kinetics of Y are actually independent of X in (15.21). Thus, one can solve the steady-state distribution for n_Y, $p_Y^{ss}(n)$. It can be shown that when putting them together, $p_{XY}^{ss}(\ell, n) =$

$p^{ss}_{X|Y}(\ell|n)p^{ss}_Y(n)$. $p^{ss}_Y(\ell)$ is called the marginal probability of n_Y. Then, one obtains the landscape

$$\varphi(x,y) = \varphi_{X|Y}(x|y) + \varphi_Y(y)$$

$$\simeq x\ln\left(\frac{x(1+x^2)}{\alpha(1+yx^2)}\right) - x + 2\arctan(x)$$

$$- \frac{2}{\sqrt{y}}\arctan\left(\sqrt{y}x\right) + \frac{y}{\gamma}\ln\left(\frac{y}{\tilde{y}^*}\right) - \frac{y-\tilde{y}^*}{\gamma}, \qquad (15.22)$$

in which $\tilde{y}^* = \frac{k_{-3}c}{k_{+3}}\gamma$.

With values $\alpha = 0.1$, $\gamma = 30$, and $\tilde{y}^* = 20$, Fig. 15.9 shows the landscape with a narrow trough along $x = 0$ for $y < 15$, and a turning point facing an "open field with downhill", somewhere near $y = 20$.

This example illustrates that one can certainly design, or the evolutionary process can certainly find, a kinetic system with a landscape that leads a system from its beginning state near an equilibrium to a final state far from equilibrium, or more generally, from a simple system to a more complex one. It requires no stretch of the imagination to think that Nature has adopted such a self-organizing mechanism in the process of adaptive evolution, and thus, biological organisms have coopted such a "program" in their genome.

15.4 Irreversible Living Process on a Landscape

Developmental biology is not stochastic; the entire "differentiation program" to a sufficiently large degree is encoded in the genome of an individual biological organism. In this section, we provide an extended discussion that reconciles this view with the nonlinear stochastic kinetic paradigm of cellular biology. In particular we ask how "symmetry breaking" occurs with relative certainty.

Two clonal cells with identical genomes can exhibit very different phenotypes, best epitomized by the various cell types in the body of multicellular organisms. To a biologist, this is something that needs to be "explained", since the tacit expectation is that the same set of genes would dictate the same cellular behavior(s) and function(s). However, with the stochastic kinetic representation presented in the present book and knowing that there is only a single copy of DNA in a cell, it is the very fact that clonal cells behave the same that requires an explanation. The stochastic dynamics on the "hillside" described below offers a possible explanation. In Chapter 13, we already studied the multistability of single-cell gene expression kinetics. The ability of a single pluripotent cell to undergo differentiation is an even more remarkable biological phenomenon.

Differentiation, as previously discussed in terms of nonlinear bifurcations, can also occur in an autonomous nonlinear chemical reaction system. This is a matter of perspective, and it can be illustrated cogently in terms of the emergent global landscape.Fig. 15.9 has a remarkable resemblance to C. H. Waddington's epigenetic landscape for cell differentiation [267, 70]. The emergent global landscape φ rooted in biochemical network dynamics transforms the biological metaphor into a physicochemically based quantitative description [269, 270, 284].

The biology narrative focuses on major "life events" in a living system, while biochemistry usually focuses on a certain part of the intracellular molecular reaction network. The former is a "higher-level", coarse-grained view of a living system. In terms of landscapes, then, for a biologist, one should consider a global topography and neglect all minor roughnesses. From this perspective, for example, a single cell undergoing cell division to become two daughter cells is a *nonstationary behavior* that consists of a sequence of "downhill hop" events [165]. The process begins with a single cell immediately after cell division until immediately before the next cell division, the numbers of all the molecular species as well as the cell size double. In terms of the concentrations of all molecular species, however, the same process is periodically oscillating.

All interesting biological processes, from a biologist's perspective, are nonstationary; all major biological events occur on the "hillsides" of a landscape. On a hillside, as a first-order approximation, the trajectories of different individuals with the same initial condition actually do follow the same deterministic sequence of events; two saddle points with comparable height are very rare in a complex system with fine tuning [160]. The stochasticity is mostly reflected in the timing of various events, and the time for an event to occur is more heterogeneous on a more rugged landscape. Without explicitly referring to timings, therefore, biologists are able to paint rather definitive sequences of events as knowledge at the cellular level. A trajectory that crosses many consecutive barriers of approximately equal height can be considered as a path along a flat landscape: multiple hops over barriers with nearly equal heights effectively reduce the ruggedness of a landscape, ultimately transforming the latter into a landscape with a "smooth" hillside. Keep in mind that if in a sequence of hops, there exists one saddle with a significantly higher barrier, then this "rate-limiting" step overwhelms all the other steps, making the timing of a final event exponentially distributed.

Traditional biochemical studies of intracellular processes, on the other hand, either focus on the kinetics of an individual reaction, which has very little relevance to the "global" cell behavior, or on the steady state of a biochemical reaction network, neglecting the slow variation of certain external variables. Therefore, the very nature of the biochemical study already puts the focus on an "attractor" of a landscape. A phenotypic switching, then, is associated with the transition between two attractors. To the hillside perspective mentioned above, this is a very detailed, reductionistic view. On the other hand, transient kinetic studies of a regulatory network without stochasticity do not encompass the overall landscape. Therefore, such studies, although valuable with respect to quantitative details, are not capable of representing the global "flow on a hillside of a rugged landscape".

The differentiation process represented by Waddington's landscape [269] does require a balanced branching kinetics [160]. At such a kinetic branching state, an attractor state has balanced saddle points. While the fate of a single cell is highly random, the probability of sixteen cells all taking one path is only $(1/2)^{16} = 0.000015$. This way, the differentiation process can use "large numbers" to counter against the uncertainty introduced by balanced kinetics. Fig. 15.10 shows an example of how a saddle-node bifurcation might also play a role in Waddington's landscape.

Giving a sustained stationary environment and assuming that "dynamics" is the ultimate underlying description [203] of any phenomenon, sequence of events and function, if any, a landscape can be obtained in two thought experiments with two extreme scenarios: (i) A single, "eternal" individual system can be followed over an infinitely long time. By "infinitely long", we mean it is longer than all the broken symmetries and reaches an ultimate ergodicity. For a single protein, such as the T4 lysozyme at $12\,^{\circ}\mathrm{C}$, this time scale could be already more than 10^8 sec $\simeq 3$ years [35]. For a single cell, this time scale is easily already longer than the age of the universe [227]. (ii) Alternatively, one has a large population of identical, independent, and asynchronized individuals. Macroscopic protein chemistry simply takes advance of the large Avogadro number: a micromolar concentration in a 2.5 ml cuvette has 10^{15}

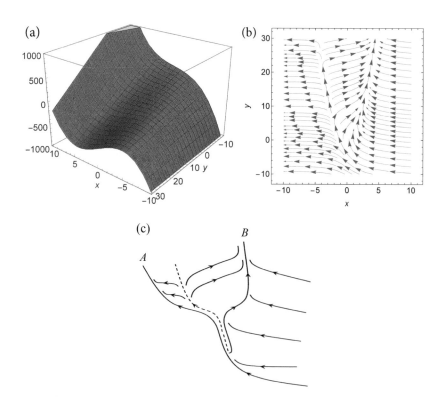

Fig. 15.10 (a) The landscape for a "differentiation program" exhibits dynamic a blue-sky bifurcation with $U(x,y) = x^3 - 2xy - 20y$: $\dot{x} = -\partial U/\partial x = -3x^2 + 2y$, $\dot{y} = \partial U/\partial y = 2x + 20$, with $x \in [-10, \infty)$. (b) The vector field for the planar system of ODEs. The nearly horizontal vectors indicate that the motion in the y-direction is much slower than that in the x-direction. For each given y, $-3x^2 + 2y = 0$ yields $x_+(y) = \sqrt{2y/3}$ and $x_-(y) = -\sqrt{2y/3}$ when $y \geq 0$. This means that for negative y near -10, x is monotonically falling to the minimum at $x = -10$ since the domain of x is restricted to $[-10, \infty)$. However, for $y > 0$, $x_+(y)$ is a "local" minimum, which is separated from $x = -10$ by the maximum at $x_-(y)$. For $y = 30$, the minimum and maximum are located at $\pm\sqrt{20} = \pm 4.47$, respectively. (c) A schematic for the "dynamic flow" in (b). This type of "blue-sky" bifurcation structure is more robust than the widely referenced "pitch-fork" bifurcation depicted by C. H. Waddington. In the presence of noise, the two types of bifurcation are nearly indistinguishable. Conceptually, however, there is a clear identification of the "old" species A and the newly appearing species B. Figure reproduced from [128] with permission.

molecules. This reduces 10^8 sec to 10^{-7} sec, if one has single-molecule detection sensitivity, e.g., observing the transition of one out of 10^{15}. The rates are expected to be Arrhenius-like, with inverse temperature replaced by the size of the population.

One never has and does not need a full landscape; most of the time, only a relatively small portion of it is relevant. This can be explored at a much shorter time scale. Current cellular studies based on cytometry with single-cell sensitivity [168] that measures millions of cells in a population at a single time point may approximate ergodicity for a given extracellular condition [140]. To capture meaningful hillside dynamics, however, one needs to have a synchronized cell population starting at unique initial states. This has been done, for example, on the differentiation path of mouse embryos from zygote to blastocyst [109].

An emergent landscape is not the "cause" of biological dynamics. It is a summary of all its "potential transient behaviors" in statistical terms. Here, we emphasize two insights from the mathematical theory of probability. First, statistical certainty concerning a population of individuals gives very little predictive power on anyone in the population; there is an uncomfortable dualism between "statistical truth" and individual reality. Biochemical kinetics in a test tube has a deterministic description, but single enzyme molecules fluctuate with "individualism". Second, stochastic dynamics with small fluctuations can undergo movements that have large deviations away from its average behavior. Interestingly, while such movements are rare, when it occurs, the sequential events leading to the rare outcome are nearly deterministic, since any other possible sequence of events is much less probable, relatively speaking. In other words, retrospective reconstruction of the history of a rare event that occurred can be made with almost 100% confidence in the context of NSDS.

Chapter 16
Nonlinear Stochastic Dynamics of Complex Individuals and Systems: New Paradigm and Syntheses

What is living matter? Our current working definition is that living matter consists of complex, energy-consuming individuals who can be further classified into *kinetic species*. If the population of each species undergoes birth (autocatalytic synthesis) and death (degradation) but otherwise all the individuals are noninteracting, then the system behaves almost like an "ideal gas", or more appropriately an "ideal solution" [118, 20]. Nevertheless, even at this simplest level, the very fact of continuous population growth gives rise to a profound new phenomenon that R. A. Fisher (1890–1962) identified as the *fundamental theorem of natural selection* (FTNS) [74, 218]. This is still a debated subject within the theory of population genetics; see [78] for a very informative discussion. Section 16.1 contains a brief introduction of the FTNS from the perspective of the present book. When spontaneous transitions are allowed among subpopulations, e.g., phenotype switching among cells, the concept of *quasi-species* emerges [58, 283, 138].

For systems of interacting individuals, two quite different mathematical descriptions have been extensively developed in the context of stochastic models. These descriptions originated from chemical kinetics in a continuous stirred-tank reactor (CSTR) [5] and from the Ising model for ferromagnetism of solids [170]. In the former, rapid spatial movement of all the "individual molecules" in an aqueous solution leads to the assumption that every individual collides with every other individual constantly, and various "reactions" occur randomly. In the Ising model, all "individual atoms" are located at fixed lattice points in a solid, and each one only interacts with its neighbors. The macroscopic limit of such an interacting particle system as the lattice size tends to zero is known as the *hydrodynamic limit*. Because of the highly technical nature of stochastic modeling based on the mathematics of infinite-dimensional interacting particle systems or stochastic partial differential equations, the present book exclusively focuses on the former model: the finite-dimensional chemical kinetic approach to nonlinear stochastic dynamics.

In the classical physics of inanimate matter, different macroscopic states are best understood and characterized through a *phase transition* [221]. In terms of nonlinear stochastic dynamics, Chapter 8 showed us that bistability and saddle-node bifurcations with catastrophe are key signatures of a macroscopic phase transition on a short time scale for both equilibrium and nonequilibrium systems alike. Chapters 13 and 15 further suggested that the emergence of far-from-equilibrium active matter in complex systems, and even the very processes of development and speciation in biology, all share a unified nonlinear stochastic dynamical system (NSDS) principle [220, 258, 237].

In our "definition" of complex systems and the analyses we performed, we integrated thoughts and methods from several major areas of studies that are widely acknowledged in the theory of complexity. These include nonequilibrium thermodynamics, statistical physics, nonlinear dynamics, stochastic processes, and information theory. The focus of our approach is to establish a coherent narrative, in mathematical terms, that synthesizes a wide range of existing thoughts and methodologies found in others' work. In this closing chapter, we discuss a series of theories on complexity. Without being exhaustive, they are:

The Belgian–Dutch school of thermodynamics, which describes nonequilibrium phenomena in terms of chemical affinity and entropy production [190]. This includes the formulation for continuous media based on a double foundation of the local equilibrium hypothesis and Onsager's principle combined with the continuity equation for matter, energy, and entropy [46]. The theory has been expanded recently to include interfaces [150];

The stochastic process approach to nonequilibrium ensemble theory and entropy production [39, 40, 25, 161];

The theory of synergetics, which articulates the ideas of nonequilibrium potential and phase transition as well as the slaving principle as an emergent phenomenon near a critical point [107, 111];

The steady-state cycle kinetic and thermodynamic formalism for free energy transduction [116, 119];

The idea of symmetry breaking from the phase transition lore in condensed matter physics [7, 123, 160];

Catastrophe theory in connection to nonlinear bifurcation [280];

The nonlinear dynamic theory of complex systems biology based on constructive biology [141, 142];

The theory of replicator dynamics and the quasispecies model [57, 58].

In the sections below, we shall give brief outlines of each of these theories.

16.1 Fundamental Theorem of Natural Selection and Price's Equation

16.1.1 The Mathematical Ideas of the FTNS and LDT

It is now well accepted among experts that certain key ideas of the large deviations theory (LDT), which is discussed in Chapter 4, were already contained in the work of Boltzmann in 1877 [48]. What is less known is that the LDT also shares some essential mathematical ideas with Fisher's fundamental theorem of natural selection (FTNS). Here, we give an illustration of this connection according to our own interpretation.

First, according to S. R. S. Varadhan [263], the LDT, in a nutshell, is concerned with the equality between the following limits about a sequence of probability measures,

$$\lim_{n \to \infty} \frac{1}{n} \ln P_n \left(A \bigcup B \right) = \max \left\{ \lim_{n \to \infty} \frac{1}{n} \ln P_n(A), \lim_{n \to \infty} \frac{1}{n} \ln P_n(A) \right\}, \tag{16.1}$$

for any two sets A and B. Now let us consider the size of a mixed population of two kinetic species,

$$u(t) \triangleq a_1 e^{\lambda_1 t} + a_2 e^{\lambda_2 t} \tag{16.2}$$

as a function of t, $a_1, a_2 > 0$. The FTNS says that the per capita growth rate of the total population,

$$\frac{1}{u(t)} \left(\frac{du(t)}{dt} \right) = \frac{\lambda_1 x_1 + \lambda_2 x_2}{x_1 + x_2} \triangleq \overline{\lambda}(t), \tag{16.3}$$

in which $x_1(t) = a_1 e^{\lambda_1 t}$ and $x_2(t) = a_2 e^{\lambda_2 t}$ represent two noninteracting subpopulations with constant per capita growth rates λ_1 and λ_2, respectively. Then, $\overline{\lambda}$ is actually the weighted mean of the two per capita growth rates. Since $x_1, x_2 \geq 0$, one has $\overline{\lambda} \leq \max\{\lambda_1, \lambda_2\}$. Furthermore,

$$\frac{d\overline{\lambda}(t)}{dt} = \frac{d^2 u(t)}{dt^2} = \frac{(\lambda_1 - \overline{\lambda})^2 x_1 + (\lambda_2 - \overline{\lambda})^2 x_2}{x_1 + x_2}. \tag{16.4}$$

Therefore, the FTNS implies that $\overline{\lambda}(t)$ is an increasing function of t with a upper bound: it has a limit!

If we identify $x_1(t)$ and $x_2(t)$ as $P_n(A)$ and $P_n(B)$, with continuous t in place of discrete n, then the LDT in Eqn. (16.1) is intimately related to the following limit as $t \to \infty$:

$$\lim_{t \to \infty} \frac{\ln u(t)}{t} = \lim_{t \to \infty} \frac{d \ln u(t)}{dt} = \lim_{t \to \infty} \overline{\lambda}(t) = \max\{\lambda_1, \lambda_2\}. \tag{16.5}$$

We see that the ideas behind the FTNS and the LDT are intimately related: through Eqn. (16.4), the FTNS actually characterizes the *rate of convergence* of $\overline{\lambda}(t) \to \max\{\lambda_1, \lambda_2\}$, as a function of t toward its limit as $t \to \infty$. The LDT proper provides the existence of the limit of $u(t)/t$. Then, L'Hôspital's rule says that $\overline{\lambda}(t) \equiv du(t)/dt$ and $u(t)/t$ will have the same limit as $t \to \infty$.

In Fisher's [74], he used Eqns. (16.2)-(16.4) to mathematically frame Darwin's idea of evolution by natural selection: $\overline{\lambda}(t)$ is the *mean per capita growth rate weighted by the subpopulation size*, which is equal to the per capita growth rate of the total population with heterogeneous subgroups. λ, which was called the Malthusian parameter by Fisher, was considered a possible definition for *fitness*, a concept that first appeared in the work of A. R. Wallace. Eqn. (16.4) then shows beautifully that $d\overline{\lambda}(t)/dt$ is equal to the "variance" among the λ's. One also clearly sees that the FTNS excludes any interactions among individuals in the population, only variations upon which "the Nature selects".

With this newfound understanding of the deep connection between the FTNS and LDT and between the LDT and thermodynamics, one is awed by the statement on p. 36 of [74]: "It will be noticed that the fundamental theorem proved above bears some remarkable resemblances to the second law of thermodynamics."

16.1.2 The FTNS and Price's Equation

Price's equation generalizes the FTNS [207]. It shows that for more complex population dynamics with interactions and nonconstant per capita growth rates, the variance term discovered by Fisher is only one of the many elements that affects the per capita growth rate. However, while all the other factors originate from "outside" a species, Fisher's positive term originates from "within" the species!

We consider a population consisting of K subgroups with the kth subpopulation size being $y_k(t)$. The very definition of instantaneous per capita growth rates gives us a set of nonlinear kinetic equations:

$$\frac{1}{y_\ell}\frac{dy_\ell}{dt} = r_\ell(\vec{y}, t), \quad \ell = 1, 2, \cdots, K, \tag{16.6}$$

where $\vec{y} = (y_1, \cdots, y_K)$. Introducing the mean per capita growth rate for the total population:

$$\bar{r}(t) = \frac{1}{\sum_{\ell=1}^{K} y_\ell}\frac{d}{dt}\left(\sum_{\ell=1}^{K} y_\ell\right) = \frac{\sum_{\ell=1}^{K} r_\ell y_\ell}{\sum_{\ell=1}^{K} y_\ell}, \tag{16.7}$$

then,

$$\frac{d\bar{r}(t)}{dt} = \sigma_r^2 + \bar{\dot{r}}, \tag{16.8a}$$

where

$$\underbrace{\sigma_r^2 = \frac{\sum_{\ell=1}^{K} y_\ell(r_\ell - \bar{r})^2}{\sum_{\ell=1}^{K} y_\ell} \geq 0}_{\text{effect of natural selection upon variations}}, \quad \underbrace{\bar{\dot{r}} = \frac{\sum_{\ell=1}^{K} y_\ell(dr_\ell/dt)}{\sum_{\ell=1}^{K} y_\ell}}_{\text{effect of dynamic change}}. \tag{16.8b}$$

This is known as Price's equation [207]. Using (16.6) to represent population dynamics in ecology or allele frequencies in population genetics, (16.8) is necessarily true a priori. It is a law! Darwin's fundamental idea is thus contained in the dynamic formulation. In other words, as soon as one expresses population changes in terms of a set of differential equations such as (16.6), Price's result and the FTNS are automatically implied by virtue of the calculus [78]. The situation here has an uncanny resemblance to the chemical thermodynamics as we learned from Chapter 7: as soon as one expresses stochastic chemical kinetics in terms of the chemical master equation, Gibbs' theory is automatically implied by virtue of the mathematics.

16.1.3 Fundamental Mechanics of Natural Selection

Part of the confusion and difficulty with Fisher's FTNS is that the equation $d\bar{r}/dt = \overline{(r_i - \bar{r})^2} = \sigma_r^2 \geq 0$ is called a theorem. From (16.8a), we see that this equation is valid only for a constant per capita growth rate r_i; its range of validity is very restricted. Price's equation (16.8a) is a much more general result. However, the "fundamentalness" of Fisher's result is not about this equation *per se*; rather, it is about decomposing the internal vs. external causes, not very different from decomposing acceleration $a = F/m$. We shall illustrate this idea further.

Consider an infinitely large population of individuals in a species Y, with its population density $y(t)$ satisfying

$$\frac{1}{y(t)} \frac{dy(t)}{dt} = r\big(y, \vec{x}(t), \vec{z}(t)\big), \tag{16.9}$$

in which $\vec{x} = (x_1, \cdots, x_M)$ represents the populations of all the other M species that interact with Y, and $\vec{z}$ represents all the possible factors from the environment that impact the growth of Y. Eqn. (16.9) then defines the mathematical concept of the *per capita growth rate*, also called the Malthusian parameter by Fisher.

Is Eqn. (16.9) absolutely true? Did we neglect any other "mechanism" that could contribute to the function $r(y, \vec{x}, \vec{z})$? We note that both $\vec{x}$ and $\vec{z}$ are factors external to population Y, and $r(y, \vec{x}, \cdot)$ represents interspecies interactions. Could looking into more detailed compositions of Y provide additional mechanisms that might affect r?

Let us now divide species Y into K subgroups, as in Eqn. (16.6), with subpopulation densities $y_1(t)$ and $y_2(t), \cdots y_K(t)$ at time t. Then,

$$y(t) = \sum_{\ell=1}^{K} y_\ell(t),$$

$$\frac{dy_\ell}{dt} = r_\ell\big(\vec{y}, \vec{x}(t), \vec{z}(t)\big),$$

$$\frac{1}{y} \frac{dy(t)}{dt} = \frac{\sum_{\ell=1}^{K} r_\ell(\vec{y}, \vec{x}, \vec{z}) y_\ell(t)}{\sum_{i=1}^{N} y_\ell(t)} = \bar{r}\big(\vec{y}, \vec{x}, \vec{z}, t\big).$$

The last equation clearly shows that the $\bar{r}$ of the entire population **is** an explicit function of t in general. Furthermore,

$$\frac{d}{dt}r\big(\vec{y}(t),\vec{x}(t),\vec{z}(t),t\big) = \frac{\sum_{\ell=1}^{K}\big[r_\ell(\vec{y},\vec{x},\vec{z})-\bar{r}\big]^2 y_\ell(t)}{\sum_{\ell=1}^{K} y_\ell(t)} \tag{16.10a}$$

$$+ \text{ terms like } \left[\frac{\partial r_i}{\partial x_j}\right]\frac{dx_j}{dt},\ \left[\frac{\partial r_i}{\partial y_j}\right]\frac{dy_j}{dt}\ \text{ and }\ \left[\frac{\partial r_i}{\partial z_j}\right]\frac{dz_j}{dt}. \tag{16.10b}$$

The explicit dependence on t always causes $\bar{r}(\vec{y},\vec{x},\vec{z},t)$, the "fitness", to increase with time while the $\vec{x}$, $\vec{y}$, and $\vec{z}$ are fixed!

We believe this is the significance of Fisher's FTNS. In fact, the terms in (16.10b) are all "indirect", through $\vec{x}(t)$, $\vec{y}(t)$ or $\vec{z}(t)$. The term in (16.10a) is the *variance in fitness at that time*, as stated by Fisher, who identified the Malthusian parameter as Darwin's fitness. In fact, one can articulate the FTNS in terms of a *partial derivative* instead of the *total derivative*:

$$\left(\frac{\partial}{\partial t}r(\vec{y},\vec{x},\vec{z},t)\right)_{\vec{x},\vec{y},\vec{z}} = \frac{\sum_{\ell=1}^{K}\big[r_\ell(\vec{y},\vec{x},\vec{z})-\bar{r}\big]^2 y_\ell(t)}{\sum_{\ell=1}^{K} y_\ell(t)} = \sigma_r^2. \tag{16.11}$$

As a mathematical concept, the per capita growth rate of a population is an explicit function of time; the dependence is not due to any external factors but to internal heterogeneity, and the marginal function over time is nonnegative.[1]

The notions of ideal gases and ideal solutions have played important roles in the development of statistical theories of matters in physics and chemistry. The "ideal" behavior stems from i.i.d. statistics. The concept of "strictly natural" selection, we argue, should be associated with the "one-body" problem. Therefore, it exhibits an ideal behavior such as the FTNS. When there are "two-body" interactions, such as competition, predator-prey, etc., many more biological effects will change the story. One of the best known examples is diploid population, which might exhibit competition and symbiosis. Similarly, sexual selection could be understood as a special form of competition.

Fisher's theory, through the current presentation, establishes an "ideal" behavior without gene-gene interactions, upon which an understanding of more realistic biological systems can be based.

----

[1] In economics, the first-order partial derivatives of a multivariable function are called marginal functions.

16.2 Classical Nonequilibrium Thermodynamics (NET)

16.2.1 Macroscopic NET of Continuous Media

The main reference for this theory is the classic text by de Groot and Mazur [46]. It is a macroscopic formulation of nonequilibrium thermodynamics (NET) without stochasticity. Diffusion equations in three-dimensional physical space are used to describe transport phenomena. In this theory, (*i*) the existence of a special macroscopic function $s(x,t)$, the instantaneous entropy density, is hypothesized via the *local equilibrium assumption*: the fundamental equilibrium thermodynamic relation

$$dS = T^{-1}\left(dU + pdV - \sum_{i=1}^{K}\mu_i dN_i\right)$$

becomes

$$\frac{\partial s(x,t)}{\partial t} = T^{-1}(x,t)\left[\frac{\partial u(x,t)}{\partial t} - \sum_{i=1}^{K}\mu_i(x,t)\frac{\partial c_i(x,t)}{\partial t}\right], \qquad (16.12)$$

in which $u(x,t)$ is the internal energy density, $c_i(x,t)$ is the concentration of the *i*th chemical species, and T and μ_i are temperature and chemical potentials, respectively. The theory of NET then proceeds as follows:

(*ii*) Establishing continuity equations for $u(x,t)$ and $c_i(x,t)$:

$$\frac{\partial u(x,t)}{\partial t} = -\frac{\partial J_u(x,t)}{\partial x}, \qquad (16.13a)$$

$$\frac{\partial c_i(x,t)}{\partial t} = -\frac{\partial J_i(x,t)}{\partial x} + \sum_{j=1}^{M}v_{ji}r_j, \qquad (16.13b)$$

where J_u and J_i are energy and particle fluxes in space, and r_j is the rate of the *j*th reaction with stoichiometric coefficients v_{ji};

(*iii*) Substituting the *J*'s and *r*'s into (16.12);

(*iv*) Grouping appropriate terms to obtain the density of the entropy production rate, $\sigma(x,t)$, as "transport flux × thermodynamics force", *à la* Onsager's principle. The remaining part is the entropy exchange flux $J_s(x,t)$:

$$\frac{\partial s(x,t)}{\partial t} = \sigma(x,t) - \frac{\partial J_s(x,t)}{\partial x}. \qquad (16.13c)$$

The set of equations in (16.13) forms a complete system describing thermodynamics of transport, with the σ playing a central role in the entropy balance equation.

As the *constitutive relations* in continuum mechanics, the functional forms of the transport and transformation fluxes, the J's and r's respectively, have to be furnished in applications of this theory to engineering problems with domain specific knowledge.

16.2.2 Mesoscopic NET With Fluctuating Variables

Initiated by Prigogine and Mazur [209], this line of research now bears the name "mesoscopic nonequilibrium thermodynamics". See [232] for a review on the subject. The x in the macroscopic NET in Section 16.2.1 represents the real three-dimensional physical space. One can apply a similar approach to the dynamics of the internal degrees of freedom in the state space of a system [176, 177]. There is only one quantity that follows a "conservation law" in the state space: probability. This is the same starting point of Bergmann and Lebowitz's stochastic Liouville dynamics, to be discussed next in Section 16.2.3. The theory of mesoscopic nonequilibrium thermodynamics (meso-NET) proceeds as follows.

(*i*) Introducing the entropy, or free energy, as a functional of the probability distribution $p_\alpha(t)$, $\alpha \in \mathscr{S}$ for a discrete system or the probability density function $f(\mathbf{x},t)$, $\mathbf{x} \in \mathbb{R}^n$ for a continuous system. The introduction of the entropy function in mesoscopic theory does not rely on the local equilibrium assumption. Rather, one follows the fundamental idea of L. Boltzmann: the entropy is a functional characterizing the *statistics* of phase space. One such functional for stochastic processes is the Shannon entropy;

(*ii*) Establishing the continuity equation in phase space connecting the probability distribution and probability flux:

$$\frac{\mathrm{d}p_\alpha(t)}{\mathrm{d}t} = \sum_{\beta \in \mathscr{S}} \Big(J_{\beta \to \alpha}(t) - J_{\alpha \to \beta}(t) \Big), \tag{16.14a}$$

$$\frac{\partial f(\mathbf{x},t)}{\partial t} = -\sum_i \frac{\partial}{\partial x_i} J_i(\mathbf{x},t); \tag{16.14b}$$

(*iii*) Computing the time derivative of $\frac{\mathrm{d}}{\mathrm{d}t}S[p_n(t)]$, or $\frac{\mathrm{d}}{\mathrm{d}t}S[f(\mathbf{x},t)]$, following the chain rule for differentiation;

(*iv*) Setting up the entropy production rate σ in terms of the appropriately identified bilinear products of "thermodynamic fluxes" and "thermodynamic forces".

While the meso-NET, the theory of stochastic Liouville dynamics in Section 16.2.3, and the main thread of the present book are all based on the dynamics described by a distribution function in the state space, their scientific narratives are quite different. The goal of the meso-NET is to *derive* the dynamics equations for stochastic fluctuations of macroscopic quantities: more specifically, the Fokker–Planck equation for the probability density function $f(\mathbf{x}, t)$. In contrast, we take a stochastic description of dynamics in state space, and the corresponding Chapman–Kolmogorov equation, as given.

We continue with Eqn. (16.14b) in a phase space, and recall the free energy dissipation rate from Chapter 6:

$$-\frac{d}{dt}\int_{\mathbb{R}^n} f(\mathbf{x},t)\ln\left(\frac{f(\mathbf{x},t)}{\pi(\mathbf{x})}\right)d\mathbf{x} = \int_{\mathbb{R}^n} f(\mathbf{x},t)\sigma^{(na)}(\mathbf{x},t)d\mathbf{x}, \qquad (16.15a)$$

in which the *local density* of the nonadiabatic entropy production rate can be identified:

$$\sigma^{(na)}(\mathbf{x},t) = -\sum_i J_i(\mathbf{x},t)\frac{\partial}{\partial x_i}\ln\left(\frac{f(\mathbf{x},t)}{\pi(\mathbf{x})}\right), \qquad (16.15b)$$

where $J(x,t)$ is the flux and $\nabla\ln\left[f(x,t)/\pi(x)\right]$ is the thermodynamic force: the gradient of the local chemical potential function. One also notices that the thermodynamic force $\nabla\ln\left[f(\mathbf{x},t)/\pi(\mathbf{x})\right]$ can be further decomposed into $\mathbf{X} + \nabla\ln f(\mathbf{x},t)$, where $\mathbf{X} = -\nabla\ln\pi(\mathbf{x})$ is the thermodynamic variable conjugate to $\mathbf{x}$. From Chapter 2, we recognize this as $\Delta\mu = \Delta\mu^o + k_BT\ln(\text{concentration ratio})$, where $\Delta\mu^o = k_BT\ln(\text{equilibrium concentration ratio})$.

Now here comes the major point of departure for meso-NET: instead of assuming the Chapman–Kolmogorov equation for Markov dynamics, meso-NET evokes Onsager's linear force-flux relation as a postulate,

$$J_i(\mathbf{x},t) = -\sum_j D_{ij}(\mathbf{x})\frac{\partial}{\partial x_j}\ln\left(\frac{f(\mathbf{x},t)}{\pi(\mathbf{x})}\right). \qquad (16.16)$$

This relation is simply a generalized form of Fick's law. Then, it is well known that combining the continuity equation (16.14b) with Fick's law (16.16) yields the Fokker–Planck equation.

"Justification" the stochastic descriptions of mesoscopic dynamics is one of the fundamental tasks of statistical physics. In addition to this meso-NET approach, there are many others, including the Bogoliubov–Born–Green–Kirkwood–Yvon

(BBGKY) hierarchy, Mori–Zwanzig (MZ) projection, Markov partition and the Kolmogorov–Sinai entropy method.

Since the meso-NET approach is based on Onsager's linear relation, its validity is limited in the linear irreversible regime. For stochastic kinetics in a discrete state space with probability $p_i(t)$ and transition rate q_{ij}, flux $J_{ij} = p_i q_{ij} - p_j q_{ji}$ is **not** linearly related to the thermodynamic potential difference $\Delta\mu = \ln(p_i q_{ij}/p_j q_{ji})$. The same logic does not work for the stochastic dynamics of mesoscopic chemical kinetics.

16.2.3 Stochastic Liouville Dynamics

Bergmann and Lebowitz initiated a new approach to nonequilibrium processes in the phase space of a Hamiltonian mechanical system that is in contact with one or multiple heat baths [25, 161]:

$$\frac{\partial p(\mathbf{x},t)}{\partial t} + \left\{ p(\mathbf{x},t), H(\mathbf{x}) \right\}_{\mathbf{x}} = \int \left[K(\mathbf{x},\mathbf{x}')p(\mathbf{x}',t) - K(\mathbf{x}',\mathbf{x})p(\mathbf{x},t) \right] d\mathbf{x}'. \quad (16.17a)$$

in which $\{p,H\}$ is the Poisson bracket, and the rhs represents the stochastic encounter with the heat bath(s). By introducing the Helmholtz potential function

$$F[p(\mathbf{x})] = \int_{\mathbb{R}^n} p(\mathbf{x},t) \left(H(\mathbf{x}) + \beta^{-1} \ln p(\mathbf{x},t) \right) d\mathbf{x}, \quad (16.17b)$$

they showed the free energy dissipation rate $f_d = -\frac{dF}{dt} \equiv \frac{dS}{dt} - \beta \frac{dU}{dt} \geq 0$ if the heat baths have a common temperature β^{-1}. If not, then

$$f_d = \frac{dS}{dt} - \sum_{i=1}^{K} \beta_i \Phi_i, \quad (16.17c)$$

where Φ_i is the mean rate of energy flow from the ith reservoir to the system: $\frac{dU}{dt} = \sum_{i=1}^{K} \Phi_i$. Furthermore, in a nonequilibrium stationary state, $f_d = \sum_{i,j=1}^{K'} (\beta_i - \beta_j) \Phi_j$, where $K' < K$ are the numbers of linearly independent energy fluxes.

16.3 The Theory of Complex Systems Biology

The Complex Systems Biology approach [142] draws its main inspiration from the modern theory of nonlinear dynamical systems, which shows how a physically meaningful invariant probability measure arises in a deterministic dynamical system [54, 159]. With a few types of elements (e.g., very low heterogeneity among the individuals) and a few simple rules for interactions, complex behavior can arise through dynamic iterations. The objective of this school is to define "emergent universality" rather than searching for the principle of "self-organization". This is defined as follows [141]:

> The approach that should be taken will be constructive in nature. We combine several basic processes and construct a class of models and to find universal logic underlying therein. With this logic, biological systems are classified into some universality classes. [An] organism; then, [is] understood as one representative for a universal class, to which the "life as it could be" also belongs. d
>
> ...
>
> Note that the approaches for complex and complicated systems should be distinguished. Since the latter are essentially understood as a combination of simple processes, what should be done here is to search for minimal sets of local processes that can fit real data. On the other hand, for complex systems [(...)], such an approach is not effective. One has to search for a general logic why such a complex system is of necessity and universal.

The first paragraph does not give a clear distinction between self-organization and emergent universality, *per se*. The real difference resides in the line "why such... is of necessity and universal". This is the ultimate question for biological science [149]. The answer to this question can be precisely represented in terms of a "landscape", which quantifies "plausibility". Indeed, the logic of the theory of probability requires consideration of "all possible outcomes and their probabilities" [133]: the question of necessity is nothing but an overwhelming probability of 1, and the issue of universality, such as the theories of thermodynamics and phase transition, has been most cogently represented as limit theorems in probability [259, 253, 139]. In a nutshell, the "constructive" nature of a theory for an individual can be reduced (or should be expanded?) into "understanding complex systems ensembles and their stochastic dynamics", as *statistical laws*. We hasten to add that one very successful mathematical approach to even purely deterministic complex dynamics is their statistical characterization [54, 159].

Many concepts in CSB are not familiar to the readers of the present book, which is built upon a worldview based on probability and statistics.However, it will be helpful to see possible correspondences between some of the key notions in CSB and NSDS: isologous diversification $\leftrightarrow$ symmetry breaking and bifurcation; dynamic consolidation $\leftrightarrow$ attractors and multistability; itinerancy $\leftrightarrow$ emergent inter-basin Markov jumps; and minority control $\leftrightarrow$ stochasticity dictated by low copy numbers.

16.4 The Theory of Replicator Dynamics

This theory, initiated by physical chemists M. Eigen and P. Schuster in the 1970s, is essentially a chemical kinetic formalism of population dynamics with birth, death, and transmutation [57, 58, 121]. As an experimental physical chemist, Eigen received the Nobel Prize in Chemistry in 1967 "for his studies of extremely fast chemical reactions, effected by disturbing the equilibrium by means of very short pulses of energy." [56] The theory focuses on fundamental questions of the origin of life [42] and biomolecular evolution and introduces the concept of *quasispecies*. This concept has become an important theme in current research on cancer evolution [193, 110, 138].

The number of emergent, discrete phenotypes of a biochemical network should not, with sufficient robustness, be overwhelmingly large. We refer our readers to an earlier work in this vein in protein physical chemistry [167]. Indeed, even though combinations of nucleotide mutations in DNA have an astronomically large number of possibilities, if functional protein three-dimensional folds and biochemical network dynamics are taken into consideration, the relevant possible outcomes of mutations should also be limited, e.g., the genotype-to-phenotype map has a great deal of degeneracy. If we assume that the set of all possible discrete phenotypes $\mathscr{S}$, e.g., attractors, is finite, then at a coarse-grained level, the dynamics of subpopulations within an organism can be represented as

$$\frac{dx_i}{dt} = \left(A_i Q_i - \tilde{D}_i\right)x_i + \sum_{j \in \mathscr{S}, j \neq i} \left[A_j Q_{ji} x_j + \tilde{w}_{ji} x_j - \tilde{w}_{ij} x_i\right], \qquad (16.18)$$

in which A_i and $\tilde{D}_i$ are the per capita birth and death rates of the ith subpopulation, $0 \leq Q_i \leq 1$ represents the proportion that has an exact reproduction, and $0 \leq Q_{ij} \leq 1$ represents "erroneous" reproduction giving birth to a jth-type individual:

$$Q_i + \sum_{j\in\mathscr{S},j\neq i} Q_{ij} = 1.$$

$\tilde{w}_{ji}$ characterizes the rates for phenotype switching from j to i. This dynamical model, originally proposed to account for mutational dynamics, is equally applicable to epigenetic switching among attractors. Eqn. (16.18) is again based on the balance of headcounts. In general, all the A's, Q's, and D's are functions of the x's.

Eqn. (16.18) can be rewritten as the standard of the replicator dynamics [57]:

$$\frac{dx_i}{dt} = \left(A_i Q_i - D_i\right)x_i + \sum_{j\in\mathscr{S},j\neq i} w_{ji}x_j, \qquad (16.19a)$$

with

$$D_i = \tilde{D}_i + \sum_{j\in\mathscr{S},j\neq i} \tilde{w}_{ij}, \quad w_{ji} = A_j Q_{ji} + \tilde{w}_{ji}. \qquad (16.19b)$$

In (16.19), the per capita death rate D_i now contains the phenotype transitions from the ith subpopulation to all the other subpopulations $j \neq i$. Similarly, the "mutation rate" w_{ji} includes all the asymmetric divisions with "erroneous" replication as well as phenotypic transitions. Differentiating these different effects from dynamics requires high-precision, single-cell measurements.

Eqn. (16.18) can also be rewritten in a third form:

$$\frac{dx_i}{dt} = \left(A_i - \tilde{D}_i\right)x_i + \sum_{j\in\mathscr{S},j\neq i} \left(w_{ji}x_j - w_{ij}x_i\right), \qquad (16.20)$$

in which A_i and $\tilde{D}_i$ are the per capita birth and death rates, respectively, irrespective of reproduction errors, and w_{ij} contains the effects of all the asymmetric divisions and phenotypic transitions. This equation was the starting point of our earlier studies [283, 218]. At the population level, one cannot distinguish between asymmetric division, e.g., Eqn. (16.18), and symmetric division followed by a phenotypic transition, as in Eqn. (16.20).

When stochasticity is introduced into replicator dynamics, it merges seamlessly with an NSDS [235, 121].

16.5 Outlooks

The entire Newtonian physics finds its foundation in the precise concepts of smooth trajectories and differential equations. Two essential elements are hidden in this "worldview": over a sufficiently short duration, changes are smooth and deterministic. To shake such a worldview amounts to a wholesale "change of religion"; this has been amply witnessed through the difficult and tortuous history of developing the concept of the *fractal* as a contradistinction to smoothness, as testified by B. B. Mandelbrot (1924–2010) [174]. Even though the theory of probability and statistics has already been widely employed in solving scientific and engineering problems, we are still learning to think about science in terms of the set of new abstract mathematical concepts from the theory of probability [133]. Information theory, another branch of studies that is based on the same mathematics, fascinates many. At this juncture, it will be illuminating to quote an excerpt from E. T. Jaynes, the master of information theory, on the depth of the mathematics [133]:

> From many years of experience with its applications in hundreds of real problems, our views on the foundations of probability theory have evolved into something quite complex, which cannot be described in any such simplistic terms as 'pro-this' or 'anti-that'. For example, our system of probability could hardly be more different from that of Kolmogorov in style, philosophy, and purpose. What we consider to be fully half of probability theory as it is needed in current applications — the principles for assigning probabilities by logical analysis of incomplete information is not present at all in the Kolmogorov system.
>
> However, when all is said and done, we find ourselves, to our own surprise, in agreement with Kolmogorov and in disagreement with its critics, on nearly all technical issues.

We hope the present book provides a concrete first step for our readers to rethink complex biological systems and processes in terms of stochastic kinetics, starting from biochemical macromolecules and their conformational transitions. There are many areas that deserve to be investigated with this newfound perspective. At the very small scale, one immediately thinks of the exciting area of the *Markov state model* for molecular dynamics [210]; in connection to our brain, one can think of the *free energy principle* [82]. There is of course the natural next step beyond the present book, integrating Belgian–Dutch school's physicochemical concept of entropy and entropy production with Eigen-Schuster's approach to biological fitness and evolution in terms of modern mathematical concepts from the theory of probability and stochastic dynamics [225], as championed by P. Ao [8, 9].

As we have mentioned, the models discussed in the present book are analogous to "ordinary differential equations-based modeling". To move into stochastic spatiotemporal dynamics, one can simply compare and contrast the two volumes of J. D. Murray's magnum opus [185, 186] for inspiration, and consult [52, 30] for more.

References

---------------- A ----------------

1. Acar, M., Mettetal, J. T. and van Oudenaarden, A. (2008) Stochastic switching as a survival strategy in fluctuating environments. *Nat. Genet.*, **40**, 471–475.
2. Agazzi, A., Dembo, A. and Eckmann, J.P. (2018) Large Deviations Theory for Markov Jump Models of Chemical Reaction Networks. *Ann. Appl. Prob.* **28** (3), 1821–1855
3. Alberty, R. A. (2003) *Thermodynamics of Biochemical Reactions*, Wiley Interscience, New York.
4. Anderson, D. F. (2011) A proof of the global attractor conjecture in the single linkage class case. *SIAM J. Appl. Math.* **71**, 1487–1508.
5. Anderson, D. F. and Kurtz, T. G. (2015) *Stochastic Analysis of Biochemical Systems*, Springer, New York.
6. Anderson, D. F., Craciun, G., Gopalkrishnan, M. and Wiuf, C. (2015) Lyapunov functions, stationary distributions, and non-equilibrium potential for reaction networks. *Bullet. Math. Biol.*, **77**, 1744–1767.
7. Anderson, P. W. (1972) More is different: Broken symmetry and the nature of the hierarchical structure of science. *Science* **177**, 393–396.
8. Ao, P. (2005) Laws in Darwinian evolutionary theory. *Phys. Life Rev.* **2**, 117–156.
9. Ao, P. (2008) Emerging of stochastic dynamical equalities and steady state thermodynamics from Darwinian dynamics. *Commun. Theor. Phys.* **49**, 1073–1090.
10. Aris, R., Aronson, D. G. and Swinney, H. L. (eds.) (1991) *Patterns and Dynamics in Reactive Media*, Springer, New York.
11. Axelrod, D., Koppel, D. E., Schlessinger, J., Elson, E. L. and Webb, W. W. (1976) Mobility measurement by analysis of fluorescence photobleaching recovery kinetics. *Biophys. J.* **16**, 1055–1069.
12. Axelrod, D., Elson, E. L., Schlessinger, J. and Koppel, D. E. (2018) Reminiscences on the "classic" 1976 FRAP article. *Biophys. J.* **115**, 1156–1159.

---------------- B ----------------

13. Bai, S., Ge, H. and Qian, H. (2018) Structure for energy cycle: a unique status of the Second Law of thermodynamics for living systems. *Sci. China Life Sci.* **61**, 1266–1273.
14. Barrick, D. E. (2018) *Biomolecular Thermodynamics: From Theory to Application*, CRC Press, Boca Raton.
15. Bak, P. and Sneppen, K. (1993) Punctuated equilibrium and criticality in a simple model of evolution. *Phys. Rev. Lett.* **71**, 4083–4086.
16. Beard, D. A. and Qian, H. (2007) Relationship between thermodynamic driving force and one-way fluxes in reversible chemical reactions. *PLoS ONE*, **2**, e144.

H. Qian, H. Ge, *Stochastic Chemical Reaction Systems in Biology*,
Lecture Notes on Mathematical Modelling in the Life Sciences,
https://doi.org/10.1007/978-3-030-86252-7

17. Beard, D. A. and Qian, H. (2008) *Chemical Biophysics: Quantitative Analysis of Cellular Systems*, Cambridge Univ. Press, U.K.
18. Beard, D. A., Babson, E., Curtis, E. and Qian, H. (2004) Thermodynamic constraints for biochemical neworks. *J. Theoret. Biol.* **228**, 327–333.
19. Ben-Naim, A. (1987) Mixing and assimilation in systems of interaction particles. *Am. J. Phys.* **55**, 1105–1109.
20. Ben-Naim, A. (1992) *Statistical Thermodynamics for Chemists and Biochemists*, Plenum, New York.
21. Ben-Naim, A. (2008) *Statistical Thermodynamics Based on Information: A Farewell to Entropy*, World Scientific, Singapore.
22. Bender, C. M. and Orszag, S. A. (1978) *Advanced Mathematical Methods for Scientists and Engineers*, McGraw-Hill, New York.
23. Bennett, C. H. (2003) Notes on Landauer's principle, reversible computation and Maxwell's demon. *Stud. Hist. Phil. Mod. Phys.* **34**, 501–510.
24. Berg, O. (1978) A model for the statistical fluctuations of protein numbers in a microbial population. *J. Theoret. Biol.* **71**, 587–603.
25. Bergmann, P. G. and Lebowitz, J. L. (1955) New approach to nonequilibrium processes. *Phys. Rev.* **99**, 578–587.
26. Bialek, W. (2012) *Biophysics: Searching for Principles*, Princeton Univ. Press, N.J.
27. Bishop, L. M. and Qian, H. (2010) Stochastic bistability and bifurcation in a mesoscopic signaling system with autocatalytic kinase. *Biophys. J.* **98**, 1–11.
28. Boltz, A. B., Kalos, M. H. and Lebowitz, J. L. (1975) A new algorithm for Monte Carlo simulation of Ising spin systems, *J. Comput. Phys.*, **17**, 10–18.
29. Botts, J. and Morales, M. (1953) Analytical description of the effects of modifiers and of enzyme multivalency upon the steady state catalyzed reaction rate. *Trans. Faraday Soc.* **49**, 696–707.
30. Bressloff, P. C. (2014) *Stochastic Processes in Cell Biology*, Springer, New York.
31. Briggs, G. E. and Haldane, J. B. (1925) A note on the kinetics of enzyme action. *Biochem J* **19**, 338–339.

——————— **C** ———————

32. Cai, L., Friedman, N. and Xie, X. S. (2006) Stochastic protein expression in individual cells at the single molecule level. *Nature* **440**, 358–36.
33. Carpinteri, A. and Paggi, M. (2014) Langrange and his Mécanique Analytique: from Kantian noumenon to present applications. *Meccanica* **49**, 1–11.
34. Chang, H. H., Hemberg, M., Barahona, M., Ingber, D. E. and Huang, S. (2008) Transcriptome-wide noise controls lineage choice in mammalian progenitor cells. *Nature*, **453**, 544–547.
35. Chen, B.-L. Baase, W. A., Nicholson, H. and Schellman, J. A. (1992) Folding kinetics of T4 lysozyme and nine mutants at 12 °C. *Biochem.* **31**, 1464–1476.
36. Chibbaro, S., Rondoni, L. and Vulpiani, A. (2014) *Reductionism, Emergence and Levels of Reality*, Springer, New York.
37. Chong, S., Chen, C., Ge, H. and Xie, X.S. (2014) Mechanism of transcriptional bursting in bacteria. *Cell* **158**, 314–326.
38. Chow, S.-N., Huang, W., Li, Y. and Zhou, H. (2012) Fokker-Planck equations for a free energy functional or Markov process on a graph. *Arch. Rational Mech. Anal.* **203**, 969–1008.
39. Cox, R. T. (1950) The statistical method of Gibbs in irreversible change. *Rev. Mod. Phys.* **22**, 238–248.
40. Cox, R. T. (1952) Brownian motion in the theory of irreversible processes. *Rev. Mod. Phys.* **24**, 312–320.
41. Creighton, T. E. (ed.) (1992) *Protein Folding*. W. H. Freeman, New York.
42. Cronin, L. and Walker, S. I. (2016) Beyond prebiotic chemistry: What dynamic network properties allow the emergence of life? *Science*, **352**, 1174–1175.
43. Crooks G. E. (1999) Entropy production fluctuation theorem and the nonequilibrium work relation for free energy differences. *Phys. Rev. E* **60**, 2721–2726.

——————— **D** ———————

44. Day, M.V. (1983) On the exponential exit law on the small parameter exit problem. *Stochastics* **8**, 297–323.

45. de Gennes, P. G. (1997) Molecular individualism. *Science*, **276**, 1999–2000.

46. de Groot, S. R. and Mazur, P. (1984) *Nonequilibrium Thermodynamics.* Dover, New York.

47. Delbrück, M. (1940) Statistical fluctuations in autocatalytic reactions. *J. Chem. Phys.* **8**, 120–124.

48. Dembo, A. and Zeitouni, O. (2014) *Large Deviations Techniques and Applications*, 2nd ed. Springer, New York.

49. Dill, K. A. and Bromberg, S. (2010) *Molecular Driving Forces: Statistical Thermodynamics in Biology, Chemistry, Physics, and Nanoscience*, 2nd ed., Garland Science, New York.

50. Doob, J. L. (1945) Markoff chains – denumerable case. *Trans. Am. Math. Soc.* **58**, 455–473.

51. Doyle, P. G. and Snell, J. L. (1984) *Random Walks and Electric Networks*, Math. Asso. Am. Pub., Washington D.C.

52. Durrett, R. and Levin, S. A. (1994) Stochastic spatial models: A user's guide to ecological applications. *Phil. Trans. Roy. Soc. B.* **343**, 329–350.

53. Dyson, H. J. and Wright, P. E. (2005) Intrinsically unstructured proteins and their functions. *Nature Rev. Mol. Cell Biol.* **6**, 197–208.

——————— E ———————

54. Eckmann, J.-P. and D. Ruelle, D. (1985) Ergodic theory of chaos and strange attractors. *Rev. Mod. Phys.* **57**, 617–656.

55. Edman, L. (2000) Theory of fluorescence correlation spectroscopy on single molecules. *J. Phys. Chem. A* **104**, 6165–6170.

56. Eigen, M. (1972) Immeasurably fast reactions. In *Nobel Lectures in Chemistry 1963-1970*, Elsevier, Amsterdam, pp. 170-203.

57. Eigen, M. and Schuster, P. (1979) *Hypercycle: A Principle of Natural Self-Organization.* Springer-Verlag, Berlin.

58. Eigen, M., McCaskill, J. and Schuster, P. (1988) Molecular quasi-species. *J. Phys. Chem.* **92**, 6881–6891.

59. Eldar, A. and Elowitz, M. (2010) Functional roles for noise in genetic circuits. *Nature* **467**, 167–173.

60. Elson, E. L. (1988) Cellular mechanics as an indicator of cytoskeletal structure and function. *Annu. Rev. Biophys. Biophys. Chem.* **17**, 397–430.

61. Epstein, I. R. and Pojman, J. A. (1998) *An Introduction to Nonlinear Chemical Dynamics: Oscillations, Waves, Patterns, and Chaos*, Oxford Univ. Press, U.K.

62. Érdi, P. and Lente, G. (2014) *Stochastic Chemical Kinetics: Theory and (Mostly) Systems Biological Applications.* Springer-Verlag, New York.

63. Esposito, M., Harbola, U. and Mukamel, S. (2007) Entropy fluctuation theorems in driven open systems: Application to electron counting statistics. *Phys. Rev. E.* **76**, 031132.

64. Esposito, M. and van den Broeck, C. (2010) Three detailed fluctuation theorems. *Phys. Rev. Lett.* **104**, 090601.

65. Ethier, S. N. and Kurtz, T. G. (2009) *Markov Processes: Characterization and Convergnece*, John Wiley & Sons, New York

66. Ewens, W. J. (2010) *Mathematical Population Genetics 1: Theoretical Introduction*, 2nd Ed., Springer, New York.

——————— F ———————

67. Feinberg, M. (1972) Complex balancing in general kinetic systems. *Arch. Rational Mech. Anal.* **49**, 187–194.

68. Feinberg, M. (1991) Some recent results in chemical reaction network theory. In *Patterns and Dynamics in Reactive Media*, Aris, R., Aronson, D. G. and Swinney, H. L. eds., Springer-Verlag, New York, pp. 43–70.

69. Feng, J. and Kurtz, T. G. (2006) *Large Deviation for Stochastic Processes*, Math. Surveys and Monographs vol. 131, AMS Press, Providence.

70. Ferrell, J. E. (2012) Bistability, bifurcations, and Waddington's epigenetic landscape. *Curr. Biol.* **22**, R458-R466.

71. Fersht, A. (1985) *Enzyme Structure and Mechanism.* 2nd ed., W. H. Freeman, New York.

72. Feynman, R. (1965) *The Character of Physical Law. Modern Library*, The MIT Press, MA.

73. Fisher, M. E. (1998) Renormalization group theory: Its basis and formulation in statistical physics. *Rev. Mod. Phys.* **70**, 653–681.

74. Fisher, R. A. (1930) *The Genetical Theory of Natural Selection.* Oxford Univ. Press, London.

75. Florin, E. L., Moy, V.T. and Gaub, H.E. (1994) Adhesion between individual ligand receptor pair. *Science* **264**, 415–417.

76. Fontana, W. and Schuster, P. (1998) Continuity in evolution: On the nature of transitions. *Science* **280**, 1451–1455.

77. Fox, R. F. (1999) Joel E. Keizer, 1942-1999: Nonequilibrium statistical thermodynamics. http://www.fefox.com/ARTICLES/JEKNST.pdf.

78. Frank, S. A. (2011) Wright's adaptive landscape versus Fisher's fundamental theorem. In *The Adaptive Landscape in Evolutionary Biology*, Svensson, E. and Calsbeek, R. eds. Oxford Univ. Press, U.K., pp. 41–56.

79. Frauenfelder, H., Sligar, S. and Wolynes, P. G. (1991) The energy landscapes and motions of proteins. *Science* **254**, 1598–1603.

80. Freidlin, M. I. and Wentzell, A. D. (1984) *Random Perturbations of Dynamical Systems.* Spinger-Verlag, New York.

81. Frieden, C. (1970) Slow transitions and hysteretic behavior in enzyme. *Ann. Rev. Biochem.* **48**, 471–489.

82. Friston, K. (2010) The free-energy principle: a unified brain theory? *Nature Rev. Neurosci.* **11**, 127–138.

——————   G   ——————

83. Gadgil, C., Lee, C. H. and Othmer, H. G. (2005) A stochastic analysis of first-order reaction networks. *Bull. Math. Biol.* **67** 901–946.

84. Galstyan, V. and Saakian, D. B. (2012) Dynamics of the chemical master equation, a strip of chains of equations in d-dimensional space. *Phys. Rev. E.* **86**, 011125.

85. Gardiner, C. (2009) *Stochastic Methods: A Handbook for the Natural and Social Sciences*, 4th ed., Series in Synergetics, Springer, New York.

86. Ge, H. (2008) Waiting cycle times and generalized Haldane equality in the steady-state cycle kinetics of single enzymes. *J. Phys. Chem. B* **112**, 61–70.

87. Ge, H. (2009) Extended forms of the second law for general time-dependent stochastic processes. *Phys. Rev. E.* **80**, 021137.

88. Ge, H. and Jiang, D.-Q. (2007) The transient fluctuation theorem of sample entropy production for general stochastic processes. *J. Phys. A. Math. Th.* **40**, F713–F723.

89. Ge, H. and Qian, H. (2009) Thermodynamic limit of a nonequilibrium steady-state: Maxwell-type construction for a bistable biochemical system. *Phys. Rev. Lett.* **103**, 148103.

90. Ge, H. and Qian, H. (2010) The physical origins of entropy production, free energy dissipation and their mathematical representations. *Phys. Rev. E* **81**, 051133.

91. Ge, H. and Qian, H. (2011) Nonequilibrium phase transition in mesoscopic biochemical systems: From stochastic to nonlinear dynamics and beyond. *J. R. Soc. Interf.* **8**, 107–116.

92. Ge, H. and Qian, H. (2011) Maximum entropy principle, equal probability *a priori* and Gibbs paradox. arXiv:1105:4118.

93. Ge, H. and Qian, H. (2012) Analytical mechanics in stochastic dynamics: Most probable path, large-deviation rate function and Hamilton-Jacobi equation. *Intern. J. Mod. Phys.* **26**, 1230012.

94. Ge, H. and Qian, H. (2012) Landscapes of non-gradient dynamics without detailed balance: Stable limit cycles and multiple attractors. *Chaos*, **22**, 023140.

95. Ge, H. and Qian, H. (2013) Heat dissipation and nonequilibrium thermodynamics of quasi-steady states and open driven steady state. *Phys. Rev. E.* **87**, 062125.

96. Ge, H. and Qian, H. (2016) Nonequilibrium thermodynamic formalism of nonlinear chemical reaction systems with Waage–Guldberg's law of mass action. *Chem. Phys.* **472**, 241–248.

97. Ge, H. and Qian, H. (2016) Mesoscopic kinetic basis of macroscopic chemical thermodynamics: A mathematical theory. *Phys. Rev. E* **94**, 052150.
98. Ge, H. and Qian, H. (2017) Mathematical formalism of nonequilibrium thermodynamics for nonlinear chemical reaction systems with general rate law. *J. Stat. Phys.* **166**, 190–209.
99. Ge, H., Jia C. and Jiang, D.-Q. (2017) Cycle symmetry, limit theorems, and fluctuation theorems for diffusion processes on the circle. *Stoc. Proc. Appl.* **127**, 1897–1925.
100. Ge, H., Qian, M. and Qian, H. (2012) Stochastic theory of nonequilibrium steady states (Part II): Applications in chemical biophysics. *Phys. Rep.* **510**, 87–118.
101. Ge, H., Qian, H. and Xie, X.S. (2015) Stochastic phenotype transition of a single cell in an intermediate region of gene-state switching. *Phys. Rev. Lett.* **114**, 078101.
102. Ge, H., Wu, P., Qian, H. and Xie, X.S. (2018) Relatively slow stochastic gene-state switching in the presence of positive feedback significantly broadens the region of bimodality through stabilizing the uninduced phenotypic state. *PLoS Comput. Biol.* **14**, e1006051.
103. Gillespie, D. T. (1976) A general method for numerically simulating the stochastic time evolution of coupled chemical reactions. *J. Phys. Chem.* **22(4)**, 403–434.
104. Gillespie, D. T. (2007) Stochastic simulation of chemical kinetics. *Annu. Rev. Phys. Chem.* **58**, 35–55.
105. Golding, I., Paulsson, J., Zawilski, S. M., and Cox, E. C. (2005) Real-time kinetics of gene activity in individual bacteria. *Cell* **123**, 1025–1036.
106. Gopalkrishnan, M., Miller, E. and Shiu, A. (2014) A geometric approach to the global attractor conjecture. *SIAM J. Appl. Dyn. Syst.* **13**, 758–797.
107. Graham, R. and Haken, H. (1971) Generalized thermodynamic potential for Markoff systems in detailed balance and far from thermal equilibrium. *Zeitschrift für Physik* **243**, 289–302.
108. Gray, P. and Scott, S. K. (1994) *Chemical Oscillations and Instabilities: Non-linear Chemical Kinetics*, Clarendon Press, U. K.
109. Guo, G., Huss, M., Tong, G. Q., Wang, C., Sun, L. L., Clarke, N. D. and Robson, P. (2010) Resolution of cell fate decisions revealed by single-cell gene expression analysis from zygote to blastocyst. *Developmental Cell* **18**, 675–685.
110. Gupta, P. B., Fillmore, C. M., Jiang, G., ..., Tao, K., Kuperwasser, C. and Lander, E. S. (2011) Stochastic state transitions give rise to phenotypic equilibrium in populations of cancer cells. *Cell*, **146**, 633–644.

————— **H** —————

111. Haken, H. (1983) *Synergetics, An Introduction: Nonequilibrium Phase Transitions and Self-Organization in Physics, Chemistry, and Biology*, 3rd rev. enl. ed., pringer-Verlag, New York.
112. Harrington, K. J., Laughlin, R. B. and Liang, S. (2001) Balanced branching in transcription termination. *Proc. Natl. Acad. Sci. USA* **98**, 5019–5024.
113. Hartwell, L. H., Hopfield, J. J., Leibler, S. and Murray, A. W. (1999) From molecular to modular cell biology. *Nature*, **402**, C47–C52.
114. Hatano, T. and Sasa, S. (2001) Steady-state thermodynamics of Langevin systems. *Phys. Rev. Lett.* **86**, 3463–4366.
115. Heuett, W. J. and Qian, H. (2006) Grand canonical Markov model: A stochastic theory for open nonequilibrium biochemical networks. *J. Chem. Phys.* **124**, 044110.
116. Hill, T. L. (1977) *Free Energy Transduction in Biology: The Steady-State Kinetic and Thermodynamic Formalism.* Academic Press, New York.
117. Hill, T. L. (1983) Some general principles in free energy transduction. *Proc. Natl. Acad. Sci. USA* **80**, 2922–2925.
118. Hill, T. L. (1985) *Cooperativity Theory in Biochemistry: Steady-State and Equilbrium Systems*, Springer-Verlag, New York.
119. Hill, T. L. (1989) *Free Energy Transduction and Biochemical Cycle Kinetics*, Springer-Verlag, New York.
120. Hill, T. L. and Plesner, I. W. (1965) Studies in irreversible thermodynamics. II. a simple class of lattice models for open systems. *J. Chem. Phys.* **43**, 267–285.
121. Hofbauer, J. and Sigmund, K (1998) *Evolutionary Games and Population Dynamics.* Cambridge Univ. Press, U.K.

122. Hopfield, J. J. (1974) Kinetic proofreading: a new mechanism for reducing errors in biosynthetic processes requiring high specificity. *Proc. Natl. Acad. Sci. USA*, **71**, 4135–4139.

123. Hopfield, J. J. (1994) Physics, computation, and why biology looks so different. *J. Theret. Biol.* **171**, 53–60.

124. Horn, F. (1972) Necessary and sufficient conditions for complex balancing in chemical kinetics. *Arch. Rational Mech. Anal.* **49**, 172–186.

125. Horn, F. and Jackson, R. (1972) General mass action kinetics. *Arch. Rational Mech. Anal.* **47**, 81–116.

126. Hu, G. (1986) Lyapounov function and stationary probability distributions. *Zeit. Phys. B Condensed Matter* **65**, 103–106.

127. Huang, S., Eichler, G., Bar-Yam, Y. and Ingber, D. E. (2005) Cell fates as high-dimensional attractor states of a complex gene regulatory network. *Phys. Rev. Lett.* **94**, 128701.

128. Huang, S., Li, F., Zhou, J. X. and Qian, H. (2017) Processes on the emergent landscapes of biochemical reaction networks and heterogeneous cell population dynamics: differentiation in living matters. *J. R. Soc. Interface* **14**, 20170097.

———— **I, J** ————

129. Ikeda, N. and Watanabe, S. (1989) *Stochastic Differential Equations and Diffusion Processes, second ed.* North-Holland Publishing Company, New York.

130. Ishii, H. and Mitake, H. (2007) Representation formulas for solutions of Hamilton-Jacobi equations with convex Hamiltonians. *Indiana Univ. Math. J.* **56**, 2159–2184.

131. Iskratsch, T., Wolfenson, H. and Sheetz, M. P. (2014) Appreciating force and shape — the rise of mechanotransduction in cell biology. *Nature Rev. Mol. Cell Biol.* **15**, 825–833.

132. Jarzynski, C. (2000) Hamiltonian derivation of a detailed fluctuation theorem. *J. Stat. Phys.* **98**, 77–102.

133. Jaynes, E. T. (2003) *Probability Theory: The Logic of Science.* Cambridge Univ. Press, U.K.

134. Jia, C., Liu, X.-F., Qian, M.-P., Jiang, D.-Q. and Zhang, Y.-P. (2012) Kinetic behavior of the general modifier mechanism of Botts and Morales with non-equilibrium binding. *J. Theore. Biol.* **296**, 13–20.

135. Jia, C., Jiang, D.-Q. and Qian, M.-P. (2016) Cycle symmetries and circulation fluctuations for discrete-time and continuous-time Markov chains. *Annals Appl. Prob.* **26**, 2454–2493.

136. Jia, C., Liu, X.-F., Qian, M.-P., Jiang, D.-Q. and Zhang, Y.-P. (2012) Kinetic behavior of the general modifier mechanism of Botts and Morales with non-equilibrium binding. *J. Theoret. Biol.* **296**, 13–20.

137. Jiang, D. Q., Qian, M. and Qian, M.-P. (2004) *Mathematical Theory of Nonequilibrium Steady States*, Lecture Notes in Mathematics Vol. 1833. Springer-Verlag, Berlin.

138. Jiang, D. Q., Wang, Y. and Zhou, D. (2017) Phenotypic equilibrium as probabilistic convergence in multi-phenotype cell population dynamics. *PLoS ONE* **12**, e0170916.

139. Jona-Lasinio, G. (2001) Renormalization group and probability theory. *Phys. Rep.* **352**, 439–458.

———— **K** ————

140. Kafri, R., Levy, J., Ginzberg, M. B., Oh, S., Lahav, G. and Kirschner, M. W. (2013) Dynamics extracted from fixed cells reveal feedback linking cell growth to cell cycle. *Nature*, **494**, 480–483.

141. Kaneko, K. (1998) Life as complex systems: Viewpoint from intra-inter dynamics. *Complexity*, **3**, 53–60.

142. Kaneko, K. (2006) *Life: An Introduction to Complex Systems Biology.* Springer, New York.

143. Karatzas, I. and Shreve, S.E. (1998) *Brownian Motion and Stochastic Calculus, second ed.* Springer, New York.

144. Keener, J. and Sneyd, J. (1998) *Mathematical Physiology*, Springer, New York.

145. Keizer, J. (1987) *Statistical Thermodynamics of Nonequilibrium Processes*, Springer-Verlag, New York.

146. Kendall, D. G. (1949) Stochastic processes and population growth. *J. R. Stat. Soc. B* **11**, 230–264.

147. Khintchine, A. Ya. (1960) *Mathematical Methods in the Theory of Queueing*, (translated from Russian) Griffin, London.
148. Kirkwood, J. G. (1935) Statistical mechanics of fluid mixtures. *J. Chem. Phys.* **3**, 300–313.
149. Kirschner, M. W. and Gerhart, J. C. (2006) *The Plausibility of Life: Resolving Darwin's Dilemma.* Yale Univ. Press, New Haven.
150. Kjelstrup, S. and Bedeaux, D. (2008) *Non-Equilibrium Thermodynamics of Heterogeneous Systems*, World Scentific, Singapore.
151. Kot, M. (2001) *Elements of Mathematical Ecology*, Cambridge Univ. Press, U.K.
152. Kramers, H. A. (1940) Brownian motion in a field of force and the diffusion model of chemical reactions. *Physica* **7**, 284–304.
153. Kubo, R., Matsuo, K. and Kitahara, K. (1973) Fluctuation and relaxation of macrovariables. *J. Stat. Phys.* **9**, 51–96.
154. Kurtz, T. G. (1972) The relationship between stochastic and deterministic models for chemical reactions. *J. Chem. Phys.* **57**, 2976–2978.
155. Kurtz, T. G. (1978) Strong approximation theorems for density dependent Markov chains. *Stoc. Proc. Appl.* 6, 223-240
156. Kurtz, T. G. (1980) Representations of Markov processes as multiparameter time changes. *Ann. Prob.* **8**, 682–715.

——————— **L** ———————

157. Landauer, R. (1961) Irreversibility and heat generation in the computing process. *IBM J. Res.* **5**, 183–191.
158. LaSalle, J. P. (1986) *The Stability and Control of Discrete Processes.* Springer-Verlag, New York.
159. Lasota, A. and Mackey, M. C. (1994) *Chaos, Fractals, and Noise: Stochastic Aspects of Dynamics.* Springer, New York.
160. Laughlin, R. B., Pines, D., Schmalian, J., Stojković, B. P., and Wolynes, P. G. (2000) The middle way. *Proc. Natl. Acad. Sci. USA*, **97**, 32–37.
161. Lebowitz, J. L. and Bergmann, P. G. (1957) Irreversible Gibbsian ensemble. *Annals Phys.* **1**, 1–23.
162. Leontovich, M. A. (1935) Basic equations of kinetic gas theory from the viewpoint of the theory of random processes. *J. Exp. Theoret. Phys.* **5**, 211–231.
163. Levitt, M. (2001) The birth of computational structural biology. *Nature Struct. Biol.* **8**, 392–393.
164. Lewis, G. N. (1925) A new principle of equilibrium. *Proc. Natl. Acad. Sci. USA* **11** 179–183.
165. Li, F., Long, T., Lu, Y., Ouyang, Q. and Tang, C. (2004) The yeast cell-cycle network is robustly designed. *Proc. Natl. Acad. Sci. USA* **101**, 4781–4786.
166. Li, G.-W. and Xie, X. S. (2011) Central dogma at the single-molecule level in living cells. *Nature* **475**, 308–315.
167. Li, H., Helling, R., Tang, C. and Wingreen, N. (1996) Emergence of preferred structures in a simple model of protein folding. *Science* **273**, 666–669.
168. Li, Q., Wennborg, A., Aurell, E., Dekel, E., Zou, J.-Z., Xu, Y., Huang, S. and Ernberg, I. (2016) Dynamics inside the cancer cell attractor reveal cell heterogeneity, limits of stability, and escape. *Proc. Natl. Acad. Sci. USA* **113**, 2672–2677.
169. Liang, J. and Qian, H. (2010) Computational cellular dynamics based on the chemical master equation: A challenge for understanding complexity (Survey). *J. Comp. Sci. Tech.* **25**, 154–168.
170. Liggett, T. M. (1985) *Interacting Particle Systems.* Springer-Verlag, New York.
171. Lin, C. C. and Segel, L. A. (1974) *Mathematics Applied to Deterministic Problems in the Natural Sciences.* Macmillan, New York.
172. Lin, C. C. (1976) On the role of applied mathematics. *Advances in Mathematics*, **19**, 267–288.
173. Lu, H. P., Xun, L. and Xie, X. S. (1998) Single molecule enzymatic dynamics. *Science*, **282**, 1877–1882.

——————— **M** ———————

174. Mandelbrot, B. B. (1982) *The Fractal Geometry of Nature*. Times Books, New York.
175. Mazo, R. M. (2002) *Brownian Motion: Fluctuations, Dynamics, and Applications*. Oxford University Press, Oxford, U. K.
176. Mazur, P. (1998) Fluctuations and nonequilibrium thermodynamics. *Physica A* **261**, 451–457.
177. Mazur, P. (1999) Mesoscopic nonequilibrium thermodynamics; irreversible processes and fluctuations. *Physica A* **274**, 491–504.
178. Michaelis, L. and Menten, M. (1913) Die kinetik der invertinwirkung, *Biochemistry Zeitung* **49**, 333-369.
179. Mino, H., Rubinstein, J. T. and White, J. A. (2002) Comparison of algorithms for the simulation of action potentials with stochastic sodium channels. *Annals Biomed. Engr.* **30**, 578–587.
180. Mirowski, P. (1992) Do economists suffer from physics envy? *Finnish Econ. Papers*, **5**, 61–68.
181. Moerner, W. E. (2015) Single-molecule spectroscopy, imaging, and photocontrol: Foundations for super-resolution microscopy (Nobel lecture). *Angew. Chem. Int. Ed.*, **54**, 8067–8093.
182. Monod, J. (1972) *Chance and Necessity: An Essay on the Natural Philosophy of Modern Biology*. Vintage Books, New York.
183. Moyal, J. E. (1962) The general theory of stochastic population processes. *Acta Math.* **108**, 1–31.
184. Murray, J. D. (1984) *Asymptotic Analysis*, Springer-Verlag, New York.
185. Murray, J. D. (2007) *Mathematical Biology: I. An Introduction*, 3rd ed., Springer, New York.
186. Murray, J. D. (2003) *Mathematical Biology: II. Spatial Models and Biomedical Applications*, 3rd ed., Springer, New York.

—————— **N, O** ——————

187. Nelson, P. (2013) *Biological Physics*, Updated ed., W. H. Freeman, New York.
188. Nicholls, D. and Ferguson, S. (2002) *Bioenergetics*, 3rd ed., Academic Press, New York.
189. Nicolis, G. and Lefever, R. (1977) Comment on the kinetic potential and the Maxwell construction in non-equilibrium chemical phase transitions. *Phys. Lett. A* **62**, 469–471.
190. Nicolis, G. and Prigogine, I. (1977) *Self-Organization in Nonequilibrium Systems: From dissipative structures to order through fluctuations*. Wiley-Interscience, New York.
191. Ninio, J. (1975) Kinetic amplification of enzyme discrimination, *Biochimie* **57**, 587-95.
192. Norman, E. B., Gazes, S. B., Crane, S. G. and Bennett, D. A. (1988) Tests of the exponential decay law at short and long times. *Phys. Rev. Lett.* **60**, 2246–2249.
193. Nowak, M. A. (1992) What is a quasispecies? *Trends Ecol. Evol.* **7**, 118–121.
194. Oksendal, B. (2003) *Stochastic Differential Equations: An Introduction with Applications*. 5th Ed., Springer-Verlag, Berlin Heidelberg.
195. Olivieri, E. and Vares, M.E. (2005) *Large Deviations and Metastability*. Cambridge University Press.
196. O'Malley, R. E. (1991) *Singular Perturbation Methods for Ordinary Differential Equations*, Springer-Verlag, New York.
197. Oono, Y. (1998) Large deviation and statistical physics. *Progr. Theoret. Phys. Supp.* **99**, 165–205.
198. Othmer, H. G. (1976) Nonuniqueness of equilibria in closed reaction systems. *Chem. Engng. Sci.* **31**, 993–1003.
199. Othmer, H. G. (2003) *Analysis of Complex Reaction Networks*, Lecture Notes, School of Mathematics, University of Minnesota, Minneapolis.

—————— **P** ——————

200. Paulsson, J. (2004): Summing up the noise in gene networks. *Nature* **427**, 415-418.
201. Paulsson, J. and Ehrenberg, M. (2000) Random signal fluctuations can reduce random fluctuations in regulated components of chemical regulatory networks. *Phys. Rev. Lett.* **84**, 5447–5450.
202. Pearson, H. (2008) Cell biology: The cellular hullaboloo. *Nature* **453**, 150–153.

203. Pearson, K. (1943) *The Grammar of Science*. J. M. Dent & Sons, London, UK.
204. Peng, Y., Qian, H., Beard, D. A. and Ge, H. (2020) Universal relation between thermodynamic driving force and one-way fluxes in a nonequilibrium chemical reaction with complex mecahnism. *Phys. Rev. Res.* **2**, 033089.
205. Polettini, M. and Esposito, M. (2014) Irreversible thermodynamics of open chemical networks. I. Emergent cycles and broken conservation laws. *J. Chem. Phys.* **141**, 024117.
206. Polettini, M., Wachtel, A. and Esposito, M. (2015) Dissipation in noisy chemical networks: The role of deficiency. *J. Chem. Phys.* **143**, 184103.
207. Price, G. R. (1972) Fisher's "fundamental theorem" made clear. *Annals of Human Genet.* **36**, 129–140.
208. Price, N. C. and Stevens, L. (1981) *Fundamentals of Enzymology*. Oxford Univ. Press, U.K.
209. Prigogine, I. and Mazur, P. (1953) Sur l'extension de la thermodynamique aux phénomènes irréversibles liés aux degrés de liberté internes. *Physica* **19**, 241–254.
210. Prinz, J.-H., Wu, H., Sarich, M., Keller, B., Senne, M., Held, M., Chodera, J. D., Schütte, C. and Noé, F. (2011) Markov models of molecular kinetics: Generation and validation *J. Chem. Phys.* **134**, 174105.

——————— **Q** ———————

211. Qian, H. (2002) From discrete protein kinetics to continuous Brownian dynamics: A new perspective (Review). *Protein Sci.* **11**, 1–5.
212. Qian, H. (2002) Equations for stochastic macromolecular mechanics of single proteins: Equilibrium fluctuations, transient kinetics and nonequilibrium steady-state. *J. Phys. Chem. B* **106**, 2065–2073.
213. Qian, H. (2006) Open-system nonequilibrium steady-state: Statistical thermodynamics, fluctuations and chemical oscillations (Feature article). *J. Phys. Chem. B* **110**, 15063–15074.
214. Qian, H. (2008) Cooperativity and specificity in enzyme kinetics: A single-molecule time-based perspective (Mini review). *Biophys. J.* **95**, 10–17.
215. Qian, H. (2009) Entropy demystified: The "thermodynamics" of stochastically fluctuating systems. *Methods Enzym.* **467**, 111–134.
216. Qian, H. (2010) Cyclic conformational modification of an enzyme: Serial engagement, energy relay, hysteretic enzyme, and Fischer's hypothesis. *J. Phys. Chem. B*, **114**, 16105–16111.
217. Qian, H. (2012) Cooperativity in cellular biochemical processes: Noise-enhanced sensitivity, fluctuating enzyme, bistability with nonlinear feedback, and other mechanisms for sigmoidal responses. *Annu. Rev. Biophys.* **41**, 179–204.
218. Qian, H. (2014) Fitness and entropy production in a cell population dynamics with epigenetic phenotype switching. *Quantitative Biology*, **2**, 47–53.
219. Qian, H. (2014) The zeroth law of thermodynamics and volume-preserving conservative system in equilibrium with stochastic damping. *Phys. Lett. A*, **378**, 609–616.
220. Qian, H. (2016) Nonlinear stochastic dynamics of complex systems, I: A chemical reaction kinetic perspective with mesoscopic nonequilibrium thermodynamics. arXiv:1605.08070.
221. Qian, H., Ao, P., Tu, Y. and Wang, J. (2016) A framework towards understanding mesoscopic phenomena: Emergent unpredictability, symmetry breaking and dynamics across scales. *Chem. Phys. Lett.* **665**, 153–161.
222. Qian, H. and Beard, D. A. (2005) Thermodynamics of stoichiometric biochemical networks in living systems far from equilibrium. *Biophys. Chem.* **114**, 213–220.
223. Qian, H. and Beard, D. A. (2006) Metabolic futile cycles and their functions: A systems analysis of energy and control. *IET Proc. Sys. Biol.* **153**, 192–200.
224. Qian, H. and Bishop, L. M. (2010) The chemical master equation approach to nonequilibrium steady-state of open biochemical systems: Linear single-molecule enzyme kinetics and nonlinear biochemical reaction networks (Review). *Intern. J. Mole. Sci.* **11**, 3472–3500.
225. Qian, H., Cheng, Y.-C. and Thompson, L. F. (2019) Ternary representation of stochastic change and the origin of entropy and its fluctuations. arXiv:1902.09536
226. Qian, H. and Elson, E. L. (2004) Fluorescence correlation spectroscopy with high-order and dual-color correlation to probe nonequilibrium steady-states. *Proc. Natl. Acad. Sci. U.S.A.* **101**, 2828–2833.

227. Qian, H. and Ge, H. (2012) Mesoscopic biochemical basis of isogenetic inheritance and canalization: Stochasticity, nonlinearity, and emergent landscape. *MCB: Mole. Cellu. Biomech.* **9**, 1–30.

228. Qian, H., Kjelstrup, S., Kolomeisky, A. B. and Bedeaux, D. (2016) Entropy production in mesoscopic stochastic thermodynamics: nonequilibrium kinetic cycles driven by chemical potentials, temperatures, and mechanical forces (Topical review). *J. Phys. Cond. Matt.* **28**, 153004.

229. Qian, H. and Reluga, T. C. (2005) Nonequilibrium thermodynamics and nonlinear kinetics in a cellular signaling switch. *Phys. Rev. Lett.* **94**, 028101.

230. Qian, H. and Xie, X. S. (2006) Generalized Haldane equation and fluctuation theorem in the steady-state cycle kinetics of single enzymes. *Phys. Rev. E* **74**, 010902(R).

———————— **R** ————————

231. Rao, R. and Esposito, M. (2016) Nonequilibrium thermodynamics of chemical reaction networks: wisdom from stochastic thermodynamics. *Phys. Rev. X* **6**, 041064.

232. Reguera, D., Rubí, J. M. and Vilar, J. M. G. (2005) The mesoscopic dynamics of thermodynamic systems. *J. Phys. Chem. B* **109**, 21502–21515.

233. Reidys, C. M. and Stadler, P. F. (2002) Combinatorial landscapes. *SIAM Rev.* **44**, 3–54.

234. Reiss, H. and Mirabel, P. (1987) Fluctuations in the concentration of the nucleus in classical nucleation theory. *J. Phys. Chem.* **91**, 1–3.

235. Rodriguez-Vargas, A. M. and Schuster, P. (1984) The dynamics of catalytic hypercycles | A stochastic simulation. In *Stochastic Phenomena and Chaotic Behavior in Complex Systems*, Schuster, P. ed., Springer-Verlag, Berlin, pp. 208–219.

236. Ross, J. (2008) *Thermodynamics and Fluctuations Far From Equilibrium*, Springer, New York.

———————— **S** ————————

237. Saakian, D. B. and Qian, H. (2016) Nonlinear stochastic dynamics of complex systems, III: Noneqilibrium thermodynamics of self-replication kinetics. arXiv:1606.02391.

238. Santillán, M. and Qian, H. (2011) Irreversible thermodynamics in multiscale stochastic dynamical systems. *Phys. Rev. E* **83**, 041130.

239. Schellman, J. A. (1997) Thermodynamics, molecules and the Gibbs conference. *Biophys. Chem.* **64**, 1–13.

240. Schlick, T. (2002) *Molecular Modeling and Simulation: An Interdisciplinary Guide*, Springer, New York.

241. Schlögl, F. (1972) Chemical reaction models for nonequilibrium phase transition. *Z. Physik.* **253**, 147–161.

242. Schrödinger, E. (1944) *What is Life? The Physical Aspect of the Living Cell*, Cambridge Univ. Press, U.K.

243. Schulz, G. E. and Schirmer, R. H. (1979) *Principles of Protein Structure.* Springer-Verlag, New York.

244. Schuss, Z. (2013) *Brownian Dynamics at Boundaries and Interfaces. In Physics, Chemistry, and Biology.* Springer-Verlag, New York.

245. Schuster, S. and Schuster, R. (1989) A generalization of Wegscheider's condition: Implications for properties of steady states and for quasi-steady-state approximation. *J. Math. Chem.* **3**, 25–42.

246. Sepúlveda, L.A., Xu, H., Zhang, J., Wang, M. and Golding, I. (2016) Measurements of gene regulation in individual cells reveals rapid switching between propomter states. *Science* **351** 1218–1222.

247. Shahrezaei, V. and Swain, P. S. (2008) Analytical distributions for stochastic gene expression. *Proc. Natl. Acad. Sci. USA* **105**, 17256–17261.

248. Shapiro, B. E. and Qian, H. (1997) A quantitative analysis of single protein-ligand complex separation with the atomic force microscope. *Biophys. Chem.* **67**, 211–219.

249. Shapiro, N. Z. and Shapley, L. S. (1965) Mass action laws and the Gibbs free energy function. *J. S.I.A.M.* **13**, 353–375.

250. Shear, D. B. (1967) An analog of the Boltzmann H-theorem (a Lyapunov function) for systems of coupled chemical reactions. *J. Theoret. Biol.* **16**, 212–228.
251. Shear, D. B. (1968) Stability and uniqueness of the equilibrium point in chemical reaction systems. *J. Chem. Phys.* **48**, 4144–4147.
252. Shwartz, A. and Weiss, A.: (1995) *Large Deviations for Performance Analysis*, Chapman & Hall, New York.
253. Sinai, Ya. G. (1976) Self-similar probability distribution. *Theory Probab. Appl.* **21**, 64–80.
254. Smith, E. (2011) Large-deviation principles, stochastic effective actions, path entropies, and the structure and meaning of thermodynamic descriptions. *Pre. Progr. Phys.* **74**, 046601.
255. Strogatz, S. H. (1994) *Nonlinear Dynamics and Chaos*. Perseus Books, Reading, MA.
256. Suter, D., Molina, N., Gatfield, D., Schneider, K., Schibler, U. and Naef, F. (2011) Mammalian genes are transcribed with widely different bursting kinetics. *Science* **332** 472–474.

──────── **T, U, V** ────────

257. Taniguchi, Y., Choi, P. J., Li, G.-W., Chen, H., Babu, M., Hearn, J., A. Emili and Xie, X. S. (2010). Quantifying E. coli proteome and transcriptome with single-molecule sensitivity in single cells. *Science*, **329**, 533–538.
258. Thompson, L. F. and Qian, H. (2016) Nonlinear stochastic dynamics of complex systems, II: Potential of entropic force in Markov systems with nonequilibrium steady state, generalized Gibbs function and criticality. arXiv:1605.08071.
259. Touchette, H. (2009) The large deviation approach to statistical mechanics. *Phys. Rep.* **478**, 1–69.
260. Turner, J. H., Beeghley, L. and Powers, C. H. (2011) *The Emergence of Sociological Theory*, Pine Forge Press, Newbury Park, CA.
261. van Kampen, N. G. (1961) A power series expansion of the master equations. *Can. J. Phys.* **39**, 551–567.
262. van Kampen, N. G. (2007) *Stochastic Processes in Physics and Chemistry*, 3rd ed., North Holland, Amsterdam.
263. Varadhan, S. R. S. (2012) Book review: Large Deviations for Stochastic Processes by Jin Feng and Thomas G. Kurtz. *Bull. Amer. Math. Soc. (N.S.)* **49**, 597–601.
264. Vellela, M. and Qian, H. (2007) A quasistationary analysis of a stochastic chemical reaction: Keizer's paradox. *Bull. Math. Biol.* **69**, 1727–1746.
265. Vellela, M. and Qian, H. (2009) Stochastic dynamics and nonequilibrium thermodynamics of a bistable chemical system: The Schlögl model revisited. *J. R. Soc. Interf.* **6**, 925–940.
266. Verdugo, A., Vinod, P. K., Tyson, J. J., and Novak, B. (2013) Molecular mechanisms creating bi-stable switches at cell cycle transitions. *Open Biol.* **3**, 120179.

──────── **W** ────────

267. Waddington, C. H. (1953) *The Epigenetics of Birds*. Cambridge Univ. Press.
268. Wang, J., Xu, L. and Wang, E. (2008) Potential landscape and flux framework of nonequilibrium networks: Robustness, dissipation, and coherence of biochemical oscillations. *Proc. Natl. Acad. Sci. USA* **105**, 12271–12276.
269. Wang, J., Xu, L., Wang, E. and Huang, S. (2010) The potential landscape of genetic circuits imposes the arrow of time in stem cell differentiation. *Biophys. J.* **99**, 29–39.
270. Wang, J., Zhang, K., Xu, L. and Wang, E. (2011) Quantifying the Waddington landscape and biological paths for development and differentiation. *Proc. Natl. Acad. Sci. USA* **108**, 8257–8262.
271. Wang, T., Zhao, J., Ouyang, Q., Qian, H., Fu, Y. V. and Li, F. (2016) Phosphorylation energy and nonlinear kinetics as key determinants for G2/M transition in fission yeast cell cycle. arXiv:1610.09637.
272. Wegscheider, R. (1902) Uber simultane gleichgewichte und die beziehungen zwischen thermodynamik und reaktionskinetik homogener systeme. *Z. Phys. Chem.* **39** 257–303.
273. Wei, J. and Prater, C. D. (1962) The structure and analysis of complex reaction systems. *Adv. Catalysis* **13**, 203–392.
274. Welch, G. R. (ed.) (1986) *The Fluctuating Enzyme*. John Wiley & Sons, New York.

275. Wyman, J. and Gill, S. (1990) *Binding and Linkage: Functional Chemistry of Biological Macromolecules*, Univ. Sci. Book, Mill Valley, CA.

─────────── **X, Y, Z** ───────────

276. Xie, X. S. (2013) Enzyme kinetics, past and present. *Science* **342**, 1457–1459.
277. Yang, C. N. and Lee, T. D. (1952) Statistical theory of equations of state and phase transitions. I. Theory of condensation. *Phys. Rev.* **87**, 404–409.
278. Yang, S. X. and Ge, H. (2018) Decomposition of the entropy production rate and nonequilibrium thermodynamics of switching diffusion processes. *Phys. Rev. E.* **98**, 012418.
279. Yule, G. U. (1925) A mathematical theory of evolution, based on the conclusions of Dr. J. C. Wills, F.R.S. *Philos. Tran. R. Soc. B.* **213**, 402–410.
280. Zeeman, E. C. (1988) Stability of dynamical systems. *Nonlinearity* **1**, 115–155.
281. Zhang, X.-J., Qian, H. and Qian, M. (2012) Stochastic theory of nonequilibrium steady states and its applications (Part I) *Phys. Rep.* **510**, 1–86.
282. Zhao, Y., Wang, D., Zhang, Z., Lu, Y., Yang, X., Ouyang, Q., Tang, C. and Li, F. (2020) Critical slowing down and attractive manifold: A mechanism for dynamic robustness in the yeast cell-cycle process. *Phys. Rev. E* **101**, 042405.
283. Zhou, J. X., Pisco, A. O., Qian, H. and Huang, S. (2014) Nonequilibrium population dynamics of phenotype conversion of cancer cells. *PLoS ONE*, **9**, e110714.
284. Zhou, P. and Li, T. (2016) Construction of the landscape for multi-stable systems: Potential landscape, quasi-potential, A-type integral and beyond. *J. Chem. Phys.* **144**, 094109.
285. Zhou, P., Gao, X., Li, X., Li, L., Niu, C., Ouyang, Q., Lou, H., Li, T. and Li, F. (2021) Stochasticity triggers activation of the S-phase checkpoint pathway in budding yeast. *Phys. Rev. X*, **11**, 011004.

Index

adenosine triphosphate (ATP), 9, 283
   hydrolysis free energy, 283
allosteric cooperativity, 169
asymptotic expansion, 164
attractors, viii, 297
autocatalysis, 36
autocorrelation function, 213

bifurcation
   blue sky, 152
   Hopf, 39
   pitch-fork, 149, 247
   saddle-node, 149, 246
   transcritical, 246
binding
   constant, 169
   positive cooperativity, 171
   protein-substrate, 169
binding polynomial, 171
biochemistry, 161
biological cell, 5
biological diversity, viii
birth-death process, 60
bistability, 36, 146
   Schlögl model, 251
   nonlinear, 251
   stochastic, 256
Boltzmann's law, 21
Briggs–Haldane theory, 168

canonical ensemble
   semigrand canonical, 222
central limit theorem, 71
chemical equilibrium, 25
chemical kinetic potential condition, 34
chemical master equation, 118, 254, 261
chemical oscillation, 38
chemical potential

of reaction, 23
of species, 24
chemostat, 24
complex balance, 32, 122
conformational selection, 306
cooperativity, 172, 199
cusp catastrophe, 251
cycle fluxes, 179

Delbrück–Gillespie process, 96, 98, 154, 208, 242
detail balance
   thermodynamic "box", 178, 222
detailed balance, 25, 108
differentiation program, 316
diffusion process, 63
dimerization reaction, 19
dissipative, viii
Doob, Boltz, Kalos, Lebowitz, and Gillespie, method of, 274
dynamic cooperativity, 196
dynamic disorder, 196
dynamic symmetry breaking, 156
dynamics, vii

emergent phenomenon, 153
entropy, 65
   balance equation, 68, 106
   production, 107
   Shannon's, 65, 106
equilibrium
   association reaction, 244
expected value, 44

far-from-equilibrium, 307
   inanimate branch, 312
fluctuating enzyme, 196
fluctuating-rate model, 274

© The Author(s), under exclusive license to Springer Nature Switzerland AG 2021   349
H. Qian, H. Ge, *Stochastic Chemical Reaction Systems in Biology*,
Lecture Notes on Mathematical Modelling in the Life Sciences,
https://doi.org/10.1007/978-3-030-86252-7

Printed in the United States
by Baker & Taylor Publisher Services